Biochemistry and Mode of Action of Hormones II

Publisher's Note

The *International Review of Biochemistry* remains a major force in the education of established scientists and advanced students of biochemistry throughout the world. It continues to present accurate, timely, and thorough reviews of key topics by distinguished authors charged with the responsibility of selecting and critically analyzing new facts and concepts important to the progress of biochemistry from the mass of information in their respective fields.

Following the successful format established by the earlier volumes in this series, new volumes of the *International Review of Biochemistry* will concentrate on current developments in the major areas of biochemical research and study. New volumes on a given subject generally appear at two-year intervals, or according to the demand created by new developments in the field. The scope of the series is flexible, however, so that future volumes may cover areas not included earlier.

University Park Press is honored to continue publication of the *International Review of Biochemistry* under its sole sponsorship beginning with Volume 13. The following is a list of volumes published and currently in preparation for the series:

Volume 1: **CHEMISTRY OF MACROMOLECULES** (H.Gutfreund)
Volume 2: **BIOCHEMISTRY OF CELL WALLS AND MEMBRANES** (C.F. Fox)
Volume 3: **ENERGY TRANSDUCING MECHANISMS** (E. Racker)
Volume 4: **BIOCHEMISTRY OF LIPIDS** (T.W. Goodwin)
Volume 5: **BIOCHEMISTRY OF CARBOHYDRATES** (W.J. Whelan)
Volume 6: **BIOCHEMISTRY OF NUCLEIC ACIDS** (K. Burton)
Volume 7: **SYNTHESIS OF AMINO ACIDS AND PROTEINS** (H.R.V. Arnstein)
Volume 8: **BIOCHEMISTRY OF HORMONES** (H.V. Rickenberg)
Volume 9: **BIOCHEMISTRY OF CELL DIFFERENTIATION** (J. Paul)
Volume 10: **DEFENSE AND RECOGNITION** (R.R. Porter)
Volume 11: **PLANT BIOCHEMISTRY** (D.H. Northcote)
Volume 12: **PHYSIOLOGICAL AND PHARMACOLOGICAL BIOCHEMISTRY** (H. Blaschko)
Volume 13: **PLANT BIOCHEMISTRY II** (D.H. Northcote)
Volume 14: **BIOCHEMISTRY OF LIPIDS II** (T.W. Goodwin)
Volume 15: **BIOCHEMISTRY OF CELL DIFFERENTIATION II** (J. Paul)
Volume 16: **BIOCHEMISTRY OF CARBOHYDRATES II** (D.J. Manners)
Volume 17: **BIOCHEMISTRY OF NUCLEIC ACIDS II** (B.F.C. Clark)
Volume 18: **AMINO ACID AND PROTEIN BIOSYNTHESIS II** (H.R.V. Arnstein)
Volume 19: **BIOCHEMISTRY OF CELL WALLS AND MEMBRANES II** (J.C. Metcalfe)
Volume 20: **BIOCHEMISTRY AND MODE OF ACTION OF HORMONES II** (H.V. Rickenberg)

(Series numbers for the following volumes will be assigned in order of publication)
DEFENSE AND RECOGNITION II (E.S. Lennox)
BIOCHEMISTRY OF NUTRITION (A. Neuberger)
MICROBIAL BIOCHEMISTRY (J.R. Quayle)

INTERNATIONAL REVIEW OF BIOCHEMISTRY

Volume 20

Biochemistry and Mode of Action of Hormones II

Edited by

H. V. Rickenberg, Ph.D.

Division of Molecular and Cellular Biology
National Jewish Hospital and Research Center
and The University of Colorado
School of Medicine

UNIVERSITY PARK PRESS

Baltimore

UNIVERSITY PARK PRESS
International Publishers in Science and Medicine
233 East Redwood Street
Baltimore, Maryland 21202

Typeset by Action Comp. Co., Inc.

Manufactured in the United States of America by
The Maple Press Company.

Library of Congress Cataloging in Publication Data

Main entry under title:

Biochemistry and mode of action of hormones II.

(International review of biochemistry; v. 20)
"The title of the volume in series one corresponding to this edition was Biochemisty of Hormones."
Bibliography: p.
Includes index.
1. Hormones. 2. Hormones—Metabolism. I. Rickenberg, H. V. II. Rickenberg, H. V. Biochemistry of hormones. III. Series. [DNLM: 1. Hormones—Metabolism. 2. Hormones—Pharmacodynamics. 3. Receptors, Hormone. W1 IN8296 v. 2 / WK102.3 B615]
QP501.B527 vol. 20 [QP571] 574.1'92'08s
[599'.01'927] 78-988

Consultant Editors' Note

The MTP *International Review of Biochemistry* was launched to provide a critical and continuing survey of progress in biochemical research. In order to embrace even barely adequately so vast a subject as "progress in biochemical research," twelve volumes were prepared. They range in subject matter from the classical preserves of biochemistry—the structure and function of macromolecules and energy transduction—through topics such as defense and recognition and cell differentiation, in which biochemical work is still a relatively new factor, to those territories that are shared by physiology and biochemistry. In dividing up so pervasive a discipline, we realized that biochemistry cannot be confined to twelve neat slices of biology, even if those slices are cut generously: every scientist who attempts to discern the molecular events that underlie the phenomena of life can legitimately parody the cry of *Le Bourgeois Gentilhomme,* "Par ma foi! Il y a plus de quarante ans que je dis de la Biochimie sans que j'en susse rien!" We therefore make no apologies for encroaching even further, in this second series, on areas in which the biochemical component has, until recently, not predominated.

However, we repeat our apology for being forced to omit again in the present collection of articles many important matters, and we also echo our hope that the authority and distinction of the contributions will compensate for our shortcomings of thematic selection. We certainly welcome criticism—we thank the many readers and reviewers who have so helpfully criticized our first series of volumes—and we solicit suggestions for future reviews.

It is a particular pleasure to thank the volume editors, the chapter authors, and the publishers for their ready cooperation in this venture. If it succeeds, the credit must go to them.

H. L. Kornberg
D. C. Phillips

Contents

Preface

The title of the volume in Series One corresponding to this volume is *Biochemistry of Hormones*; the present volume is entitled *Biochemistry and Mode of Action of Hormones*. The change in title is meant to emphasize the current preoccupation of workers in the field with the molecular mechanisms by which hormones exert their effects.

During the four years that have elapsed between the publication of the two volumes, there has been significant progress in our understanding of how hormones act. In the case of a majority of the polypeptide and catecholamine hormones, and to a lesser extent in the case of prostaglandins, receptors for the hormones have been identified and, in some instances, isolated; the discreteness of membranal receptor and nucleotide cyclase has been established. The precise nature of the coupling between receptor and cyclase, though, has proved elusive. It appears that many, possibly most, of the effects of hormones, which are mediated by changes in the cellular concentrations of cyclic nucleotides, find their penultimate expression in alterations in the state of phosphorylation of a variety of proteins that presumably play key roles in the regulation of permeability, the maintenance and transitions of cellular morphology, chromosomal behavior, etc.

There have been important advances also in the elucidation of the mechanism of cell-hormone interaction in the case of the other major class of hormones, the steroids. There is evidence that the intranuclear steroid-proteinaceous receptor complex affects the transcription of genetic information. The detailed mechanism of this modulation of transcription still awaits clarification.

It may well be that the ultimate elucidation, i.e., at the molecular level, of the workings of both steroid hormones and those hormones mediated in their effects by cyclic nucleotides will come from studies with insects and with the lower eukaryotes. Viable mutants, abnormal in either the production of, or response to, hormones can be obtained and offer a logical approach to the investigation of the role and mode of action of hormones, especially in early development.

The use of cells cultured in isolation from the organism and tissue of origin has proven invaluable as a tool for the study of hormone action, both at the physiological and at the molecular level. However, not only have experiments with cultured cells aided the investigation of known hormones, but they have also led to the discovery of a variety of growth factors which, although not necessarily active at sites remote from the sites of synthesis, appear to meet the other criteria inherent in the definition of a hormone.

The one common denominator that characterizes the miscellany of compounds classified as plant hormones is that they act, indeed, on plants. Little is known about the mechanisms of action of these compounds. They exist, their structures are known, and their effects on a variety of processes have been described. Perhaps in a third edition of this volume we shall be told how plant hormones act and what new principles of biological regulation might be discovered from their study.

H. V. Rickenberg

Biochemistry and Mode of Action of Hormones II

International Review of Biochemistry
Biochemistry and Mode of Action of Hormones II, Volume 20
Edited by H. V. Rickenberg

1
Membrane-active Hormones: Receptors and Receptor Regulation

P. A. INSEL
University of California, San Francisco

The author's work was funded by Grants GM-16496, GM-00001, and HL-06285 from the United States Public Health Service, National Institutes of Health, Bethesda, Maryland.

Recently, hormones have been defined (1) as substances carrying "information from 'sensor' cells in direct contact with environmental signals to more sequestered responder cells" (1). In this formulation Tomkins has suggested that hormones evolved as a means of communication: altered metabolic states in the "sensor" cells were "encoded" by the synthesis and secretion of molecules that circulate to responder cells, in which the metabolic message is "decoded" into appropriate intracellular symbols. By such a mechanism, organs may respond in a coordinate fashion to external perturbations.

In this review (covering literature published before September, 1976), some of the factors underlying the qualitative and quantitative specificity of hormonal response are examined. The scope of this analysis has been arbitrarily restricted to membrane-active hormones, i.e., those interacting with membrane-bound receptors and thereby triggering specific intracellular events. This analysis, therefore, eliminates the gene-active hormones, including glucocorticoids, mineralocorticoids, sex steroids, thyroid hormone, and vitamin D. Many recent articles in this series and elsewhere review the rapid developments underlying our understanding of receptors for steroid and thyroid hormones and mechanisms of the hormone-receptor complex interaction with the nucleus (2–10).

In contrast, the membrane-active hormones (Table 1) encompass polypeptides, vasoactive amines, prostaglandins, acetylcholine, and other neurotransmitters, perhaps including naturally occurring opiates (endorphin, enkephalin). Clearly, it would be impossible in this chapter to present details on receptors and functions of that long list of compounds. Instead, certain aspects of hormone action are examined primarily to understand how cells respond to such agents in a unique and programmed fashion. Recent studies of membrane structure and function, measurements of receptors, coupling of receptors to effectors, regulation of hormone-receptor interactions, and the consequences of those interactions for cellular responses are reviewed. Studies on insulin and catecholamines as prototypes for polypeptide hormones and vasoactive amines are emphasized. Reviews of developments on cholinergic receptors (11–15), opiate receptors (16–18), and prostaglandins (19–22) have appeared recently.

PLASMA MEMBRANE STRUCTURE AND FUNCTION

Structural and functional studies of plasma membranes have been performed most extensively with erythrocytes (23–27), primarily because of the ease with which one can obtain purified preparations of cells, hemoglobin-free ghosts, and plasma membrane vesicles. In spite of the fact that erythrocytes in mammalian species are enucleate, with a programmed lifespan (about 120 days for human erythrocytes), and are, therefore, highly differentiated, observations on erythrocytes are generally assumed to apply to other actively mitotic cells.

Table 1. Membrane-active hormones

Polypeptide hormones
Adrenocorticotropin (ACTH), angiotensin, calcitonin, epidermal growth factor (EGF), exophthalmogenic factor, follicle-stimulating hormone (FSH), glucagon, human chorionic gonadotropin (hCG), luteinizing hormone (LH), LH-releasing factor, melanocyte-stimulating hormone (MSH), nerve growth factor (NGF), oxytocin, parathyroid hormone, prolactin, secretin, somatomedin, thyrotropin (TSH), TSH-releasing factor, vasoactive intestinal peptide (VIP), vasopressin (antidiuretic hormone, ADH), insulin
Vasoactive amines
Epinephrine, norepinephrine, dopamine, histamine, serotonin
Prostaglandins
Acetylcholine?
Enkephalins ? (endorphins)

Cellular plasma membranes are composed of lipid (about 40% by weight) and protein (about 50%), with the remainder being carbohydrate. The notion that the lipid matrix is arranged in a bimolecular leaflet (28, 29) with hydrophobic bonding of membrane proteins to the nonpolar lipids has led to numerous proposed models of membrane organization (30–34).

Lipids in the membrane are principally of three groups: phospholipids, glycolipids, and sterols. The proportion of each of these classes varies minimally with functional activity of a particular cell. It has been suggested (24–27) that the phospholipids are asymmetrically distributed in specific halves of the lipid bilayer so that choline-containing phospholipids (phosphatidylcholine and sphingomyelin) are on the outer half and amine-containing phospholipids (phosphatidylserine and phosphatidylethanolamine) are on the cytoplasmic side. The nature of the fatty acid residues of phospholipids determines the temperatures of phase (melting) transition in artificial lipid bilayers (35) and thus, it is presumed, in plasma membranes; membranes containing unsaturated phospholipids have low melting temperatures ($-10\,°C$ for dioleyl lecithin), whereas those with saturated phospholipids have much higher melting temperatures (61 °C for distearoyl lecithin). Below the melting temperature the hydrocarbon chains are fixed, whereas above it they are free to move. This movement, molecular lateral diffusion, has been estimated to have diffusion constants on the order of $10^{-8}\,cm^2\,s^{-1}$, a rate 10^{10} times more rapid than the passage of phospholipid molecules from one side of a bilayer to the other (flip-flop) (29, 36–38). In addition to those two types of molecular movements, at least three others have been described in membranes (39–41): 1) gauche-*trans* (twisting) transition; 2) segmental motion (flexing of fatty acid chains and their oscillation about an axis perpendicular to the plane of the bilayer, usually at rates less than $10^{-9}\,s^{-1}$); and 3) rotational motion, i.e., rotation of molecules about their axes perpendicular to the plane of the bilayer at rates of 10^{-6}

to 10^{-9} s^{-1}. The interaction of cholesterol, the primary membrane sterol, with membrane phospholipids produces a "condensing" effect that hinders the free movement of mobile fatty acid chains (42). "Microviscosity" denotes the Newtonian flow characteristics of fluidity (i.e., the thermal motion of the lipid bilayer) of both artificial liposomes and intact plasma membranes (43–46); microviscosity is largely determined by the molar ratio of cholesterol to phospholipids in both artificial and physiological membranes (46). As the cholesterol to phospholipid ratio increases, microviscosity and the degree of order in the system increase, and, as a result, phase transitions are gradually abolished (47, 48).

The extent to which the lipid bilayer is penetrated by proteins and the contribution of such penetration to membrane structure are still open to question (24, 49, 50). It has been proposed (31, 51) that membrane proteins can be classified as either extrinsic (peripheral), i.e., water soluble and easily removed, or intrinsic (integral), i.e., water insoluble, lipid associated, and removed with difficulty. Roughly 75% of membrane proteins are intrinsic, and this includes "most membrane-bound enzymes, antigenic proteins, transport proteins, drug and hormone receptors, and receptors for lectins" (49). Coleman (52) has reviewed the critical role of the membrane microenvironment (e.g., lipids) in determining the properties of membrane-bound enzymes. Perhaps this is best demonstrated by the ability of exogenously added lipid to reactivate inactive, solubilized enzymes such as Na-K-ATPase. The use of sodium dodecyl sulfate (SDS) combined with gel electrophoresis (53–55) and, more recently, two-dimensional gel electrophoresis (56) has assisted greatly in the resolution of protein components of erythrocyte membranes. In addition, electron microscopic techniques such as freeze-fracture and the use of membrane probes that permit selective destruction, alteration, or radioactive labeling of proteins on the surface of the cell membrane, as opposed to those entirely inside the lipid bilayer, have also facilitated localization of membrane proteins (26, 29, 49, 50). From such studies, it appears that the extrinsic proteins are located at least partially inside the lipid bilayer and that at least two of the major intrinsic proteins in erythrocytes extend across the plasma membrane from the external surface, where their carbohydrate portions are detectable, through the lipid bilayer into the cytoplasm itself (26, 51). Attempts to purify intrinsic proteins have been actively pursued, and recently the major intrinsic protein of human erythrocytes, Band 3 (named for its SDS-gel position), has been purified in an electrophoretically homogeneous form (57).

As discussed above with respect to lipids, proteins also demonstrate rapid lateral diffusion (diffusion constant 10^{-10} cm^2 s^{-1}) in the plane of the membrane. This was first shown by Frye and Edidin (58), who measured the rate of intermixing of surface antigens by using heterokaryons (cells formed by fusion of two parent cells by addition of virus). This diffusion was temperature-dependent but insensitive to inhibitors of protein synthesis or uncouplers of oxidative phosphorylation (39, 41). In addition, movement of fluorescently labeled antimembrane antibody (59), rhodopsin diffusion in rod outer segments (60, 61), and binding of divalent (but not monovalent) antibodies to cell surface antigens (39, 41, 62–64) are thought to represent lateral movement of proteins. In experiments with fluorescein- or ferritin-coupled antibodies directed against surface antigens, aggregation of antigens results in so-called "patch formation." This process, also temperature-dependent but independent of protein synthesis and oxidative phosphorylation, is followed by "cap formation," i.e., aggregation of patches at one cellular pole. In contrast with patch formation, cap formation depends on cellular metabolism. The cap, once formed, may then undergo endocytosis or be cast off from the membrane. Similar aggregation effects are also observed when lectins or bacterial toxins bind to cell surfaces (29, 65–68). Further studies of capping by Edelman and his colleagues (reviewed by Edelman (64)) have shown that treatment of lymphocytes with the plant lectin concanavalin A prevents subsequent patching and capping of cell surface receptors. This phenomenon, called anchorage modulation, propagates to involve the entire cell surface after only a small fraction of the receptors are bound and may involve activation of a cellular cytoskeleton that includes actin-like microfilaments, myosin, and microtubules (63, 64).

At the present time, fluid mosaic or liquid crystalline models (31, 34, 49, 65, 69) (Figure 1) of membrane structure seem to be those most compatible with the majority of available data on membrane structure. These models propose that the globular intrinsic proteins and glycoproteins are distributed within and stabilized by the fluid lipid bilayer. The proteins are thought to be amphipathic, with their highly polar ends (ionic amino acid residues and covalently bound saccharide regions) in contact with the aqueous phase and their nonpolar regions embedded in the hydrophobic core of the membrane. Extrinsic proteins are partially bound to the lipid bilayer by interaction with polar groups of phospholipids or intrinsic proteins. Some of the protein-glycoprotein complexes traverse the lipid bilayer as transmembrane structural linkages. The fundamental feature of this model is its "mosaic" character; i.e., substantial portions of the lipids and proteins are organized independently of each other and are capable of movement in the membrane. An additional feature may be the interaction of the inner membrane surface with cellular cytoskeletal elements (microtubules, microfilaments) (70). This latter feature may be especially critical for the aggregation-endocytosis phenomena described above (39, 65).

One important aspect of plasma membrane structure and function is the heterogeneous organization (polarity) that is found on the surface of certain cells so that morphological features, enzymic activities, and transport systems are selectively distributed. For example, studies on renal proximal tubular and intestinal epithelial cells (71–74) indicate that luminal

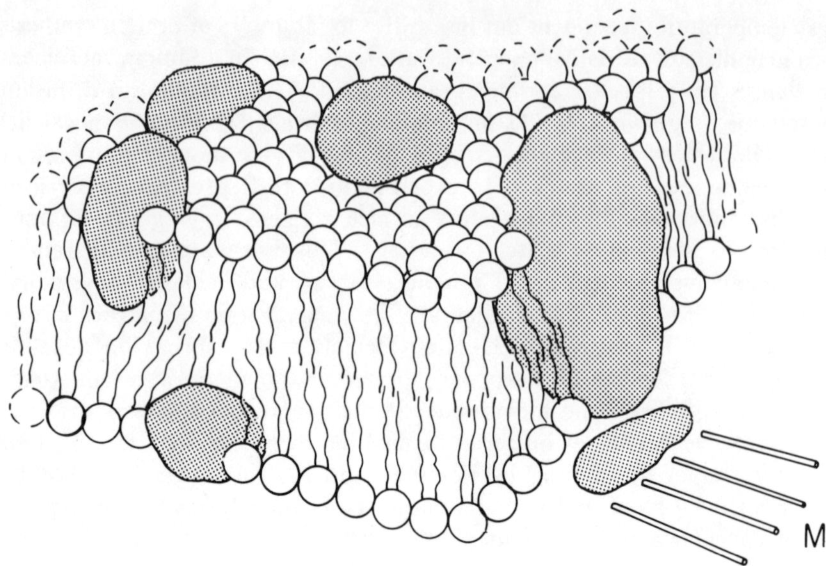

Figure 1. The fluid mosaic model of plasma membrane structure. Schematic three-dimensional and cross-sectional views. The *solid bodies* with stippled surfaces represent the globular intrinsic proteins that are distributed unevenly in the plane of the membrane. The *circles* represent the polar heads and the *wavy lines* the fatty acid chains of membrane phospholipids. Below the bilayer, M refers to the cellular cytoskeletal elements, microtubules, and microfilaments oriented near a membrane protein. Adapted from Singer and Nicolson (31), Singer (49), and Nicolson (65).

(brush border) membranes have well developed microvilli, prominent alkaline phosphatase and disaccharidase activities, and sodium-dependent active transport systems for glucose and amino acids. In contrast, antiluminal (basal) membranes from the same cells have prominent infoldings, striking Na-K-ATPase activity, and quite different transport systems for sugars. Additionally, in the kidney the catalytic and binding activities of cyclic AMP-dependent protein kinase are preferentially located on the luminal surface, whereas hormone receptors (for parathyroid hormone, calcitonin, or vasopressin) and adenylyl cyclase are located on the antiluminal surface (75–78). Although this functional organization is well described in broken cell preparations, the nature of the interaction of the two surfaces in the intact cell remains inadequately explained.

STUDY OF HORMONE RECEPTORS

In the context of this chapter, membrane receptors may be defined as the recognition sites at which cell surfaces interact with drugs or hormones. An important aspect of this recognition process is that a message is ultimately decoded (1) and that cellular function is perturbed by this interaction between hormone and receptor. (Birnbaumer et al. (79) have distinguished between receptors, which function as just defined, and acceptors, which are sites of hormonal interaction that produce no observable alteration in cellular function; this distinction may only reflect the investigator's inability to recognize a particularly labile tissue response.) Differentiation of a particular cell is reflected by that cell's complement of receptors and receptor-mediated events. The fundamental premise of studies of hormone binding to target tissues, cells, or subcellular particles is that specificity of binding is detectable and related to target cell responses. Because the detailed chemical structure of receptors remains a mystery, the term "receptor" is an operational one, describing observations of ligand interaction with cells. The problem lies, therefore, in distinguishing "nonspecific" binding (semifacetiously described by Roth (80) as "binding that is not of interest to me now") from "specific" binding. Although no single criterion completely separates "interesting" from "uninteresting" binding, a series of features (81) taken together greatly strengthen the likelihood that one is examining specific binding.

1. Strict structural and steric specificity
2. Saturability, i.e., a definable number of binding sites
3. Tissue specificity that parallels target cell responsiveness
4. Affinity that corresponds to physiologically active hormone concentrations
5. Reversibility that corresponds to reversal of physiological effects when hormone is removed from the preparation

Most studies of receptor binding are performed with ligands that are radioactively labeled, usually with ^{125}I, ^{131}I, or ^{3}H. It is important to show that such radioactive derivatives behave as do the parent compounds. Insulin, for example, if radioiodinated to less than 1 atom of iodine per molecule, maintains full biological activity, but with increased iodination activity decreases rapidly (82). In contrast, the biological activity of glucagon may increase with increased iodination (83). Binding studies are generally performed by incubation of the labeled hormone with cellular material under equilibrium conditions. When cells or particulate preparations are used, a rapid and efficient separation of membrane-associated ligand is accomplished by centrifugation or by filtration over synthetic polymer or porous glass filters. Nonspecific binding is usually measured by incubating parallel samples in the presence of excess unlabeled ligand or other agonists or antagonists for the receptor. Cuatrecasas et al. (81), Cuatrecasas and Hollenberg (84), and Kahn (85) have recently reviewed many of the techniques, pitfalls, and problems inherent in hormone binding experiments.

The results obtained in recent years in several laboratories on ^{3}H agonist binding to presumed β-adrenergic catecholamine receptors are illustrative of the problems that can be encountered in ligand binding studies (86). A variety of agents ([^{3}H]isoproterenol, [^{3}H]epinephrine, and [^{3}H]norepinephrine) was used to examine β-adrenergic receptors in liver (87–89), heart and myocardial cells (90–93), adipocytes and adiposic tissues (94, 95), uterus (96), avian erythrocytes (97, 98), and skeletal muscle (99). Although interaction of these ligands with cells and subcellular material was generally time- and temperature-dependent, saturable, and reversible for some preparations, certain of the above criteria were not satisfied in all instances. In particular, typical β-adrenergic stereoselectivity ((–)-isomers more potent than (+)-isomers) was not found; nonbiologically active catechol compounds competed for binding sites with the radioligands; some biologically active compounds did not inhibit binding; and, in some systems, binding was irreversible (86–101). Cuatrecasas and co-workers proposed that such binding represented interaction with the membrane-bound form of the metabolizing enzyme catechol-*O*-methyltransferase; others have debated this contention (102–105). Wolfe et al. (102) and Maguire et al. (106) provided strong evidence that [^{3}H]epinephrine and [^{3}H]norepinephrine binding involved primarily agonist oxidation and subsequent reaction with tissue macromolecules rather than interactions with β-adrenergic receptors. It has been suggested in preliminary studies that the agent pyrocatechol may be useful in blocking components of this binding that are not related to true β-adrenergic receptors (107), but further evidence is required before this approach is validated.

An important aspect of studies of ligand binding is the comparison of binding with experiments testing the ligand's biological activity. Such comparisons serve not only to validate a particular binding assay but also to assist in understanding how the recognition process is coupled to activation of cellular events. It is, therefore, desirable that binding and biological studies be carried out with the use of homogeneous preparations from viable cells. For this reason, many investigators have used hormonally responsive cells in tissue culture (108–120) as easily manipulable systems with which to study receptors and hormone action. A particularly promising use of cultured cells seems to be the generation of genetically altered clones with definable lesions in their pathway of hormonal responses (112, 113). One theoretical drawback to the use of cultured cells is that most arose as tumors in animals and may, therefore, have properties different from their normal tissue counterparts (118). Although this may be the case, most of the conclusions drawn from studies of cultured cells have tended to confirm and extend, rather than contradict, previous results obtained in other systems. In addition, studies with cultured cells avoid many of the theoretical and technical difficulties inherent in experiments with intact tissues, isolated organ perfusion, tissue slices, isolated cells, tissue homogenates, or even plasma membrane preparations. The heterogeneity of cellular populations comprising most tissues constitutes a major objection to studies in which tissues or organs rather than cultured cells are employed. In addition, the connective tissue and basement membrane may be barriers to diffusion in intact tissues and organs. Many of the techniques (collagenase or hyaluronidase treatment, tissue homogenizers) used to disrupt those barriers may themselves alter membrane structure and function. The marked lability and changing properties (e.g., acquisition of NaF stimulability) of adenylyl cyclase upon cellular disruption are examples of this (121).

EXAMPLES OF STUDIES ON MEMBRANE RECEPTORS

Insulin

Because studies on insulin receptors have been reviewed extensively in several recent articles (81, 84, 108, 109, 122, 123), only a brief summary of the findings is presented here. Cuatrecasas and Hollenberg (84), in particular, have discussed quantitative aspects of hormone-receptor binding studies, and the interested reader may consult that reference for the mathematical derivation of both kinetic constants and graphic analysis. Kahn's recent review (122) is also an excellent summary of current data on the characterization of hormone receptors.

Until a few years ago, only indirect studies had suggested that receptors for polypeptide hormones were located on the plasma membrane. Experiments in which hormones were bound to cellulose, Sepharose, or other inert supports (124–129) strongly implicated cell surface receptors as the sites of initial hormone recognition. However, problems related to leakage of the hormone from the support (130–133) and to the estimation of the concentrations of the insoluble hormones (134, 135) have made such data difficult to interpret. Beginning in 1970, however, direct binding studies in which labeled polypeptides were employed proved successful in meeting criteria for the demonstration of receptor binding (136, 137) (see above); studies of [^{125}I]insulin binding to cellular receptors ensued rapidly thereafter (138, 139).

Receptors for insulin are located on the outside of the cell (139–142) and are separable from sites of insulin degradation (143–146). (Terris and Steiner (147), however, have recently suggested that in the liver membrane-bound insulin may be the substrate for insulin degradation.) Jarett and Smith (141, 142) have conjugated insulin to ferritin and have shown by electron microscopy that the ferritin-insulin conjugates bind specifically and in clusters at the external side of adipocyte plasma membranes. However, some of the ferritin-insulin complex is located inside a network of membrane invaginations termed surface-connected vesicles; the significance of these is ill defined.

Binding of [^{125}I]insulin to the plasma membrane is rapid, reversible, and saturable; insulin analogs with decreased biological potency compete with the radioligand with reduced potency; and occupancy of the high affinity (dissociation constant, $K_d = 10^{-10}$ M) saturable binding sites (about 10,000 sites per intact fat cell (138) parallels closely the biological effects of the [^{125}I]insulin preparation in the same cells (81, 148). For comparison, basal levels of insulin concentrations in human serum are roughly 10^{-10} M (15 μU/ml), suggesting that the measured K_d is an appropriate range for physiological response (149). Certain cells (e.g., circulating human monocytes (146, 150), cultured human fibroblasts (81, 109), and rat thymus lymphocytes (151)) have K_d values for insulin binding that are one or two orders of magnitude higher, yet those values still correspond to the potency of insulin for the physiological regulation of these cells. The Scatchard (152) analysis (plotting of bound counts/free counts versus bound counts) (Figure 2) of insulin binding to plasma membrane preparations, or to whole cells, yields a characteristically curvilinear relationship. This observation, which has been interpreted both as indicating multiple classes of binding sites and as negative cooperativity is discussed subsequently.

In spite of the evidence that insulin binds to cell surfaces, findings emerging from several studies suggest binding to other cellular organelles, including isolated nuclei in vitro (153) and intracellular membranes, Golgi, and endoplasmic reticulum (85, 121, 154). The physiological role of such intracellular receptors has not been determined.

The positive results obtained in studies of [^{125}I]insulin binding to intact cells and plasma membrane preparations facilitated attempts to

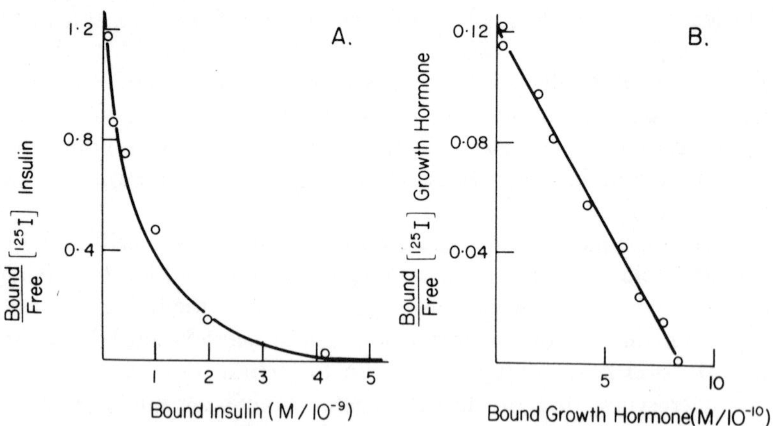

Figure 2. Scatchard (152) analysis of binding of [^{125}I]porcine insulin (*A*) and [^{125}I]human growth hormone (*B*) to cultured human lymphocytes. The growth hormone data are thought to represent binding to a single class of binding sites, whereas the curvilinear pattern for insulin has been interpreted as reflecting either 1) multiple classes of binding sites, 2) negative cooperativity, or 3) a single class of receptor sites with differing affinity for effector depending on hormonal occupancy of the receptor. See text for details. Adapted from De Meyts et al. (287).

solubilize insulin receptors. Hormone receptors seem to be intrinsic membrane proteins (49), and, therefore, detergents (or organic solvents) are required for their removal from the lipid matrix. Cuatrecasas reported successful solubilization of insulin receptors from fat and liver cell membranes (155–157) with the use of the detergent Triton X-100; he estimated a K_d of about 10^{-10} M, a molecular diameter of about 70 Å, a sucrose sedimentation coefficient of about 11 S, and a molecular weight of about 300,000. By using the detergent NP40 and also by culturing cells in serum-free media, Gavin et al. (158, 159) solubilized from human lymphocytes insulin receptors having similar properties. Until this time, no further work has appeared regarding solubilized insulin receptors. This may be a reflection of the technical problems (160) in handling the large amounts of material required for further characterization.

By using techniques that include x-ray analyses, circular dichroism, and receptor binding of analogs, Pullen et al. (161) have attempted to define the region of the insulin molecule bound to the receptor. Their findings indicate that both hydrophobic residues and polar regions of the molecule seem to be involved and that receptor binding may be analogous to the formation of insulin dimers.

Catecholamines

As discussed above, earlier studies of β-adrenergic receptor binding in which ^{3}H agonists were used subsequently proved to be primarily measurements of oxidation of these agents, followed by binding of the oxidized compounds to tissue macromolecules (102, 106). However, since late in 1974, three separate radioactive antagonists ([^{3}H](−)-dihydroalprenolol, [^{3}H](±)-propranolol, and [^{125}I](±)-iodohydroxybenzylpindolol) have become available, and all seem to fulfill criteria expected of ligands for β-adrenergic receptors. More recently, four ligands ([^{3}H]dihydroergocryptine, [^{3}H]dihydroazapetine, [^{3}H]2 ((2′,6′-dimethoxy)phenoxyethylamino)methylbenzodioxan (WB 4101), [^{3}H]clonidine) have been suggested as agents for the identification of α-adrenergic receptors (Table 2).

The distinction between β- and α-adrenergic receptors was orginally established in the experiments of Ahlquist (162), who defined the relative potencies of norepinephrine, epinephrine, isoproterenol, and other adrenergic agonists in producing a variety of pharmacological responses. Responses such as vasoconstriction that were more sensitive to epinephrine and norepinephrine and less sensitive to isoproterenol were termed α-adrenergic, whereas those (e.g., vasodilatation, myocardial stimulation) more sensitive to isoproterenol than to norepinephrine, with epinephrine being intermediate, were classified as β-adrenergic. Subsequent experiments in which specific α- and β-adrenergic antagonists have been examined reinforced and extended (163, 164) this functional classification by differentiating two

Table 2. Radiolabeled ligands used to assay adrenergic receptors [a]

Compound	Structure	Isotope	Specific Activity (Ci/mmole)
BETA-ADRENERGIC RECEPTORS			
(−) Dihydroalprenolol	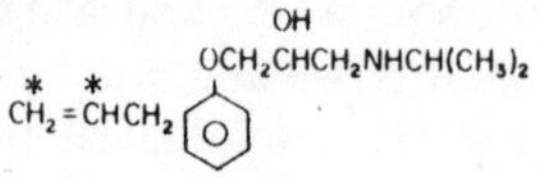 $CH_2{=}CHCH_2$–(ring)–$OCH_2CH(OH)CH_2NHCH(CH_3)_2$ (* * on $CH_2{=}CH$)	[^{3}H]	17-33
(±) Propranolol	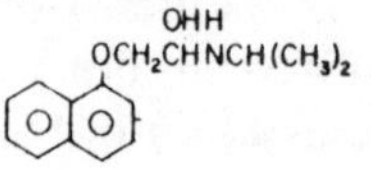 (naphthyl)–$OCH_2CH(OH)CH_2NHCH(CH_3)_2$	[^{3}H] Generally labeled	4-10
(±) Hydroxy-benzylpindolol	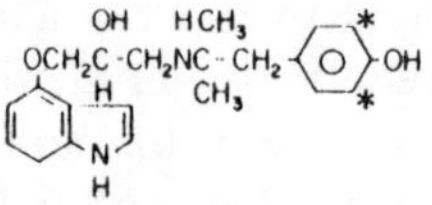 (indole)–$OCH_2C(OH)H$–$CH_2NHC(CH_3)_2$–CH_2–(ring, * *)–OH	[^{125}I]	2200
ALPHA-ADRENERGIC RECEPTORS			
(±) Dihydro-ergocryptine	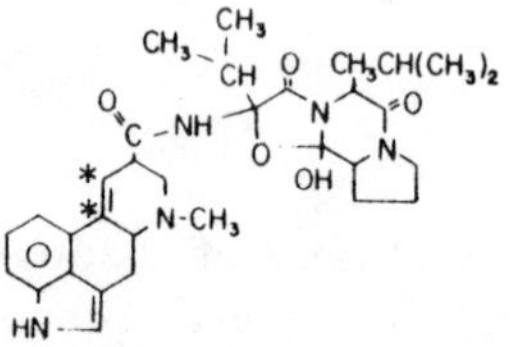 CH_3, CH_3, CH, O, $CH_2CH(CH_3)_2$, C, NH, N, O, N, O, OH, *, *, $N{-}CH_3$, HN	[^{3}H]	25
(±) Dihydroazapetine	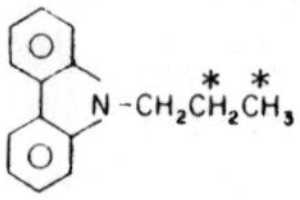 $N{-}CH_2CH_2CH_3$ (* *)	[^{3}H]	50
(±) Clonidine	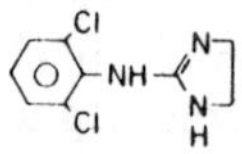 Cl, NH, N, Cl, N, H	[^{3}H]	1-6
(±) WB 4101	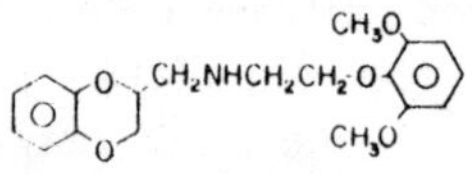 O, O, $CH_2NHCH_2CH_2{-}O$, CH_3O, CH_3O	[^{3}H]	7

[a]Taken from references 169, 174, 199–201.
* Site of radioactive labeling.

subclasses of β receptors: β_1, present in adipose and cardiac tissue in which the agonist potency of isoproterenol is about 4-fold greater than that of norepinephrine and epinephrine, and β_2, present in bronchial, vascular, and uterine smooth muscle in which isoproterenol is about 100-fold more potent than norepinephrine, which in turn is less potent than epinephrine.

Although the primary biochemical event that is induced by α-adrenergic agonists remains ill-defined, the interaction of β-adrenergic agonists and receptors is thought to activate the enzyme adenylyl (adenyl, adenylate) cyclase (EC 4.6.1.1) and thereby to increase cellular levels of adenosine 3′,5′-monophosphate (cyclic AMP, cAMP) (165, 166) (Figure 3). (Sutherland's initial discovery of cAMP had actually originated from his studies on the mechanism of epinephrine action (167).) The cAMP generated by the cyclase is then thought to activate the enzyme cAMP-dependent protein kinase (EC 2.7.a.37). Activation of protein kinase stimulates phosphorylation of protein substrates, and these phosphorylated substrates ultimately produce the observed biological effect of the hormone (168). Epinephrine produces increased levels of cAMP via interaction with β-adrenergic receptors in at least 25 different tissues (169).

Although many of those β-adrenergic receptors were identified and examined by activation of adenylyl cyclase, direct analysis of receptor-amine interactions required studies of hormone binding. Using [^{3}H]propranolol, Levitzki, Steer, and co-workers (170–172) studied binding to turkey erythrocyte ghosts. They showed that such binding closely paralleled effects on β-adrenergic-stimulable adenylyl cyclase. At approximately the

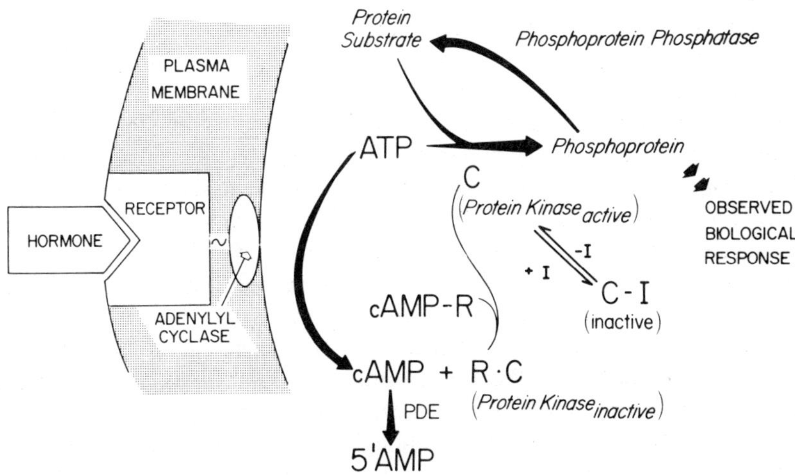

Figure 3. The membrane-active hormone response system. Hormone interacts with receptor initially, activating the catalytic adenylyl cyclase component via a coupling or transducing element (~). The cAMP generated from ATP is either degraded by phosphodiesterase (PDE) to 5′-AMP or activates protein kinase(s) by binding to the R (regulatory) subunit of the enzyme, thus freeing the C (catalytic) subunit. The active C subunits phosphorylate protein substrates (with ATP as substrate) or are bound and inactivated by a protein kinase inhibitor (396). The phosphorylated protein substrate produces the observed biological response, although this may require many intermediate steps. The phosphoprotein is dephosphorylated (inactivated) by phosphoprotein phosphatase.

same time, Aurbach and co-workers (173–175) used [^{125}I]iodohydroxybenzylpindolol ([^{125}I]HYP) and Lefkowitz and co-workers (176–178) used [^{3}H]dihydroalprenolol to study β-receptors on turkey and frog erythrocytes, respectively. Subsequently, those latter two ligands have been used to identify β-adrenergic receptors on cultured cells (179–182), human lymphocytes (183), canine heart (184), rabbit uterus (185), and rat heart (186), brain (187, 188), adipocytes (189), liver (190), erythrocytes (191), and pineal gland (192).

Of the various ligands available (Table 2), [^{3}H]propranolol shows considerable (more than 40% total counts) nonspecific binding, i.e., binding in the presence of high concentrations of unlabeled propranolol or isoproterenol (170–172), as does [^{3}H]dihydroalprenolol under certain conditions (184). This may be partially attributable to the high concentration of ligand required for binding experiments, as well as to the inherent solubility of the hydrophobic antagonists in the lipid matrix of the membrane (193). The high specific activity of [^{125}I]HYP (2,200 Ci/mmol) and its high affinity (K_d values of 10^{-11}–5×10^{-10} M) in most (174, 179, 180) systems permit the use of less radioactivity and consequently lower nonspecific binding. On the other hand, because of the relatively short half-life of ^{125}I (60 days), [^{125}I]HYP must be prepared frequently and the purified compound must be protected from radiolysis.

Binding equilibrium is achieved fairly rapidly in all tissues examined thus far (3–60 min at 37 °C), and the measured binding (at least for [^{125}I] HYP) does not seem to involve receptor or ligand degradation (174, 179, 186). Binding of all three of the ligands is reversible, with half-times of reversal ranging from less than 30 s (177) to about 2 hr (180). Although receptor affinities vary depending on the ligand, estimates of the density of binding sites with the use of different ligands in the same tissue are quite close (170, 174, 187, 188). Estimates of the density of receptor sites per cell range from 400 to 1,800 for erythrocytes (170, 174, 177) and from 300 to 4,000 for cultured cells (179, 180), corresponding to 0.05–1.0 pmol of receptor per mg of tissue protein in preparations of widely varying purity.

Correlations between the dissociation constants of inhibition of radioligand binding and the effects (activation and inhibition) of β-adrenergic agonists and antagonists on adenylyl cyclase have been quite close generally (171, 175, 177–180, 183, 192), although this has not always been seen (especially in experiments with β-adrenergic agonists (175, 186, 188, 193, 194–197)). This latter observation may relate to the coupling mechanism between receptors and cyclase involving guanyl nucleotides (175, 195–197), which is discussed below.

Progress on the characterization of the α-adrenergic receptor has been inhibited both by the lack of suitable ligands to study binding and by the absence of a recognized second messenger. Attempts to use the haloalkylamines (e.g., phenoxybenzamine), classical α-adrenergic blocking agents, as radioligands were unsuccessful (198). However, preliminary reports on four ligands that may be useful markers for the α-adrenergic receptors (199–201) have recently appeared. [^{3}H]Dihydroergocryptine, a potent α-adrenergic antagonist (199), binds to rabbit uterine membranes rapidly, stereoselectively, in a saturable manner (0.13 pmol/mg of protein at saturation), and with the pharmacological specificity of an α-adrenergic receptor. [^{3}H]Dihydroazapetine (200), another α-adrenergic blocking agent, was studied in the rat vas deferens and, although similar features were noted (binding capacity, however, was 40 pmol/mg protein), some α-adrenergic agonists actually produced an increase in binding rather than the expected competition with the ligand. In a brief report, Greenberg et al. (201) compared a radioactively labeled α-adrenergic agonist ([^{3}H]clonidine) with an antagonist ([^{3}H]WB 4101) in binding studies with rat brain particulates. α-Adrenergic agonists seemed to have higher affinities for [^{3}H]clonidine sites than for [^{3}H]WB 4101 sites, and the reverse was true for α-antagonists. All of these preliminary reports require further documentation.

MECHANISMS OF COUPLING OF HORMONE RECEPTORS TO EFFECTOR RESPONSES

A fundamental premise underlying studies of hormone-receptor binding is that such studies will increase our understanding of how cells are activated by membrane-active hormones. As discussed above, the studies of Sutherland and co-workers (165–168) were seminal in formulating the notion of a second messenger, i.e., of cAMP as the mediator by which target cells initially respond to those hormones. Numerous factors have been suggested as coupling or transducing agents between hormone-receptor binding and activation of adenylyl cyclase. In this section, some of these are briefly reviewed. An extensive review of this subject has recently appeared (202).

Insulin Receptors

Much has been learned regarding insulin binding to its receptor, but the relationship between insulin-receptor binding and subsequent metabolic changes remains poorly understood. Although some evidence has been marshaled in favor of the inhibition by insulin of adenylyl cyclase (203, 204), activation of phosphodiesterase (205–209), and stimulation of cyclic GMP formation (210), none of those mechanisms has been universally accepted (211). Insulin, therefore, would seem to be a hormone still in search of a second messenger.

In contrast, the response of target tissues to most other polypeptide hormones and to β-adrenergic amines seems to be mediated by activation of adenylyl cyclase (168) (Figure 3). Guanyl nucleotides, membrane lipids, divalent cations, and phosphorylation have all been implicated as regulators of cyclase activation by the hormone-receptor complex.

Guanyl Nucleotides

Rodbell and his co-workers discovered the key role of guanyl nucleotides in the stimulation of hepatic adenylyl cyclase activity by glucagon (212, 213) and later extended these observations to the stimulation of cyclase in a wide variety of eukaryotic cells (214). Although the effect of GTP was emphasized in early work, subsequent experiments by Rodbell's group and others (194, 196, 214–226) have been performed with synthetic analogs of GTP, such as 5′-guanylylimidodiphosphate (Gpp(NH)p) and 5′-guanylylmethylene diphosphate (Gpp(CH_2)p). Gpp(NH)p and Gpp(CH_2)p, in which the terminal phosphate group cannot be used in transferase and hydrolase reactions, stimulate adenylyl cyclase activity in a manner that appears to be synergistic with the hormonal activation of cyclase (201, 215, 227). Activation by the analogs is time- and temperature-dependent, but once attained it is essentially irreversible. In this irreversibly activated state, the enzyme activity is very stable and refractory to further stimulation by either hormones or sodium fluoride, a nonspecific activator of cyclase (215–227).

Reported findings on the effects of guanyl nucleotides on hormone binding are somewhat inconsistent. It was noted initially (213) that guanyl nucleotides increased the rate and degree of dissociation of [^{125}I]glucagon, decreased uptake of glucagon, and decreased the affinity of binding sites for glucagon. Recent studies with catecholamines indicate that although binding of the labeled antagonists used for binding studies is unaltered by Gpp(NH)p (171, 174, 177, 216), the ability of agonists to compete with radioligand may be decreased (195–197); i.e., the agonists have a decreased apparent affinity for the receptors. In contrast, the potency of agonists for stimulation of the cyclase activity is often (195–197, 216, 221, 226), but not always (213, 227), increased. Part of the variability among different studies may reflect the time dependence of activation by guanyl nucleotides (217, 222) and the irreversible activation by Gpp(NH)p.

The apparently paradoxical result, on the one hand, of decreased receptor affinity and, on the other hand, increased potency of the hormones in stimulating cyclase activity in the presence of guanyl nucleotides, the time- and temperature-dependent lags in guanyl nucleotide activation, and the recent finding (216) that the hormone and ATP may actually convert high activity cyclase to a lower activity form are difficult to resolve by any simple model of allosteric regulation. Many investigators (175, 195, 196, 202, 213, 216, 217, 220) have proposed models of the system, most of which have in common the notion that the cyclase-hormone-receptor-guanyl nucleotide complex can exist in more than one active state. Hammes and Rodbell (228) have recently attempted to simplify some of the more complicated models by suggesting that a two-state system exists in which only one state is active and able to bind activating ligands. Clearly, the absence of more detailed structural information about each of the putative components of the system makes detailed model building a difficult exercise. The work of Neer (229), who has reported that solubilized renal medullary cyclase can exist in two forms that are resolved by gel filtration and activated at differing rates by Gpp(NH)p, may offer new insights into mechanisms of regulation of cyclase by nucleotides.

Pfeuffer and Helmreich (218) have reported the isolation of a guanyl nucleotide binding protein from pigeon erythrocyte membranes. This protein fraction, containing 90% of the [^{14}C]Gpp(NH)p originally bound to the membranes, was purified 40–80-fold and was estimated to have a molecular weight of 230,000 by gel electrophoresis. This observation requires further substantiation; the significance of radioactive guanyl nucleotide bindings (218, 225, 230) has been questioned by Jacobs and Cuatracasas (231), who have suggested that no more than 2% of such binding might be related to adenyl cyclase.

Membrane Lipids

As discussed under "Plasma Membrane Structure and Function," lipids constitute about 40% of plasma membranes by weight, seem to be heterogeneously dispersed in the membrane, and are actively mobile. It has been suggested that hormone and drug receptors may be intrinsic membrane proteins, primarily because detergents usually (155–157, 232–235), but not always (158), are required for their extraction. As intrinsic membrane proteins, receptors would presumably be closely associated with membrane lipids. (In *Escherichia coli* membranes, it has been estimated (236) that each intrinsic membrane protein is surrounded by about 600 lipid molecules, of which 130 are closely linked to the protein molecule.) In addition, perturbation of the pigeon erythrocyte membrane cholesterol with polyene antibiotics (e.g., filipin, nystatin, amphotericin), which perhaps alter the hydrophobic core of the lipid matrix, produces a loss of catecholamine-stimulable, but unaltered fluoride-stimulable cyclase and unchanged [^{3}H]isoproterenol binding activities (237). Similar findings have been reported with frog erythrocytes (238) in which filipin treatment selectively lowers catecholamine-stimulable cyclase activity without a change in [^{3}H]dihydroalprenolol binding.

A similar approach has been the use of phospholipases (85, 238–242). Treatment with phospholipases inhibits catecholamine binding to frog erythrocytes (238), [^{125}I]glucagon binding to liver membranes, and glucagon-stimulated, but not sodium fluoride-stimulated, cyclase activity (239). Further studies (243) in which liver membranes are treated with phospholipases that specifically hydrolyze subclasses of phospholipids suggest that acidic phospholipids (phosphatidylserine, phosphatidylinositol) are important for activation of cyclase of glucagon (perhaps by liganding of the NH_2-terminal histidine to a regulatory site) and for the increased rate of dissociation of glucagon from its receptor produced by guanyl nucleotides.

Two recent studies emphasize the role of lipids in hormone-responsive adenylyl cyclase. Orly and Schramm (244) have shown that incubation of turkey erythrocyte membranes with fatty acids alters the behavior with respect to temperature of the catecholamine-stimulated cyclase, perhaps by modifying the GTP binding site. Shattil and co-workers (245, 246) have used human platelets incubated with liposomes containing increased cholesterol-phospholipid ratios and have shown that cholesterol uptake by the platelets produces 1) elevation of basal cyclase activity and loss of hormone-stimulable enzyme activity, and 2) decreased membrane fluidity and increased order within the hydrophobic core of platelet membranes.

This latter finding may be related to the role of membrane mobility in the regulation of cellular function. Although lateral mobility in the plane of the membrane has been studied extensively for lectins (64–66), little information is available on hormone-receptor mobility. Craig and Cuatrecasas (67) studied the binding of fluorescein-labeled cholera toxin to rat mesenteric lymph nodes and demonstrated both patch and cap formation that preceded production of cAMP. This process could be inhibited by anticholera toxin IgG, by inhibitors of the assembly of microtubules and microfilaments (e.g., cytochalasin B, colchicine, vinblastine), by metabolic poisons (NaN_3, NaF, dinitrophenol), and by concanavalin A, which inhibits motility of surface antigens. However, these inhibitory effects on mobility were not consistently reflected by changes in toxin-stimulated cyclase activity. In addition, a biologically inactive competitive antagonist of toxin action produced redistribution of receptors but no cyclase stimulation. It has been demonstrated that membrane ganglioside GM_1 is the cholera toxin receptor (84, 247, 248) (other gangliosides may be receptors for thyroid-stimulating hormone (TSH), luteinizing hormone, (LH), and human chorionic gonadotropin (hCG) (249)), and the role that ganglioside redistribution plays in cyclase activation is somewhat mysterious at this time. Helmreich (202) has reported that microtubules may be involved in guanyl nucleotide- and catecholamine-stimulable cyclase activity, and others have shown that adenylyl cyclase is inhibited by lectins such as concanavalin A (250–252). However, the lack of specificity of lectin (253, 254) treatment makes it impossible to draw any firm conclusions from such results.

Levey et al. (233–235) have reported solubilization of cat heart hormone-sensitive adenylyl cyclase activity with Lubrol PX and restoration of the lost hormonal responses by the addition of specific purified phospholipids, i.e., phosphatidylinositol for catecholamine and phosphatidylserine for glucagon and histamine-stimulable cyclase. Cat heart is the only system in which specific lipids have been used to reconstitute the hormone-responsive system.

Divalent Cations

Although magnesium or other alkaline earth metals (Mn^{2+}, Co^{2+}) are well known co-factors for adenylyl cyclase, the role of calcium as a coupling agent between receptors and cyclase is a subject of much current interest. Because Berridge has recently reviewed cellular regulation by calcium in detail (255, 256), this topic is not considered extensively here.

Rasmussen (257) was the first to stress that the interaction between calcium and cAMP was important to understanding the action of polypeptide hormones. Although some systems require low (submicromolar) concentrations of calcium for adenylyl cyclase activity, at higher concentrations (0.1–5 mM) Ca^{2+} (and also Cu^{2+} and Zn^{2+}) inhibits enzyme activity (basal, NaF, or hormone-stimulable) in many tissues (258, 259). In turkey erythrocytes, this inhibition produces a decrease in V_{max} of the enzyme without affecting the affinity for the hormone or for ATP (171, 260, 261), and at a site distinct from the Mg^{2+} binding site. This last result may not be true for cardiac tissue (262). Recent work with intact and homogenized human platelets (263) has led to the suggestion that Ca^{2+} concurrently inhibits adenylyl cyclase and stimulates guanylyl cyclase.

In early studies (264), chelators were used to assess the role of divalent cations on hormone receptors and adenylate cyclase. In such studies concentrations of the chelator EGTA (ethyleneglycol-bis-(β-aminoethyl ether) N,N'-tetra-acetic acid) that had no effect on binding markedly inhibited ACTH-stimulable cyclase activity. This indicated that calcium, although not required in ACTH-receptor interaction, seemed to be involved in subsequent activation of the cyclase. More recent studies (164, 171, 177, 179, 215) of catecholamine binding have also emphasized the relative lack of effect of either Ca^{2+} itself or of Ca^{2+} chelators on ligand binding. This discrepancy between effects on binding and effects on cyclase has suggested that calcium might function as a coupler between receptor and cyclase. However, most of the evidence is indirect and circumstantial. Because EGTA may have effects on cyclase independent of its chelating action (265) and because calcium is bound in intracellular stores, a meaningful delineation of the function of calcium as a regulator of hormone response requires further study. In this regard, Levitzki (193) has shown that a solubilized, partially purified adenylyl cyclase preparation retains its calcium sensitivity in a manner resembling the membrane-bound enzyme. Studies in which solubilized preparations are employed should help clarify how calcium regulates cyclase.

Phosphorylation-Dephosphorylation

Constantopoulos and Najjar (266, 267) have postulated that adenylyl cyclase might be regulated through a series of reactions involving phosphorylation (deactivation) and dephosphorylation (activation). This theory is based on a similar model that had been proposed for the regulation of phosphoglucomutase. It was suggested that hormones or NaF would promote nucleophilic displacement of the phosphoryl group of a presumed phospho-(inactivated) adenylate cyclase and that protein kinase would produce phosphorylation and inactivation of the enzyme (266, 267). To date, no definitive evidence is available to substantiate or to refute this hypothesis.

Fluid Mosaic Model and Models of Cyclase Regulation by Hormones

The fluid mosaic model (Figure 1) (31, 34, 49, 69) encompasses much of what is currently accepted regarding membrane structure. Some investigators (81, 84, 121, 160, 202, 259, 268–271) have attempted to incorporate the ideas regarding membrane fluidity into models that explain the activation of cyclase by hormones. Cuatrecasas and co-workers (81, 84, 160) have suggested the term "mobile receptor hypothesis" for their version of such a model. In this model, hormone receptors and cyclase are discrete and separate plasma membrane structures located on the external surface (i.e., exposed to the extracellular environment) and on the inner surface (i.e., facing the cytoplasm), respectively. Occupancy of receptors by hormone is thought to alter their specificity and affinity for complex formation with cyclase. An important feature of this model is the obligatory requirement for hormone receptors and cyclase to diffuse in the plane of the membrane. The notion that hormone occupancy of receptors alters the ability of receptor and cyclase to interact is quite analogous to the changes in steroid receptors observed upon binding of steroids (5, 56). A similar model, "the non-stoichiometric floating receptor model for hormone sensitive adenylyl cyclase," has been suggested recently by De Haën (271). In addition, he proposes: 1) two conformational states (one active and one inactive) for hormone receptors and adenylyl cyclase; 2) that only receptors are mobile, and 3) that, although no fixed stoichiometric relationship exists between the number of receptor units and cyclase units, they can only form 1:1 complexes with each other. These two strikingly similar hypotheses offer useful frameworks within which to devise experiments that may begin to explain not only the mechanism of cyclase activation by hormone receptor complexes, but also the role of guanyl nucleotides, divalent cations, and lipids in the regulation of hormonal response.

The conclusion that receptors and cyclase are independent entities is based largely on indirect evidence: perturbations in one function do not perturb the other. Although definitive proof of the separate nature of receptors and cyclase requires purification of the individual components and reconstitution of the intact system, this author's laboratory (112, 180, 272, 273) has addressed the question by using a genetic approach with cultured S49 lymphoma cells. When isoproterenol is used as a selecting agent in a single-step selection procedure, the resistant clones that are isolated lack not only β-adrenergic amine-stimulable cyclase activity but also enzymic activity responsive to other effectors (prostaglandin E, cholera toxin, NaF, or guanyl nucleotides). This suggests (272, 273) that one type of cyclase molecule responds to different effectors. In addition, the fact that this variant cell possesses β-adrenergic receptors that are similar to those of wild type cells, indicates that β-adrenergic receptors and adenylyl cyclase are the products of separate genes (180). These results offer additional evidence for the independence of catecholamine receptors and the catalytic component of adenylyl cyclase. In addition, the occurrence of normal cyclase activity in a mutant clone lacking cAMP-dependent protein kinase argues against a cAMP-dependent protein kinase-mediated phosphorylation-dephosphorylation model of cyclase activity (266, 267).

REGULATION OF HORMONE RECEPTORS

Spare Receptors

Clark's original theory (274) proposing that the response to a drug or hormone is directly related to the occupancy of receptor sites in a target tissue (the occupancy theory) presumed that a maximum response is obtained only when all receptors are occupied. This theory predicts that dose-response curves of hormone (or drug) concentration versus amount bound or versus biological response should be identical. Because this was not the case under certain circumstances, the occupancy theory was further refined by Ariëns and de Groot (275), who introduced the concept of partial agonists, i.e., agents that are able to bind but unable to maximally activate a particular response system. Similarly, Stephenson (276) hypothesized that drugs might produce a maximum response with only partial occupancy of the total pool of available receptors, predicting a nonlinear relationship between receptor occupancy and response. Such a relationship has been observed for many hormones, including vasopressin (277), hCG (278, 279), epidermal growth factor (81, 84), ACTH (136, 280), TSH (281), glucagon (234, 282), and insulin under certain conditions (81, 84, 122, 283, 284). This observation, leading to the concept of spare receptors, is not a universal finding and is not seen with certain other hormone receptors, including uterine receptors for oxytocin (285) and β-adrenergic receptors in cultured cells (179).

It is important to distinguish between two types of observations in the literature, both of which have been interpreted in terms of spare receptors. In one, receptors seem to be in excess of functional catalytic adenylyl cyclase moieties (e.g., solubilized preparations of cat myocardial adenylyl cyclase (234)), whereas in other systems (e.g., in the responses to the tropic hormones hCG (278, 279), ACTH (280), and TSH (281)), the dose-response curves for certain hormone-stimulated events (e.g., steroidogenesis in response to ACTH or hCG) seem to be shifted to the left of curves of radioactive hormone binding or of hormone-stimulated cAMP synthesis. These latter experiments have been interpreted as showing that only a small fraction of receptor sites have to be occupied in order to produce the observed biological response. Such conclusions ignore the role that other components of the cAMP system (Figure 3) may play in determining the magnitude of target tissue responses to hormones. Specifically, if factors such as activities of cAMP-dependent protein kinase (286) and of phosphodiesterase are not considered, it seems unrealistic to compare results of hormone binding of the plasma membrane to cellular responses (e.g., steroidogenesis) that require several intervening steps.

Because breaking cells and preparing plasma membranes may alter recovery of receptors and cyclase to a different extent (194, 234), and probably because cyclase activity tends to be more sensitive to cellular disruption than does hormone binding activity, analyses that attempt to relate occupancy of receptors to adenylyl cyclase activation in membrane preparations may be difficult to interpret. It is likely that in such membrane preparations cyclase activity is less than 10% of the activity observed in unbroken cells (194). Because few data exist that relate the kinetics of binding of biologically active hormones to stimulation of cAMP synthesis in intact cells (in which one can ignore or correct for the activity of phosphodiesterase), it is not yet possible to draw any conclusions regarding the role of spare receptors in hormone binding and cyclase activation.

Negative Cooperativity

De Meyts and co-workers (271, 287–289) were the first to suggest that the curvilinear pattern of insulin binding observed on Scatchard analysis (Figure 2) might not reflect the presence of multiple "classes" of binding sites. Instead they believe that progressive saturation of insulin receptors by insulin reduces their affinity for the hormone. By calling this phenomenon negative cooperativity, they have introduced into the lexicon on membrane-bound receptors a term previously used to describe negative homotropic interactions of soluble proteins (290). The principal experimental data that are presumed to reflect negative cooperativity for hormone binding are studies of the dissociation of bound radioactively labeled hormone either by large dilution in buffer alone or in buffer to which unlabeled hormone is added. The presence of the unlabeled ligand seems to accelerate dissociation of the bound hormone. Initially demonstrated for insulin (271), this phenomenon has now been observed with numerous other hormones—nerve growth factor (291), epidermal growth factor (292), TSH (293, 294), and β-adrenergic antagonist binding by frog erythrocytes (295–297) (but not with the β-adrenergic receptors of turkey erythrocytes (174, 297)). Attempts have been made to consider this a true site-site interaction and to eliminate some possible explanations, such as rebinding of released ligand in the bulk solution or from unstirred layers, polymerization of bound ligand (debated by Cuatrecasas and Hollenberg (292)), and effects of nonspecific binding (271, 289, 297). Studies with insulin receptors (271) and β-adrenergic receptors (297) indicate that less than 10% receptor occupancy is required to observe this response. Concanavalin A, which does not itself bind to the insulin receptor, inhibits the accelerated dissociation from insulin receptors (271), but inhibitors of microfilament formation (e.g., cytochalasin), which produce decreases in receptor number, do not (298). Gpp(NH)p has no effect on the cooperative response of β-adrenergic receptors (297). Most experiments that demonstrate negative cooperativity have been performed at relatively low temperatures (25°C), and the contribution of such cooperative interactions to physiological responses is not yet defined. De Meyts (299) has recently reported that a more detailed quantitative analysis of insulin binding data suggests a model to explain cooperativity; in this model, the receptor has four interacting subunits that depolymerize upon hormone binding. Other investigations, however, have found recently that negative cooperativity is not a significant factor in insulin binding to solubilized insulin receptors (300). Similarly, a kinetic formulation of the mobile receptor hypothesis (301) that assumes greater affinity between cyclase (or its counterpart for hormones such as insulin) and receptor upon occupancy of the receptor by hormone predicts that a single homogeneous receptor may seem to have two classes of high and low affinity binding sites, cooperativity, and even spare receptors. The resolution between the opposing views regarding cooperativity must await further studies in both solubilized and membrane systems.

Self-regulation of Hormone Receptor Concentration

A well-known characteristic of some membrane-active hormones is the acute spike response observed in target tissues (Figure 4), in which the rapid onset of action after hormone treatment is often followed shortly thereafter by a decrease in response. In addition, chronically elevated levels of a hormone seem to be capable of reducing tissue response to the hormone. These phenomena have been termed refractoriness, desensitization, down regulation, tolerance, or tachyphylaxis.

Regulation of the concentration of insulin receptors has been actively investigated in recent years, and numerous animal and human cell types have been studied (see review by Roth et al. (108)). For example, the decreased number of insulin receptors detected on tissues in obese hyperglycemic mice, which have marked hyperinsulinemia and insulin resistance, is considered an animal analog of the decreased number of insulin receptors observed in human obesity. Recently, Chang et al. (302) have suggested that the decreased insulin binding in the obese hyperglycemic mice may result from a generalized alteration in membrane glycoproteins; this suggestion, however, has been debated (122). Similar findings, i.e., hormone-specific decreases in receptor number or function, have been observed in studies of intact animals and of other hormones such as calcitonin (303), β-adrenergic amines (304–306), gonadotropins (307), and glucagon (308). In contrast, animals with decreased levels of circulating insulin (309) or β-adrenergic amines (306) have increased concentrations of receptors on target tissues.

Refractoriness has been observed with many hormones, including

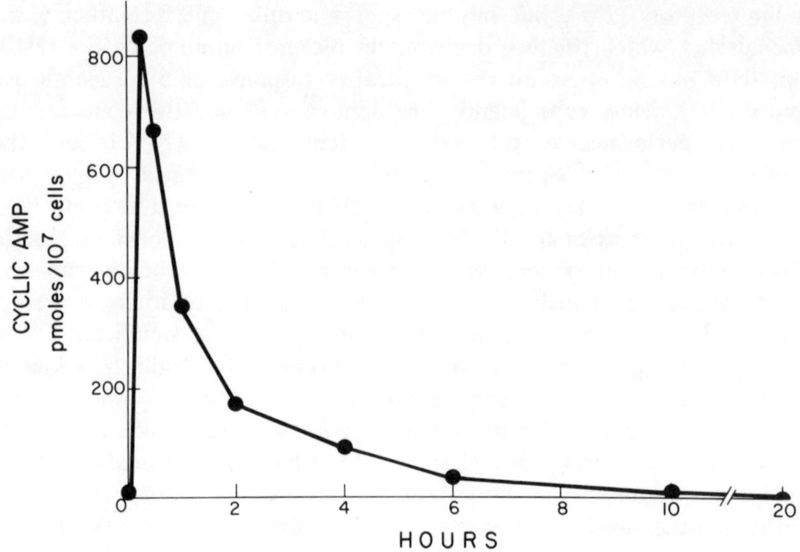

Figure 4. A typical time course of the response to isoproterenol during prolonged incubation of S49 lymphoma cells to isoproterenol (10^{-6} M). Note the early peak followed by the decline in cellular cAMP levels. Adapted from Shear et al. (325).

insulin (310), growth hormone (311), TSH (312), glucagon (308), TSH-releasing hormone (313), ACTH (314), β-adrenergic catecholamines (111, 315–327), prostaglandins (111, 319, 323, 325, 326), acetylcholine (328), melanocyte-stimulating hormone (MSH) (329) in experiments in which isolated organ perfusions, tissue slices, and isolated and cultured cells were used. Although refractoriness is generally agonist-specific, i.e., it occurs only for the initial agent to which a particular cell was exposed, this is not always the case (111, 307, 330). When cells that are refractory are placed in medium free of the hormone, loss of refractoriness (resensitization) occurs (305, 310, 311, 313, 319, 321, 322, 328, 330).

Many mechanisms have been suggested to explain cellular refractoriness to hormones. These include the production of nonspecific antagonists of adenylyl cyclase activity (317, 331), cellular extrusion of cAMP (323, 332), increased degradation of cAMP by the enzyme phosphodiesterase (which may be induced as cAMP levels increase (111, 321, 324, 326, 333–335), and decreased numbers of hormone receptors (304–306, 310, 311, 313, 319, 325, 327, 328). If a loss of hormone receptors is observed, it is accompanied by a decrease in hormone-stimulable adenylyl cyclase activity (for hormones whose second messenger is cAMP (111, 304–308, 312, 314, 315, 318, 327, 329, 330)). By regulating receptor number after exposure to a particular hormone, cells that respond to several hormones are able to selectively modulate their response to the initial agent while maintaining responsiveness to the other hormones.

In cells that show a loss of hormone receptors after hormonal exposure, the alteration seems to be a decrease in the number of receptors; the remaining receptors have unaltered affinity (304–306, 308, 310, 313, 322, 325, 327). Two types of experiments suggest that receptor occupancy is required for the loss in receptors to occur: 1) the dose-response relationship for hormonal occupancy of receptor compared with the hormone-induced loss in receptor number, and 2) experiments in which hormone analogs or agonists of differing potency were used (121, 305, 310, 311, 318, 319, 323, 330, 336, 337). However, occupancy alone is not sufficient to produce refractoriness. This has been demonstrated unequivocally with β-adrenergic catecholamines; β-adrenergic antagonists that occupy the receptor but do not produce cyclase stimulation fail to produce refractoriness themselves and block refractoriness produced by β-adrenergic agonists (305, 319, 330, 336, 338).

The author's laboratory (325) has studied refractoriness to β-adrenergic catecholamines in the S49 mouse lymphoma cell, a tissue culture cell line in which mutant clones have been selected with specific lesions in the pathway of cAMP generation and function. With these cells, it has been demonstrated that refractoriness is agonist-specific and associated with a decrease in amount, but not affinity, of β-adrenergic-stimulable adenylyl cyclase activity and β-adrenergic receptor binding activity. Because similar responses were found in a clone lacking cAMP-dependent protein kinase, it seems that refractoriness and the decreases in receptor number and cyclase activity are not mediated by protein kinase. In addition, when a clone that had no effector-stimulable adenylyl cyclase activity but still had β-adrenergic receptors was used, initial incubation with isoproterenol did not produce a subsequent decrease in receptor number. Besides indicating that receptor occupancy alone is insufficient to produce refractoriness, this finding suggests that adenylyl cyclase may itself regulate β-adrenergic receptors.

It has been suggested that self-regulation of membrane receptors may be an especially "important homeostatic regulatory mechanism in cell communication" (339) and that the mechanisms that produce modulation of antigenic receptors (see under "Plasma Membrane Structure and Function") may be similar to those that regulate hormone receptors (64, 339). Possible mechanisms would include decreased receptor synthesis, increased degradation, clustering (patching? capping?) with decrease in the ability to bind the hormone, release from the cell, translocation into an intracellular site, or irreversibility of the binding of the hormone. Some of these hypotheses have been tested and no one mechanism explains all the results.

The rate of loss of receptors is usually slower than the onset of hormonal response, and generally the recovery from refractoriness is even slower than the rate of loss (311, 313, 330, 336). Some investigators (313, 321, 340) have found, although others (305, 311, 330, 337) have not, that the decreases in receptor number are blocked by treatment with cycloheximide. Discrepant results are also observed when inhibitors of protein synthesis are used to block the recovery phase (305, 311, 330, 336, 337). For insulin and growth hormone receptors, the loss in receptors seems to result from increased rates of degradation of receptors (122, 310, 311, 340). It has been suggested (341) that the ability of insulin to affect insulin receptors results from a catalytic activity that is associated with the insulin molecule itself (perhaps on the B chain) and that is independent of the ability of insulin to occupy physiological receptors. Although this may be true for insulin, it seems unlikely that such a mechanism would explain the decreases in receptors observed with so many other hormones.

Two groups (342, 343), both working with broken cell preparations, have recently demonstrated refractoriness in vitro. Bockaert et al. (342), studying adenylyl cyclase responses to LH in plasma membranes prepared from Graafian follicles, found that desensitization developed as a function of time and that it required magnesium, ATP, and higher concentrations of LH than did activation of cyclase. Because Gpp(NH)p and GTP were not able to cause desensitization, phosphorylation was suggested as a possible mechanism. Using frog erythrocyte membranes, Mukherjee and Lefkowitz (343) reported that refractoriness to β-adrenergic amines developed more rapidly in membranes than in whole cells, required occupancy of the receptors by agonists, and did not occur with solubilized β receptors. With frog membranes, recovery from refractoriness was promoted by purine nucleotides, most effectively by Gpp(NH)p, suggesting that nucleotides are not only involved in coupling in cyclase and receptors (see under "Guanyl Nucleotides"), but also, perhaps, in the modulation of receptor number.

The possibility that microtubules or microfilaments (298, 340) or that changes in membrane fluidity leading to aggregation of receptors (64) regulate refractoriness remains largely unexplored. It is also too early to decide if results obtained with erythrocytes will be applicable to actively mitotic cells. The possibility that guanyl nucleotides are involved both in coupling hormone receptors to cyclase and in the regulation of the subsequent loss of hormone receptors (343) emphasizes the importance of learning more about the structural and functional relationships between hormone receptor, nucleotide binding, and catalytic cyclase components. For β-adrenergic receptors, this goal may be facilitated by affinity labels which have become available recently (344–355) and by fluorescent probes (346–347) for the receptor, as well as by solubilized preparations of receptors (348). The use of immunofluorescence (349) for detection of peptide hormones in isolated target cells may yield similar information for polypeptides.

Other Mechanisms

Although decreases in the binding affinity and binding capacity of receptors have been studied in great detail, less complete evidence suggests that hormones may be able to alter their receptors in the opposite direction.

Indirect studies indicate that the number of prolactin receptors may be induced by prolactin itself (350) and may also be influenced by other hormones, in particular estrogen (see Posner (351)). Other examples of receptor regulation by hormones that act on separate receptors include the increased numbers of ovarian LH receptors observed after animals are injected with follicle-stimulating hormone (FSH) (352), the increased number and affinity of uterine oxytocin receptors after animals are treated with estrogen (353), and the changes in membrane receptors for β-adrenergic amines (190) and vasopressin (354) observed after adrenalectomy. Douglas and Catt (355) have reported that changes in sodium or potassium balance as well as changes in angiotensin II levels can produce changes in the number and affinity of angiotensin II receptors of rat adrenal cells. In the pineal gland, the light-dark cycle with its concomitant altered levels of sympathetic stimulation produces alternate states of subsensitivity and supersensitivity which are reflected by changes both in β-adrenergic receptors (306) and in the activity of cAMP-dependent protein kinase (356).

An additional regulatory feature controlling receptors might be the stage of an individual cell in the cell cycle. Mouse melanoma cells in culture have MSH receptors that are clustered (357) on the cell surface, perhaps overlying the Golgi area (358). These receptors are detected only during the G_2 phase of the cell cycle (359). Whether a similar situation exists with respect to other hormone receptors remains to be seen.

REGULATION OF HORMONAL RESPONSE BEYOND HORMONE RECEPTORS

The regulation of cellular response by hormones is largely determined by the presence or absence of specific hormone receptors, the manner in which those receptors are coupled to adenylyl cyclase (or other second messenger systems), and the various factors that may be involved in the regulation of the behavior or number of the receptors. This final section briefly discusses some of the mechanisms which modulate the cellular responses to hormones and which are subsequent to the generation of cAMP. For the reasons discussed previously, cells maintained in tissue culture have been particularly useful for examining mechanisms of cAMP action (see reviews 111 and 114). Studies both in tissue culture cells and other systems have strongly implicated calcium (see under "Mechanisms of Coupling of Hormone Receptors to Effector Responses") as an important regulator of adenylyl cyclase and of other cellular activities. The role of calcium is discussed in great detail in recent reviews (255, 256) and thus is not considered further here. Hormonal regulation of phosphoprotein phosphatase (Figure 3) also is not discussed.

cAMP-dependent Protein Kinases

It has been proposed that the cAMP generated by adenylyl cyclase produces its cellular effects by subsequent activation of cAMP-dependent protein kinases (Figure 3) (360, 361). Protein kinases have been studied most extensively in soluble cytosolic fractions; at the same time, cAMP-dependent membrane phosphorylation and dephosphorylation have been demonstrated in many cells (361, 362). For the regulation of membrane-associated events such as transport (363–367), membrane-bound enzymes (368, 369), or the generation of electrical potentials (370), it might be particularly advantageous for a cell to have membrane-bound protein kinase in close proximity to the site of cAMP generation. The extent to which membrane-bound or nuclear (371, 372) protein kinases rather than soluble kinases mediate hormonal responses remains unknown.

Although the hypothesis (360) that the effects of cAMP are mediated by protein kinases has been clearly established for certain enzymes regulated by phosphorylation-dephosphorylation (see ref. review 362) such as phosphorylase kinase (373), glycogen synthetase (373), and hormone-sensitive lipase (374), the general applicability of the hypothesis to other cellular events regulated by cAMP is not established conclusively. This author and his colleagues have used the S49 mouse lymphoma cell line to examine this hypothesis (112). Intact wild type S49 cells show at least five responses to treatment with cAMP (in its dibutyryl form) or to agents whose effects are mediated by cAMP (β-adrenergic amines, prostaglandins of the E series, cholera toxin): 1) induction of phosphodiesterase (334); 2) cell cycle-specific (G_1) arrest of growth (375); 3) changes in membrane transport (376); 4) decreases in the activities of ornithine decarboxylase and *S*-adenosylmethionine decarboxylase (377); and 5) eventual cytolysis (378–381). By growing large numbers of cells in the presence of dibutyryl-cAMP, our group has isolated mutant clones that are resistant to cytolysis. Each of the resistant clones fell into one of three general classes (381): those with no response to cyclic AMP, those in which cyclic AMP had a decreased potency compared with wild type cells (K_a mutants), and those in which cyclic AMP had a similar potency compared with wild type but in which the maximum responses to the nucleotide were decreased (V_{max} mutants) (Figure 5). When the dose-response pattern of cytosolic cAMP-dependent protein kinase activity was examined, strikingly similar (no response, K_a, V_{max}) relationships between the in vitro kinase assays and the patterns of response observed in the intact mutant and wild type cells were found (Figure 5). These results suggest that effects of cAMP on growth, transport, and enzyme induction are mediated by cAMP-dependent protein kinase. In addition, the findings demonstrate that cellular response to cyclic AMP is not solely determined by the absolute level of cyclic AMP in the cell, because cells with altered protein kinase activity respond to a given level of cyclic AMP differently from wild type cells. Of particular

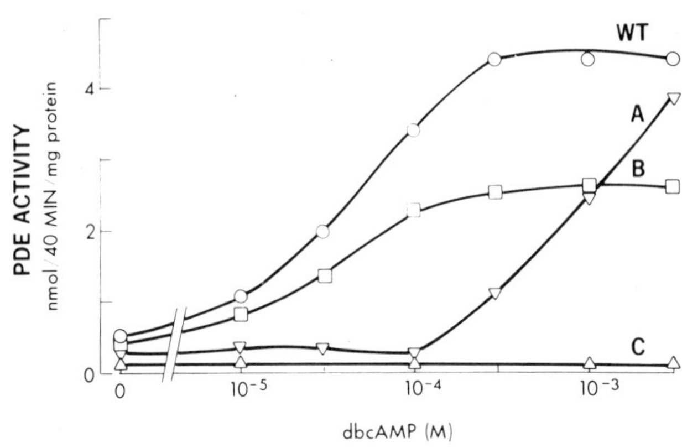

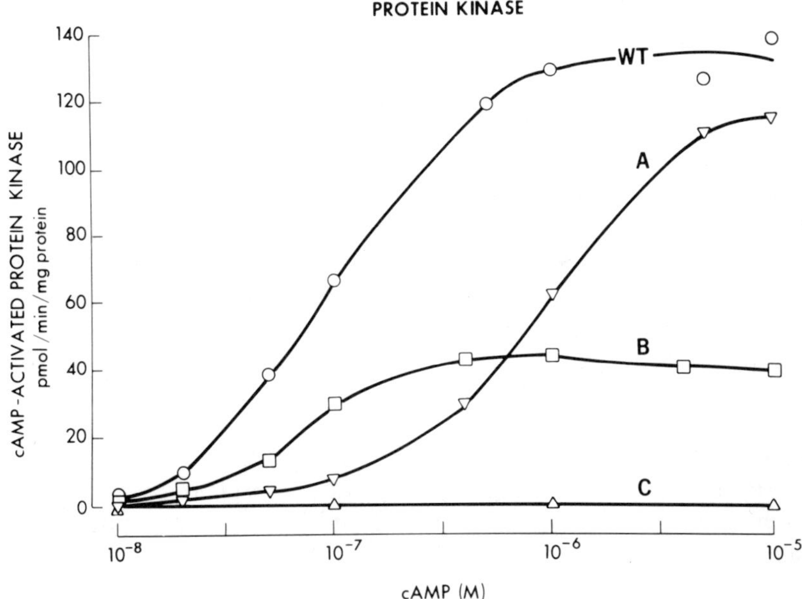

Figure 5. Dose-response patterns for cAMP phosphodiesterase induction (*top*) in intact S49 mouse lymphona cells and of cytosolic cAMP-dependent protein kinase activity (*bottom*) from wild type (o—o, WT) and mutant clones (A, B, and C designate K_a, V_{max}, and kinase-negative clones, respectively). Reproduced from Insel et al. (381) with permission from the American Association for the Advancement of Science.

interest, results with the V_{max} mutants seem to indicate that the absolute amount of cAMP-dependent protein kinase activity in a cell determines the extent to which a cell is capable of responding to elevated levels of cAMP. This latter result may have special relevance to tissues in which higher levels of cAMP are generated than seem to be required for the maximum physiological response (378–380, 382). Perhaps, the level of protein kinase activity in a cell determines the extent to which the cell can respond to a particular hormone.

Recent studies (383–385) of cAMP-dependent protein kinase indicate that this enzyme, which is composed of two dissociable subunits (R or regulatory and C or catalytic), exists as at least two isoenzymes in a variety of tissues. These two isoenzymes, having different R subunits, are distinguishable biochemically (383, 384), immunologically (385), by their tissue distribution (383), and perhaps by their relative proportions at different stages of the cell cycle (386). Phosphorylation and dephosphorylation of one of the two types of R subunits seem to alter the ability of the dissociated R subunit to reassociate with the enzymically active C subunit and to regenerate the inactive holoenzyme (387). The extent to which tissues with differing hormonal responses may be regulated by differences between their kinase subunits remains conjectural at this time, but in S49 cells, mutational alteration in the R subunits that affects both dissociation of holoenzyme and affinity for cyclic AMP also produces changes in hormonal response of the intact S49 cell (388–390).

Compartmentalization of cAMP

Whether or not intracellular cyclic AMP exists in separate subcellular compartments is unknown at present. Such compartmentalization might explain certain superficially anomalous observations such as experiments in which total cellular levels of cAMP correlate poorly with either biological responses or with estimates of the affinity constants of cAMP-dependent protein kinase (391–394), and immunofluorescent studies in which different stimulators of adenylyl cyclase activity increase cAMP in different subcellular regions (395). However, Beavo et al. (286) have suggested that it is unnecessary to invoke cellular compartmentalization to explain discrepancies between cellular cAMP levels ($\sim 10^{-6}$ M) and independent estimates of affinity constants of kinase of cAMP ($K_a \sim 10^{-8}$ M). If the tissue concentrations of protein kinase (roughly equal to cellular cAMP levels in skeletal muscle (~ 0.2 μM)) and of the heat stable inhibitor (396) of protein kinase are considered, the apparent affinity of purified kinase for cAMP decreases to a value comparable to tissue concentrations of the nucleotide. Similarly, Swillens et al. (397) have shown that the inappropriate manner in which kinetic data on cAMP-dependent protein kinase have been analyzed may explain the apparent discrepancy between tissue levels of cAMP and levels of cAMP required to activate the kinase. The extent to which hormones are able to regulate tissue levels of protein kinase or levels of the protein kinase inhibitor is unknown.

Phosphorylated Substrates

The nature of the substrates phosphorylated by protein kinase is extensively discussed in a recent review (361) and is mentioned only briefly. Olsen (398) has suggested that interaction of cAMP with filamentous proteins (i. e., microtubules, microfilaments, and actin-like proteins) may be the basis for many cAMP-mediated processes. In this context, Sloboda et al. (399) have shown that a cAMP-dependent protein kinase of a microtubule preparation from brain preferentially phosphorylates a microtubule-associated protein. As discussed earlier, microtubules play an important role both in the capping of antigenic receptors and perhaps in the regulation of hormone receptors.

Induction of cAMP phosphodiesterase (321, 324, 333–335) is a well described response of numerous tissues to increased cAMP levels. Because induction of phosphodiesterase can be a major mechanism for a regulation of hormonal responses (324) in some cells, it is likely that changing hormonal response within a differentiating or dedifferentiated tissue might reflect the acquisition or loss of the ability to induce phosphodiesterase. The induction of phosphodiesterase, which is apparently mediated by protein kinase (381), is not observed universally (321, 322).

The recent finding (361, 400) that several types of steroid hormones and cAMP regulate the phosphorylation and dephosphorylation of a protein of molecular weight 54,000 (termed SCARP, for steroid and cyclic AMP-regulated phosphoprotein) raises the intriguing possibility that membrane-active hormones and gene-active hormones may have certain common mechanisms for cellular regulation.

ACKNOWLEDGMENTS

I should like to thank Drs. A. Levitzki, B. Sacktor, M. K. Jain, and S. J. Shattil for permission to cite their work that had not yet been published. In addition, I thank Drs. H. Bourne, P. Coffino, and G. Johnson for their critical reviews of this manuscript, Ms. V. Abe and Mr. J. Ragland for assistance with the typing, and Dr. K. L. Melmon for advice and support.

REFERENCES

1. Tomkins, G. M. (1975). Science 189:760.
2. Liao, S. (1974). *In* H. V. Rickenberg (ed.), MTP International Review of Science, Biochemistry Series I, Vol. 8, Biochemistry of Hormones, pp. 153–186. University Park Press, Baltimore.
3. O'Malley, B. W., and Means, A. R. (1974). *In* H. V. Rickenberg (ed.), MTP International Review of Science, Biochemistry Series I, Vol. 8, Biochemistry of Hormones, pp. 187–210. University Park Press, Baltimore.

4. Wicks, W. D. (1974). *In* H. V. Rickenberg (ed.), MTP International Review of Science, Biochemistry Series I, Vol. 8, Biochemistry of Hormones, pp. 211–242. University Park Press, Baltimore.
5. Jensen, E. Unpublished manuscript.
6. Yamamoto, K. R., and Alberts, B. M. (1976). Annu. Rev. Biochem. 45:721.
7. DeLuca, H. F., and Schnoes, H. K. (1976). Annu. Rev. Biochem. 45:631.
8. Tata, J. R. (1975). Nature (Lond.) 257:18.
9. Oppenheimer, J. H., Schwartz, H. L., Surks, M. I., Koerner, D., and Dillmann, W. H. (1976). Recent Progr. Horm. Res. 32:529.
10. Buller, R. E., and O'Malley, B. W. (1976). Biochem. Pharmacol. 25:1.
11. DeRobertis, E., and Schacht, J. (1974). Neurochemistry of Cholinergic Receptors. Raven Press, New York.
12. Cohen, J. B., and Changeux, J.-P. (1975). Annu. Rev. Pharmacol. 15:83.
13. Snyder, S. H., Chang, K. H., Kuhar, M. J., and Yamamura, H. I. (1975). Fed. Proc. 34:1915.
14. Raftery, M. A., Vandlen, R., Michaelson, D., Bode, J., Moody, T., Chao, Y., Reed, K., Deutsch, J., and Duguid, J. (1974). J. Supramol. Struct. 2:582.
15. Heilbronn, E. (1975). *In* P. G. Waser (ed.), Cholinergic Mechanisms, pp. 343–364. Raven Press, New York.
16. Goldstein, A. (1975). The Opiate Narcotics: Neurochemical Mechanisms in Analgesia and Dependence. Pergamon Press, New York.
17. Snyder, S. H., Pasternak, G. W., and Pert, C. B. (1975). *In* L. C. Iverson, S. D. Iverson, and S. H. Snyder (eds.), Handbook of Psychopharmacology, Vol. 5, p. 329. Plenum Press, New York.
18. Snyder, S. H. (1975). Biochem. Pharmacol. 24:1371.
19. Gorman, R. This volume.
20. Kadowitz, P. J., Joiner, P. D., and Hyman, A. L. (1975). Annu. Rev. Pharmacol. 15:285.
21. Samuelsson, B., Granström, E., Green, K., Hamberg, M., and Hammerström, S. (1975). Annu. Rev. Biochem. 44:669.
22. Samuelsson, B., and Paoletti, R. (1976). Advances in Prostaglandin and Thromboxane Research. Raven Press, New York.
23. Guidotti, G. (1972). Annu. Rev. Biochem. 41:731.
24. Bretscher, M. S. (1973). Science 181:622.
25. Juliano, R. L. (1973). Biochim. Biophys. Acta 300:341.
26. Steck, T. L. (1974). J. Cell Biol. 62:1.
27. Marchesi, V. T., Furthmayr, H., and Tomita, M. (1976). Annu. Rev. Biochem. 45:667.
28. Danielli, J. F., and Davson, H. (1935). J. Cell. Physiol. 5:495.
29. Oseroff, A., Robbins, P. W., and Burger, M. M. (1973). Annu. Rev. Biochem. 42:647.
30. Goldup, A., Ohki, S., and Danielli, J. F. (1970). Recent Progr. Surf. Sci. 3:193.
31. Singer, S. J., and Nicolson, G. L. (1972). Science 175:720.
32. Kreutz, W. (1972). Angew. Chem. (Engl.) 11:551.
33. Lenard, J., and Singer, S. J. (1966). Proc. Natl. Acad. Sci. USA 56:1828.
34. Vanderkooi, G., and Green, D. E. (1970). Proc. Natl. Acad. Sci. USA 66:615.
35. Chapman, D., and Wallach, D. F. H. (1968). *In* D. Chapman (ed.), Biological Membranes, p. 125. Academic Press, New York.
36. Kornberg, R. D., and McConnell, H. M. (1971). Proc. Natl. Acad. Sci. USA 68:2564.
37. Devaux, P., and McConnell, H. M. (1972). J. Am. Chem. Soc. 94:4475.

38. Kornberg, R. D., and McConnell, H. M. (1971). Biochemistry 10:1111.
39. Edidin, M. (1974). Annu. Rev. Biophys. Bioeng. 3:179.
40. Jain, M. K., and White, H. B. (1977). Adv. Lipid Res. 15:1.
41. Cherry, R. J. (1975). FEBS Lett. 55:1.
42. Phillips, M. D. (1972). *In* J. F. Danielli, M. D. Rosenberg, and D. A. Cadenhead (eds.), Recent Progress in Surface and Membrane Science, Vol. 5, pp. 139–221. Academic Press, New York.
43. Gitler, C. (1972). Annu. Rev. Biophys. Bioeng. 1:51.
44. Shinitzky, M., and Inbar, M. (1974). J. Mol. Biol. 85:603.
45. Fuchs, P., Parola, A., Robbins, P. W., and Blout, E. R. (1975). Proc. Natl. Acad. Sci. USA 72:3351.
46. Shinitzky, M., and Inbar, M. (1976). Biochim. Biophys. Acta 433:133.
47. Cogan, U., Shinitzky, M., Weber, G., and Nishida, T. (1973). Biochemistry 12:521.
48. De Kruyff, B., Demel, R. A., Slotbloom, A. J., Van Deenen, L. L. M., and Rosenthal, A. F. (1973). Biochim. Biophys. Acta 307:1.
49. Singer, S. J. (1974). Annu. Rev. Biochem. 43:805.
50. Lenaz, G. (1974). Subcell. Biochem. 3:167.
51. Green, D. E. (1972). Ann. N. Y. Acad. Sci. 195:150.
52. Coleman, R. (1973). Biochim. Biophys. Acta 300:1.
53. Shapiro, A. L., Vinuela, E., and Maizel, J. V. (1967). Biochem. Biophys. Res. Commun. 28:815.
54. Weber, K., and Osborn, M. (1969). J. Biol. Chem. 244:4406.
55. Fairbanks, G., Steck, T. L., and Wallach, D. F. H. (1971). Biochemistry 10:2606.
56. Conrad, M. S., and Penniston, J. T. (1976). J. Biol. Chem. 251:1253.
57. Furthmayr, H., Kahane, I., and Marchesi, V. T. (1976). J. Membr. Biol. 26:173.
58. Frye, L. D., and Edidin, M. (1970). J. Cell Sci. 7:319.
59. Edidin, M., and Fambrough, D. (1973). J. Cell. Biol. 57:27.
60. Poo, M., and Cone, R. A. (1974). Nature (Lond.) 247:438.
61. Liebman, P. A., and Entine, G. (1974). Science 185:457.
62. Taylor, R. B., Duffus, W. P. H., Raff, M. C., and de Petris, S. (1971). Nature (New Biol.) 233:225.
63. de Petris, S. (1975). J. Cell Biol. 65:123.
64. Edelman, G. M. (1976). Science 192:218.
65. Nicolson, G. L. (1976). Biochim. Biophys. Acta 457:57.
66. Nicolson, G. L., Robbins, J. C., and Hyman, R. (1976). J. Supramol. Struct. 4:15.
67. Craig, S., and Cuatrecasas, P. (1975). Proc. Natl. Acad. Sci. USA 72:3844.
68. Lis, H., and Sharon, N. (1973). Annu. Rev. Biochem. 42:541.
69. Vanderkooi, G. (1972). Ann. N. Y. Acad. Sci. 195:6.
70. Perdue, J. F. (1973). J. Cell Biol. 58:265.
71. Kinne, R. (1976). *In* F. Bronner and A. Kleinzeller (eds.), Curr. Top. Membr. Transport, Vol. 8, p. 209. Academic Press, New York.
72. Sacktor, B. (1977). *In* R. Sanadi (ed.), Curr. Top. Bioenerg, Vol. 8, p. 39. Academic Press, New York.
73. Neville, D. M. (1974). Methods Enzymol. XXXI:115.
74. Eichholz, A., and Crane, R. K. (1974). Methods Enzymol. XXXI:123.
75. Kinne, R., Schlatz, L., Kinne-Saffron, E., and Schwartz, I. L. (1975). J. Membr. Biol. 24:145.
76. Schlatz, L., Schwartz, I. L., Kinne-Saffron, E., and Kinne, R. (1975). J. Membr. Biol. 24:131.

77. Insel, P., Balakir, R., and Sacktor, B. (1975). J. Cyclic Nucl. Res. 1:107.
78. Marx, S. J., Fedak, S. A., and Aurbach, G. D. (1972). J. Biol. Chem. 247: 6913.
79. Birnbaumer, L., Pohl, S., and Kaumann, A. J. (1974). *In* P. Greengard and G. A. Robison (eds.), Advances in Cyclic Nucleotide Research, Vol. 4, pp. 239–282. Raven Press, New York.
80. Roth, J. (1975). Recent Progr. Horm. Res. 31:94.
81. Cuatrecasas, P., Hollenberg, M. D., Chang, K.-J., and Bennett, V. (1975). Recent Progr. Horm. Res. 31:37.
82. Blundell, T. L., Dodson, G. C., Hodgkin, D. C., and Mercola, D. (1972). Adv. Protein Chem. 26:279.
83. Bromer, W. W., Boucher, M. E., and Patterson, J. M. (1973). Biochem. Biophys. Res. Commun. 53:134.
84. Cuatrecasas, P., and Hollenberg, M. D. (1976). Adv. Protein Chem. 30:251.
85. Kahn, C. R. (1975). *In* E. D. Korn (ed.), Methods in Membrane Biology, Vol. 3, pp. 81–146. Plenum Press, New York.
86. Haber, E., and Wrenn, S. (1976). Physiol. Rev. 56:317.
87. Marinetti, G. V., Ray, T. K., and Tomasi, V. (1969). Biochem. Biophys. Res. Commun. 36:185.
88. Tomasi, V., Koretz, S., Ray, T. K., Dunnick, J. K., and Marinetti, G. V. (1970). Biochim. Biophys. Acta 211:31.
89. Dunnick, J. K., and Marinetti, J. V. (1971). Biochim. Biophys. Acta, 249:122.
90. Lefkowitz, R. J., and Haber, E. (1971). Proc. Natl. Acad. Sci. USA 68:1773.
91. Lefkowitz, R. J., O'Hara, D., and Haber, E. (1972). Proc. Natl. Acad. Sci. USA 69:2828.
92. Lefkowitz, R. J., Sharp, G., and Haber, E. (1973). J. Biol. Chem. 248:342.
93. Lefkowitz, R. J., O'Hara, D., and Warshaw, J. (1974). Biochim. Biophys. Acta 332:317.
94. Aprille, J. R., Lefkowitz, R. J., and Warshaw, J. B. (1974). Biochim. Biophys. Acta 373:502.
95. Jarett, L., Smith, R., and Crespin, S. (1974). Endocrinology 94:719.
96. Keller, D., and Goldfien, A. (1973). Clin. Res. 21:202.
97. Bilzekian, J. P., and Aurbach, G. D. (1973). J. Biol. Chem. 248:5575.
98. Schramm, M., Feinstein, H., Naim, E., Lang, N., and Lasser, M. (1972). Proc. Natl. Acad. Sci. USA 69:523.
99. Valieries, J., Drummond, G. I., and Severson, D. L. (1973). Fed. Proc. 32:773.
100. Cuatrecasas, P., Tell, G. P. E., Sica, V., Parikh, I., and Chang, K.-J. (1974). Nature (Lond.) 247:92.
101. Tell, G. P. E., and Cuatrecasas, P. (1974). Biochem. Biophys. Res. Commun. 57:793.
102. Wolfe, B., Zirrolli, J., and Molinoff, P. (1974). Mol. Pharmacol. 10:582.
103. Lefkowitz, R. J. (1974). Biochem. Biophys. Res. Commun. 58:1110.
104. Koretz, S. H., and Marinetti, G. V. (1974). Biochem. Biophys. Res. Commun. 61:22.
105. Aprille, J. R., and Malamud, D. F. (1975). Biochem. Biophys. Res. Commun. 64:1293.
106. Maguire, M. E., Goldmann, P. H., and Gilman, A. G. (1974). Mol. Pharmacol. 10:563.
107. Pairault, J., and Laudat, M.-H. (1975). FEBS Lett. 50:61.
108. Roth, J., Kahn, C. R., Lesniak, M. A., Gorden, P., De Meyts, P., Megyesi, K., Neville, D. M., Gavin, J. R., Soll, A. H., Freychet, P., Goldfine, I. D., Bar, R. S., and Archer, J. A. (1975). Recent Progr. Horm. Res. 31:195.

109. Hollenberg, M. D., and Cuatrecasas, P. (1975). *In* G. Litwack (ed.), Biochemical Actions of Hormones, Vol. III, pp. 41-85. Academic Press, New York.
110. Armelin, H. A. (1975). *In* G. Litwack (ed.), Biochemical Actions of Hormones, Vol. III, pp. 1-21. Academic Press, New York.
111. Chlapowski, F. S., Kelly, L. A., and Butcher, R. W. (1975). *In* P. Greengard and G. A. Robison (eds.), Adv. Cyclic Nucl. Res. Vol. 6, pp. 245-338. Raven Press, New York.
112. Coffino, P., Bourne, H. R., Freidrich, U., Hochman, J., Insel, P. A., Lemaire, I., Melmon, K. L., and Tomkins, G. M. (1976). Recent Progr. Horm. Res. 32:669.
113. Maguire, M. E., Brunton, L. L., Wiklund, R. A., Anderson, H. J., Van Arsdale, P. M., and Gilman, A. G. (1976). Recent Progr. Horm. Res. 32:633.
114. Pastan, I., Johnson, G. S., and Anderson, W. B. (1975). Annu. Rev. Biochem. 44:491.
115. Sato, G. (1974). Methods Enzymol. XXXII:557.
116. Breckenridge, B. McL. (1975). *In* B. Weiss (ed.), Cyclic Nucleotides in Disease, p. 67. University Park Press, Baltimore.
117. Perkins, J. P. (1975). *In* B. Weiss (ed.), Cyclic Nucleotides in Disease, p. 351. University Park Press, Baltimore.
118. De Meyts, P. (1976). *In* M. Blecher (ed.), Methods in Receptor Research, Part 1, pp. 301-383. Marcel Dekker, Inc., New York.
119. Rechler, M. M., and Podskalny, J. M. (1976). Diabetes 25:250.
120. Winand, R. J., and Kohn, L. D. (1975). J. Biol. Chem. 250:6534.
121. Jarett, L., Reuter, M., McKeel, D. W., and Smith, R. M. (1971). Endocrinology 89:1186.
122. Kahn, C. R. (1976). J. Cell Biol. 70:261.
123. Freychet, P. (1976). Diabetologia 12:83.
124. Cuatrecasas, P. (1969). Proc. Natl. Acad. Sci. USA 63:450.
125. Cuatrecasas, P. (1973). Science 179:1143.
126. Schimmer, B. P., Ueka, K., and Sato, G. H. (1968). Biochem. Biophys. Res. Commun. 32:806.
127. Johnson, C. B., Blecher, M., and Giorgio, N. A., Jr. (1972). Biochem. Biophys. Res. Commun. 46:1035.
128. Blatt, L. M., and Kim, K. H. (1971). J. Biol. Chem. 246:4895.
129. Oka, T., and Topper, Y. J. (1971). Proc. Natl. Acad. Sci. USA 68:2066.
130. Davidson, M. D., Van Herle, A. J., and Gerschenson, L. E. (1973). Endocrinology 92:1442.
131. Schwartz, J., Nutting, D. F., Goodman, H. M., Kostyo, J. L., and Fellows, R. E. (1973). Endocrinology 92:439.
132. Kolb, H. J., Renner, R., Hepp, K. D., Weiss, L., and Wieland, O. H. (1975). Proc. Natl. Acad. Sci. USA 72:248.
133. Garwin, J. L., and Gelherter, T. D. (1974). Arch. Biochem. Biophys. 164:52.
134. Katzen, H. M., and Vlahakes, G. J. (1973). Science 179:1142.
135. Butcher, R. W., Crofford, O. B., Gammeltoft, S., Gliemann, J., Gavin, J. R., Goldfine, I. D., Kahn, C. R., Roth, J., Jarett, L., Larner, J., Lefkowitz, R. J., Levine, R., and Marinetti, G. V. (1973). Science 182:396.
136. Lefkowitz, R. J., Roth, J., Pricer, W., and Pastan, I. (1970). Proc. Natl. Acad. Sci. USA 65:745.
137. Lin, S.-Y., and Goodfriend, T. L. (1970). Am. J. Physiol. 218:1319.
138. Cuatrecasas, P. (1971). Proc. Natl. Acad. Sci. USA 68:1264.
139. Freychet, P., Roth, J., and Neville, D. M. (1971). Proc. Natl. Acad. Sci. USA 68:1833.

140. Bennett, V., and Cuatrecasas, P. (1973). Biochim. Biophys. Acta 311:362.
141. Jarett, L., and Smith, R. M. (1974). J. Biol. Chem. 249:7024.
142. Jarett, L., and Smith, R. M. (1975). Proc. Natl. Acad. Sci. USA 72:3526.
143. Freychet, P., Kahn, C. R., Roth, J., and Neville, D. M. (1972). J. Biol. Chem. 247:3953.
144. Cuatrecasas, P. (1971). J. Biol. Chem. 246:7265.
145. Cuatrecasas, P., Desbuquois, B., and Krug, F. (1971). Biochem. Biophys. Res. Commun. 44:333.
146. Gavin, J. R., Gorden, P., Roth, J., Archer, J. A., and Buell, D. N. (1973). J. Biol. Chem. 248:2202.
147. Terris, S., and Steiner, D. F. (1975). J. Biol. Chem. 250:8382.
148. Gliemann, J., Gammeltoft, S., and Vinten, J. (1974). Acta Endocrinol. (Suppl.) 77(190):131.
149. Sherwin, R. S., Kramer, K. J., Tobin, J. D., Insel, P. A., Liljenquist, J., Berman, M., and Andres, R. (1974). J. Clin. Invest. 53:1481.
150. Schwartz, R. H., Bianco, A. R., Handwerger, B. S., and Kahn, C. R. (1975). Proc. Natl. Acad. Sci. USA 72:474.
151. Goldfine, I. D., Gardner, J. D., and Neville, D. M. (1972). J. Biol. Chem. 247:6919.
152. Scatchard, G. (1949). Ann. N. Y. Acad. Sci. 51:660.
153. Goldfine, I. D., and Smith, G. J. (1976). Proc. Natl. Acad. Sci. USA 73: 1427.
154. Horvat, A., Li, E., and Katsoyannis, P. G. (1975). Biochim. Biophys. Acta 382:609.
155. Cuatrecasas, P. (1972). J. Biol. Chem. 247:1980.
156. Cuatrecasas, P. (1972). Proc. Natl. Acad. Sci. USA 69:318.
157. Cuatrecasas, P. (1972). Proc. Natl. Acad. Sci. USA 69:1277.
158. Gavin, J. R., Buell, D. N., and Roth, J. (1972). Science 178:168.
159. Gavin, J. R., Mann, D. L., and Roth, J. (1972). Biochem. Biophys. Res. Commun. 49:870.
160. Cuatrecasas, P. (1974). Annu. Rev. Biochem. 43:169.
161. Pullen, R. A., Lindsay, D. G., Wood, S. P., Tickle, I. J., Blundell, T. L., Wollmer, A., Krail, G., Brandenburg, D., Zahn, H., Gliemann, J., and Gammeltoft, S. (1976). Nature (Lond.) 259:369.
162. Ahlquist, R. P. (1948). Am. J. Physiol. 153:536.
163. Lands, A. M., Arnold, A., McAuliff, J. P., Luduena, F. P., and Brown, T. G. (1967). Nature (Lond.) 214:597.
164. Lefkowitz, R. J. (1975). Biochem. Pharmacol. 24:583.
165. Sutherland, E. W., Øye, I., and Butcher, R. W. (1965). Recent Progr. Horm. Res. 21:623.
166. Robison, G. A., Butcher, R. W., and Sutherland, E. W. (1967). Ann. N. Y. Acad. Sci. 139:703.
167. Sutherland, E. W. (1972). Science 177:401.
168. Robison, G. A., Butcher, R. W., and Sutherland, E. W. (1971). Cyclic AMP. Academic Press, New York.
169. Lefkowitz, R. J., Limbird, L. E., Mukherjee, C., and Caron, M. G. (1976). Biochim. Biophys. Acta 457:1.
170. Levitzki, A., Atlas, D., and Steer, M. L. (1974). Proc. Natl. Acad. Sci. USA 71:2773.
171. Levitzki, A., Sevilla, N., Atlas, D., and Steer, M. L. (1974). J. Mol. Biol. 97:35.
172. Atlas, D., Steer, M. L., and Levitzki, A. (1974). Proc. Natl. Acad. Sci. USA 71:4246.

173. Aurbach, G. D., Fedak, S. A., Woodard, C. J., Palmer, J. S., Hauser, D., and Troxler, F. (1974). Science 186:1223.
174. Brown, E. M., Hauser, D., Troxler, F., and Aurbach, G. D. (1976). J. Biol. Chem. 251:1232.
175. Brown, E. M., Fedak, S. A., Woodard, C. J., Aurbach, G. D., and Rodbard, D. (1976). J. Biol. Chem. 251:1239.
176. Lefkowitz, R. J., Mukherjee, C., Coverstone, M., and Caron, M. G. (1974). Biochem. Biophys. Res. Commun. 60:703.
177. Mukherjee, C., Caron, M. G., Coverstone, M., and Lefkowitz, R. J. (1975). J. Biol. Chem. 250:4869.
178. Mukherjee, C., Caron, M. G., Mullikin, D., and Lefkowitz, R. J. (1976). Mol. Pharmacol. 12:16.
179. Maguire, M. E., Wiklund, R. A., Anderson, H. J., and Gilman, A. G. (1976). J. Biol. Chem. 251:1221.
180. Insel, P. A., Maguire, M. E., Gilman, A. G., Bourne, H. R., Coffino, P., and Melmon, K. L. (1976). Mol. Pharmacol. 12:1069.
181. Sheppard, J. R. (1976). Fed. Proc. 35:1641.
182. Insel, P. A., Stoolman, L., and Melmon, K. L. (1976). Pharmacologist 18: 220.
183. Williams, L. T., Snyderman, R., and Lefkowitz, R. J. (1976). J. Clin. Invest. 57:149.
184. Alexander, R. W., Williams, L. T., and Lefkowitz, R. J. (1975). Proc. Natl. Acad. Sci. USA 72:1564.
185. Roberts, J. M., Goldfien, A., and Insel, P. A. (1976). Pharmacologist 18:186.
186. Harden, T. K., Wolfe, B. B., and Molinoff, P. B. (1976). Mol. Pharmacol. 12:1.
187. Sporn, J. R., and Molinoff, P. B. (1976). J. Cyclic Nucl. Res. 2:149.
188. Alexander, R. W., Davis, J. N., and Lefkowitz, R. J. (1975). Nature (Lond.) 258:437.
189. Williams, L. T., Jarett, L., and Lefkowitz, R. J. (1976). J. Biol. Chem. 251:3096.
190. Wolfe, B. B., Harden, T. K., and Molinoff, P. B. (1976). Proc. Natl. Acad. Sci. USA 73:1343.
191. Charness, M. E., Bylund, D. B., Beckman, B. S., Hollenberg, M. D., and Snyder, S. H. (1976). Life Sci. 19:243.
192. Zatz, M., Kebabian, J. W., Romero, J. A., Lefkowitz, R. J., and Axelrod, J. (1976). J. Pharmacol. Exp. Ther. 196:714.
193. Levitzki, A. (1977). *In* P. Cuatrecasas and M. F. Greaves (eds.), Receptors and Recognition, Series A, Vol. 3. Chapman and Hall, London. In press.
194. Stoolman, L. S., Insel, P. A., and Melmon, K. L. (1977). Clin. Res. 25:106A.
195. Maguire, M. E., Van Arsdale, P. M., and Gilman, A. G. (1976). Mol. Pharmacol. 12:335.
196. Maguire, M. E., Ross, E. M., Biltonen, R. C., and Gilman, A. G. (1976). Fed. Proc. 35:1634.
197. Lefkowitz, R. J., Mullikin, D., and Caron, M. G. (1976). J. Biol. Chem. 251:4686.
198. Yong, M. S., and Nickerson, M. (1973). J. Pharmacol. Exp. Ther. 186:100.
199. Williams, L. T., and Lefkowski, R. J. (1976). Science 192:791.
200. Ruffolo, R. R., Fowble, J. W., Miller, D. D., and Patil, P. N. (1976). Proc. Natl. Acad. Sci. USA 73:2730.
201. Greenberg, D. A., U'Prichard, D. C., and Synder, S. H. (1976). Life Sci. 19:69.
202. Helmreich, E. J. M., Zenner, H. P., Pfeuffer, T., and Cori, C. F. (1976). *In*

B. L. Horecker and E. R. Stadtman (eds.), Current Topics in Cellular Reg. Vol. 10, pp. 41–87. Academic Press, New York.

203. Illiano, G., and Cuatrecasas, P. (1972). Science 175:906.
204. Renner, R., Kemmler, W., and Hepp, K. D. (1974). Eur. J. Biochem., 49:129.
205. House, P. D. R., Poulis, P., and Weidemann, M. J. (1972). Eur. J. Biochem. 24:429.
206. Loten, E. G., and Sneyd, J. G. T. (1970). Biochem. J. 120:187.
207. Kono, T., Robinson, F. W., and Sarver, J. A. (1975). J. Biol. Chem. 250:7826.
208. Manganiello, V., and Vaughan, M. (1973). J. Biol. Chem. 248:7164.
209. Zinman, B., and Hollenberg, C. H. (1974). J. Biol. Chem. 249:2182.
210. Illiano, G., Tell, G. P. E., Siegel, M. J., and Cuatrecasas, P. (1973). Proc. Natl. Acad. Sci. USA 70:2443.
211. Pilkis, S., Claus, T. H., Johnson, R. A., and Park, C. R. (1975). J. Biol. Chem. 250:6328.
212. Rodbell, M., Krans, H. M. J., Pohl, S. L., and Birnbaumer, L. (1971). J. Biol. Chem. 246:1872.
213. Rodbell, M., Birnbaumer, L., Pohl, S. L., and Krans, H. M. J. (1971). J. Biol. Chem. 246:1877.
214. Londos, C., Salomon, Y., Lin, M. C., Harwood, J. P., Schramm, M., Wolff, J., and Rodbell, M. (1974). Proc. Natl. Acad. Sci. USA 71:3087.
215. Spiegel, A. M., Brown, E. M., Fedak, S. A., Woodard, C. J., and Aurbach, G. D. (1976). J. Cyclic Nucl. Res. 2:47.
216. Sevilla, N., Steer, M. L., and Levitzki, A. (1976). Biochemistry 15:3493.
217. Bennett, V., and Cuatrecasas, P. (1976). J. Membr. Biol. 27:207.
218. Pfeuffer, T., and Helmreich, E. J. M. (1975). J. Biol. Chem. 250:867.
219. Salomon, Y., Lin, M. C., Londos, C., Rendell, M., and Rodbell, M. (1975). J. Biol. Chem. 250:4239.
220. Rendell, M., Salomon, Y., Lin, M. C., Rodbell, M., and Berman, M. (1975). J. Biol. Chem. 250:4253.
221. Schramm, M., and Rodbell, M. (1975). J. Biol. Chem. 250:2232.
222. Lefkowitz, R. J., and Caron, M. G. (1975). J. Biol. Chem. 250:4418.
223. Lefkowitz, R. J. (1974). J. Biol. Chem. 249:6119.
224. Johnson, R. A., Pilkis, S., and Hamet, P. (1975). J. Biol. Chem. 250:6599.
225. Spiegel, A. M., and Aurbach, G. D. (1974). J. Biol. Chem. 249:7630.
226. Salomon, Y., and Rodbell, M. (1975). J. Biol. Chem. 250:7245.
227. Hanoune, J., Lacombe, M.-L., and Pecker, F. (1975). J. Biol. Chem. 250: 4569.
228. Hammes, G. G., and Rodbell, M. (1976). Proc. Natl. Acad. Sci. USA 73: 1189.
229. Neer, E. J. (1976). J. Biol. Chem. 251:5831.
230. Lefkowitz, R. J. (1975). J. Biol. Chem. 250:1006.
231. Jacobs, S., and Cuatrecasas, P. (1976). Biochem. Biophys. Res. Commun. 70:885.
232. Dufau, M. L., Ryan, D., and Catt, K. J. (1975). J. Biol. Chem. 250:4822.
233. Levey, G. S., Fletcher, M. A., Klein, I., Ruiz, E., and Schenk, A. (1974). J. Biol. Chem. 249:2665.
234. Levey, G. S. (1974). Methods Enzymol. XXXVIII:174.
235. Levey, G. S. (1973). Recent Progr. Horm. Res. 29:361.
236. Träuble, H., and Overath, P. (1973). Biochim. Biophys. Acta 307:491.
237. Puchwein, G., Pfeuffer, T., and Helmreich, E. J. M. (1974). J. Biol. Chem. 249:3232.
238. Limbird, L., and Lefkowitz, R. J. (1976). Mol. Pharmacol. 12:559.

239. Pohl, S. L., Krans, H. M. J., Kozyreff, V., Birnbaumer, L., and Rodbell, M. (1971). J. Biol. Chem. 246:4447.
240. Rubalcava, B., and Rodbell, M. (1973). J. Biol. Chem. 248:3831.
241. Rethy, A., Tomasi, V., Trevisani, A., and Barnabei, O. (1972). Biochim. Biophys. Acta 290:58.
242. Yamashita, K., and Field, J. B. (1973). Biochim. Biophys. Acta 304:686.
243. Barden, N., and Labrie, F. (1973). J. Biol. Chem. 248:7601.
244. Orly, J., and Schramm, M. (1975). Proc. Natl. Acad. Sci. USA 72:3433.
245. Sinha, A. K., Shattil, S. J., and Colman, R. W. (1976). Fed. Proc. 35:1714.
246. Shattil, S. J., and Cooper, R. A. (1976). Biochemistry 22:4832.
247. Cuatrecasas, P. (1973). Biochemistry 12:3547.
248. Cuatrecasas, P. (1973). Biochemistry 12:3558.
249. Mullin, B. R., Fishman, P. H., Lee, G., Aloj, S. M., Ledley, F. D., Winand, R. J., Kohn, L. D., and Brady, R. (1976). Proc. Natl. Acad. Sci. USA 73:842.
250. Cuatrecasas, P., and Tell, G. P. E. (1973). Proc. Natl. Acad. Sci. USA 70: 485.
251. Majerus, P., and Brodie, G. N. (1972). J. Biol. Chem. 247:4253.
252. Michaelis, E. K., and Michaelis, M. L. (1976). Life Sci. 18:1021.
253. Sela, B. A., Wang, J. L., and Edelman, G. M. (1975). J. Biol. Chem. 250: 7535.
254. Schlesinger, J., Koppel, D. E., Axelrod, D., Jacobson, K., Webb, W. E., and Elson, E. L. (1976). Proc. Natl. Acad. Sci. USA 73:2409.
255. Berridge, M. J. (1975). *In* P. Greengard and G. A. Robison (eds.), Advances in Cyclic Nucleotide Research, Vol. 6, pp. 2-98. Raven Press, New York.
256. Berridge, M. J. (1975). J. Cyclic Nucl. Res. 1:305.
257. Rasmussen, H. (1970). Science 170:404.
258. Birnbaumer, L. (1973). Biochim. Biophys. Acta 300:129.
259. Perkins, J. P. (1973). *In* P. Greengard and G. A. Robison (eds.), Advances in Cyclic Nucleotide Research, Vol. 3, pp. 1-64. Raven Press, New York.
260. Steer, M. L., and Levitzki, A. (1975). J. Biol. Chem. 250:2080.
261. Steer, M. L., and Levitzki, A. (1975). Arch. Biochem. Biophys. 167:371.
262. Drummond, G. I., and Duncan, L. (1970). J. Biol. Chem. 245:976.
263. Rodan, G. A., and Feinstein, M. B. (1976). Proc. Natl. Acad. Sci. USA 73: 1829.
264. Lefkowitz, R. J., Roth, J., and Pastan, I. (1970). Nature (Lond.) 228:864.
265. Johnson, R. A., and Sutherland, E. W. (1973). J. Biol. Chem. 248:5114.
266. Constantopoulos, A., and Najjar, V. (1973). Biochem. Biophys. Res. Commun. 53:794.
267. Najjar, V., and Constantopoulos, A. (1973). Mol. Cell. Biochem. 2:87.
268. Levitzki, A. (1974). J. Theor. Biol. 44:367.
269. McGuire, R. F., and Barber, R. (1976). J. Supramol. Struct. 4:259.
270. De Meyts, P., Bianco, A. R., and Roth, J. (1976). J. Biol. Chem. 251:1877.
271. De Haën, C. (1976). J. Theor. Biol. 58:383.
272. Bourne, H. R., Coffino, P., Melmon, K. L., Tomkins, G. M., and Weinstein, Y. (1974). *In* G. I. Drummond, P. Greengard, and G. A. Robison (eds.), Advances in Cyclic Nucleotide Research, Vol. 5. Raven Press, New York.
273. Bourne, H. R., Coffino, P., and Tomkins, G. M. (1975). Science 187:750.
274. Clark, A. J. (1937). *In* W. Hubner and T. Schüler (eds.), Heffler's Handbuch der Experimentellen Pharmakilogie, Erganzungswerk, Bd. 4, p. 33. Springer-Verlag, Berlin.
275. Ariëns, E. J., and de Groot, W. M. (1954). Arch. Int. Pharmacodyn. Ther. 99:193.

276. Stephenson, R. P. (1956). Br. J. Pharmacol. 11:379.
277. Jard, S., Roy, C., Barth, T., Rajerison, R., and Bockaert, J. R. (1974). *In* G. I. Drummond, P. Greengard, and G. A. Robison (eds.), Advances in Cyclic Nucleotide Research, Vol. 5, p. 31. Raven Press, New York.
278. Catt, K. J., and Dufau, M. L. (1973). Nature (New Biol.) 244:719.
279. Dufau, M. L., Wantabe, K., and Catt, K. (1973). Endocrinology 92:6.
280. Beall, R. J., and Sayers, G. (1972). Arch. Biochem. Biophys. 148:70.
281. Williams, J. A. (1972). Endocrinology 91:1411.
282. Birnbaumer, L., and Pohl, S. (1973). J. Biol. Chem. 248:2056.
283. Kono, T., and Barham, F. W. (1971). J. Biol. Chem. 246:6210.
284. Gliemann, J., Gammeltoft, S., and Vinten, J. (1975). J. Biol. Chem. 250: 3368.
285. Soloff, M. S. (1976). Br. J. Pharmacol. 57:381.
286. Beavo, J. A., Bechtel, P. J., and Krebs, E. G. (1974). Proc. Natl. Acad. Sci. USA 71:3580.
287. De Meyts, P., Roth, J., Neville, D. M., Gavin, J. R., Lesniak, M. A. (1973). Biochem. Biophys. Res. Commun. 55:154.
288. De Meyts, P., and Roth, J. (1975). Biochem. Biophys. Res. Comun. 66:1118.
289. De Meyts, P. (1976). J. Supramol. Struct. 4:241.
290. Matthews, B. W., and Bernhard, S. A. (1973). Annu. Rev. Biophys. Bioeng. 2:257.
291. Frazier, W. A., Boyd, L. F., Pulliam, M. W., Szutowicz, A., and Bradshaw, R. (1974). J. Biol. Chem. 249:5918.
292. Cuatrecasas, P., and Hollenberg, M. (1975). Biochem. Biophys. Res. Commun. 62:31.
293. Tate, R. L., Schwartz, H. I., Holmes, J. M., Kohn, L. D., and Winand, R. J. (1975). J. Biol. Chem. 250:6509.
294. Tate, R. L., Holmes, J. M., Winand, R. J., and Kohn, L. D. (1975). J. Biol. Chem. 250:6527.
295. Verrier, B., Fayet, G., and Lissitzky, S. (1974). Eur. J. Biochem. 42:355.
296. Limbird, L. E., De Meyts, P., and Lefkowitz, R. J. (1975). Biochem. Biophys. Res. Commun. 64:1160.
297. Limbird, L. E., and Lefkowitz, R. J. (1976). J. Biol. Chem. 251:5007.
298. Van Obberghen, E., De Meyts, P., and Roth, J. (1976). Diabetes 25:321.
299. De Meyts, P. (1976). Endocrinology 96:68.
300. Pollet, R., Haase, B., and Standaert, M. L. (1976). Endocrinology 96:69.
301. Jacobs, S., and Cuatrecasas, P. (1976). Biochim. Biophys. Acta 433:482.
302. Chang, K.-L., Huang, D., and Cuatrecasas, P. (1975). Biochem. Biophys. Res. Commun. 64:566.
303. Sraer, J., Ardaillou, R., and Couette, S. (1974). Endocrinology 95:632.
304. Mukherjee, C., Caron, M., and Lefkowitz, R. J. (1975). Proc. Natl. Acad. Sci. USA 72:1945.
305. Mukherjee, C., Caron, M., and Lefkowitz, R. J. (1976). Endocrinology 99:347.
306. Kebabian, J. W., Zatz, M., Romero, J. A., and Axelrod, J. (1975). Proc. Natl. Acad. Sci. USA 72:3735.
307. Hunzicker-Dunn, M., and Birnbaumer, L. (1976). Endocrinology 99:211.
308. DeRobertis, F. R., and Craven, P. (1976). J. Clin. Invest. 57:435.
309. Hepp, K. D., Langley, J., Von Funcke, H. J., Renner, R., and Kemmeler, W. (1975). Nature (Lond.) 258:154.
310. Gavin, J. R., Roth, J., Neville, D. M., De Meyts, P., and Buell, D. N. (1974). Proc. Natl. Acad. Sci. USA 71:84.
311. Lesniak, M. A., and Roth, J. (1976). J. Biol. Chem. 251:3720.
312. Shuman, S., Zor, U., Chayoth, R., and Field, J. B. (1976). J. Clin. Invest. 57:1132.

313. Hinkle, P. M., and Tashjian, A. H. (1975). Biochemistry 14:3845.
314. Wishnow, R. M., and Lifrak, E. (1976). Fed. Proc. 35:1711.
315. Kakiuchi, S., and Rall, T. (1968). Mol. Pharmacol. 4:367.
316. Manganiello, V., Murad, F., and Vaughan, M. (1971). J. Biol. Chem. 246: 2195.
317. Ho. R., and Sutherland, E. W. (1971). J. Biol. Chem. 246:1822.
318. Remold-O'Donnell, E. (1974). J. Biol. Chem. 249:3615.
319. Mickey, J. V., Tate, R., Mullikin, D., and Lefkowitz, R. J. (1976). Mol. Pharmacol. 12:409.
320. Makman, M. (1971). Proc. Natl. Acad. Sci. USA 68:885.
321. de Vellis, J., and Brooker, G. (1974). Science 186:1221.
322. Franklin, T. J., and Foster, S. J. (1973). Nature (New Biol.) 246:146.
323. Clark, R. B., Su, Y.-F., Cubeddu, L. X., Johnson, G. L., and Perkins, J. P. (1975). Metabolism 24:343.
324. Browning, E. T., Brostrom, C. O., and Groppi, V. E. (1976). Mol. Pharmacol. 12:32.
325. Shear, M., Insel, P. A., Melmon, K. L., and Coffino, P. (1976). J. Biol. Chem. 251:7572.
326. Newcombe, D. S., Ciosek, C. P., Ishikawa, Y., and Fahey, J. V. (1975). Proc. Natl. Acad. Sci. USA 72:3124.
327. Mickey, J. V., Tate, R., Mullikin, D., and Lefkowitz, R. J. (1975). J. Biol. Chem. 250:5727.
328. Klein, W., Nathanson, N. M., and Nirenberg, M. (1976). Fed. Proc. 35:1576.
329. Pawelek, J., Wong, G., Sansone, M., and Morowitz, J. (1973). Yale J. Biol. Med. 46:430.
330. Su, Y.-F., Cubeddu, L. X., and Perkins, J. P. (1976). J. Cyclic Nucl. Res. 2:257.
331. Ho, R., and Sutherland, E. W. (1975). *In* G. I. Drummond, P. Greengard, and G. A. Robison (eds.), Adv. Cyclic Nucl. Res. Vol. 5, pp. 533–548. Raven Press, New York.
332. Doore, B. J., Bashor, M. M., Spitzer, N., Mawe, R. C., and Saier, M. H. (1975). J. Biol. Chem. 250:4371.
333. d'Armiento, M., Johnson, G., and Pastan, I. (1972). Proc. Natl. Acad. Sci. USA 69:459.
334. Bourne, H. R., Tomkins, G. M., and Dion, S. (1973). Science 181:952.
335. Schwartz, J. P., and Passaneau, J. V. (1974). Proc. Natl. Acad. Sci. USA 71:3844.
336. Franklin, T. J., Morris, W. P., and Twose, P. A. (1975). Mol. Pharmacol. 11:485.
337. Morishima, I., Robison, G. A., Thompson, W. J., and Strada, S. J. (1976). Pharmacologist 18:186.
338. Lauzon, G. J., Kulshrestha, S., Starr, L., and Bär, H. P. (1976). J. Cyclic Nucl. Res. 2:99.
339. Raff, M. (1976). Nature (Lond.) 259:265.
340. Kosmakos, F. C., and Roth, J. (1976). Endocrinology 98:69.
341. Huang, D., and Cuatrecasas, P. (1975). J. Biol. Chem. 250:8251.
342. Bockaert, J., Hunzicker-Dunn, M., and Birnbaumer, L. (1976). J. Biol. Chem. 251:2653.
343. Mukherjee, C., and Lefkowitz, R. J. (1976). Proc. Natl. Acad. Sci. USA 73:1494.
344. Atlas, D., and Levitzki, A. (1976). Biochem. Biophys. Res. Commun. 69:397.
345. Atlas, D., Steer, M. L., and Levitzki, A. (1976). Proc. Natl. Acad. Sci. USA 73:1921.
346. Melamed, E., Lahav, M., and Atlas, D. (1976). Nature (Lond.) 261:420.

347. Atlas, D., and Levitzki, A. Proc. Natl. Acad. Sci. USA. In press.
348. Caron, M., and Lefkowitz, R. J. (1976). J. Biol. Chem. 251:2374.
349. Hsuch, A. J., Dufau, M. L., Katz, S. I., and Catt, K. J. (1976). Nature (Lond.) 261:710.
350. Posner, B. I., Kelly, P. A., and Friesen, H. G. (1975). Science 188:57.
351. Posner, B. I. (1975). Can. J. Physiol. Pharmacol. 53:689.
352. Zeleznik, A. J., Midgley, A. R., and Reichert, L. E. (1974). Endocrinology 95:818.
353. Soloff, M. S. (1975). Biochem. Biophys. Res. Commun. 65:205.
354. Rajerison, R., Marchetti, J., Roy, C., Bockaert, J., and Jard, S. (1974). J. Biol. Chem. 249:6390.
355. Douglas, J., and Catt, K. J. (1976). J. Clin. Invest. 58:834.
356. O'Dea, R., Zatz, M., and Axelrod, J. (1976). Fed. Proc. 35:1384.
357. Varga, J. M., Saper, M. A., Lerner, A. B., and Fritsch, P. (1976). J. Supramol. Struct. 4:44.
358. Varga, J. M., Moellmann, G., Fritsch, P., Godawska, E., and Lerner, A. B. (1976). Proc. Natl. Acad. Sci. USA 73:559.
359. Varga, J. M., DiPasquale, A., Pawelek, J., McGuire, J. S., and Lerner, A. B. (1974). Proc. Natl. Acad. Sci. USA 71:1590.
360. Kuo, J. F., and Greengard, P. (1969). Proc. Natl. Acad. Sci. USA 64:1349.
361. Malkinson, A. M., Krueger, B. K., Rudolph, S. A., Casnellie, J., Haley, B., and Greengard, P. (1975). Metabolism 24:331.
362. Rubin, C. S., and Rosen, O. M. (1974). Annu. Rev. Biochem. 44:831.
363. Rea, C., and Segal, S. (1973). Biochim. Biophys. Acta 311:615.
364. Weiss, I. W., Morgan, K., and Phang, J. M. (1972). J. Biol. Chem. 247:760.
365. Chang, K.-J., Marcus, N. A., and Cuatrecasas, P. (1974). J. Biol. Chem. 249:760.
366. Kirchberger, M. A., and Chu, G. (1976). Biochim. Biophys. Acta 419:559.
367. Podevin, R. A., and Boumendil-Podevin, E. F., (1975). Biochim. Biophys. Acta 375:106.
368. Tria, E., Luly, P., Tomasi, V., Trevisani, A., and Barnabei, O. (1974). Biochim. Biophys. Acta 343:297.
369. Dowd, F., and Schwartz, A. (1975). J. Mol. Cell Cardiol. 7:483.
370. Greengard, P. (1976). Nature (Lond.) 260:101.
371. Verhaegen, M., and Sand, G. (1976). Biochim. Biophys. Acta 429:163.
372. Kish, V. M., and Kleinsmith, L. (1974). J. Biol. Chem. 249:750.
373. Soderling, T. R., Hickenbottom, J. P., Reimann, E. M., Hunkeler, F. L., Walsh, D. A., and Krebs, E. G. (1970). J. Biol. Chem. 245:637.
374. Huttunen, D. K., and Steinberg, D. (1971). Biochim. Biophys. Acta 239:411.
375. Coffino, P., Gray, J., and Tomkins, G. M. (1975). Proc. Natl. Acad. Sci. USA 72:878.
376. Daniel, V., Bourne, H. R., and Tomkins, G. M. (1973). Nature (New Biol.) 244:167.
377. Insel, P., and Fenno, J. Proc. Natl. Acad. Sci. USA. In press.
378. Bourne, H. R., Coffino, P., and Tomkins, G. M. (1975). J. Cell. Physiol. 85:603.
379. Coffino, P., Bourne, H. R., and Tomkins, G. M. (1975). J. Cell. Physiol. 85:603.
380. Daniel, V., Litwack, G., and Tomkins, G. M. (1973). Proc. Natl. Acad. Sci. USA 70:751.
381. Insel, P. A., Bourne, H. R., Coffino, P., and Tomkins, G. M., (1975). Science 190:896.
382. Robison, G. A., Butcher, R. W., and Sutherland, E. W. (1971). Cyclic AMP, p. 31. Academic Press, New York.

383. Corbin, J. D., Keely, S. L., and Park, C. R. (1975). J. Biol. Chem. 250:218.
384. Hofmann, F., Beavo, J. A., Bechtel, P. J., and Krebs, E. G. (1975). J. Biol. Chem. 250:7795.
385. Fleischer, N., Rosen, O. M., and Reichlin, M. (1976). Proc. Natl. Acad. Sci. USA 73:54.
386. Costa, M., Gerner, E. W., and Russell, D. H. (1976). J. Biol. Chem. 251: 3313.
387. Rangel-Aldao, R., and Rosen, O. M. (1976). J. Biol. Chem. 251:3375.
388. Hochman, J., Insel, P. A., Bourne, H. R., Coffino, P., and Tomkins, G. M. (1975). Proc. Natl. Acad. Sci. USA 72:5051.
389. Hochman, J., Bourne, H. R., Coffino, P., Insel, P. A., Krasny, L., and Melmon, K. L. (1977). Proc. Natl. Acad. Sci. USA 74:116.
390. Steinberg, R., O'Farrell, P., Freidrich, U., and Coffino, P. (1977). Cell 10:391.
391. Mendelson, C., Dufau, M., and Catt, K. (1975). J. Biol. Chem. 250:8818.
392. Rall, T. (1972). Pharmacol. Rev. 24:399.
393. Do Khoc, L., Harbon, S., and Clauser, H. J. (1973). Eur. J. Biochem. 40: 177.
394. Exton, J. H., Lewis, S. B., Ho, R.-J., Robison, G. A., and Park, C. R. (1971). Ann. N. Y. Acad. Sci. 185:185.
395. Parker, C. W., Sullivan, J. T., and Wedner, H. J. (1974). *In* P. Greengard and G. A. Robison (eds.), Adv. Cyclic Nucl. Res. Vol. 4, p. 26. Raven Press, New York.
396. Walsh, D. A. (1973). Recent Progr. Horm. Res. 29:329.
397. Swillens, S., Van Cauter, E., and Dumont, J. E. (1974). Biochim. Biophys. Acta 364:250.
398. Olsen, R. W. (1975). J. Theor. Biol. 49:263.
399. Sloboda, R. D., Rudolph, S. A., Rosenbaum, J. L., and Greengard, P. (1975). Proc. Natl. Acad. Sci. USA 72:177.
400. Liu, A. Y.-C., and Greengard, P. (1976). Proc. Natl. Acad. Sci. USA 73:568.

International Review of Biochemistry
Biochemistry and Mode of Action of Hormones II, Volume 20
Edited by H. V. Rickenberg

2
Assessment of the Role of Cyclic Nucleotides as Hormonal Mediators

H. SANDS[1] and H. V. RICKENBERG[2]
National Jewish Hospital and Research Center and the University of Colorado School of Medicine, Denver, Colorado

---

---

[1]Supported by a grant (HL-14964) from the National Institutes of Health, Bethesda, Maryland.

[2]Ida and Cecil Green Investigator of Developmental Biochemistry. Research supported by grants from the National Institutes of Health (AM-11046), Bethesda, Maryland, and the National Science Foundation (PCM-76-10272).

The study of cyclic nucleotides is the subject of numerous reviews. A journal, "Cyclic Nucleotide Research," is dedicated exclusively to research in that area, and "Advances in Cyclic Nucleotide Research" monitors progress in the field on a continuous basis. It seems then that there is a plethora of information on the subject of cyclic nucleotides. This contribution to the ballooning literature is justified on two grounds: on the one hand, this volume deals with the mode of action of hormones and, clearly, cyclic nucleotides play a key role in the mechanism of action of many hormones; on the other hand, there is, paradoxically, increasing uncertainty as to the precise nature of that role. This is particularly true, for example, with respect to the relationship between cyclic AMP, cyclic GMP, and inorganic ions such as calcium. In many cases, it is not clear whether the cyclic nucleotide or calcium or some yet unidentifed intermediate exerts its activity proximal to the observed physiological end effect. Or again, the assumption is generally made that the effects of cyclic nucleotides are mediated by protein kinases; there is a great deal of evidence in favor of this hypothesis, and the occurrence of cyclic nucleotide-activated protein kinases parallels that of cyclic nucleotides in eukaryotic cells. Yet it is known that in prokaryotes cyclic AMP acts by a mechanism which does not involve protein kinases, and, in fact, to date no cyclic nucleotide-activated protein kinase has been observed in prokaryotes. The possibility should not be ruled out, therefore, that in eukaryotes, also, cyclic nucleotides may exert certain effects that are not mediated by protein kinases. By the same token, there is a great deal of evidence which indicates that cyclic AMP modulates the de novo synthesis of certain proteins in eukaryotes; yet the mechanism of this crucial regulatory function of the cyclic nucleotide has remained elusive. In contrast, the mode of action of cyclic AMP in the control of prokaryotic protein synthesis is now understood; the possibility of analogies in the modes of action of cyclic AMP in the regulation of prokaryotic and eukaryotic protein synthesis remains to be explored. Ignorance also prevails with respect to the nature of the interactions between the two major classes of hormones, i.e., between those hormones which act at the cell surface and cause changes in the levels of cyclic nucleotides and the steroid hormones which enter the target cell to combine with a cytoplasmic receptor (see Chapter 4) and eventually penetrate into the nucleus. Such interactions are implicit in the so-called permissive effect of glucocorticoids on cyclic AMP action and possibly also in the fact that certain enzymes are inducible by either steroids or by cyclic nucleotides. These then are just a few areas in the study of cyclic nucleotides where considerable uncertainty prevails. It is our intention to concentrate on these lacunae in our knowledge and to refer the reader to reviews of the general subject of cyclic nucleotides wherever appropriate.

GENERAL CONSIDERATIONS

Recognition of the physiological role of cyclic AMP dates from the discovery by Sutherland and his collaborators (1) that cyclic AMP mediated the hyperglycemic effect of epinephrine and of glucagon by stimulating the conversion of inactive to active glycogen phosphorylase. Cyclic GMP was discovered somewhat more recently by Ashman et al. (2); its physiological role is still far from clear. It seems that the conceptually attractive Yin Yang hypothesis (3-6), i.e., the postulate that cyclic AMP and cyclic GMP play a generally antagonistic role in the regulation of cellular metabolism, may be too simplistic. In any case, the recognition that the actions of many hormones were mediated by changes in the cellular concentrations of cyclic nucleotides led to the study of the mechanisms by which the changes in cyclic nucleotide concentration were brought about.

The findings indicate that cells responsive to a given peptide or catecholamine hormone carry on their surfaces specific hormone receptors. Cells controlled in their metabolism by several hormones are endowed with a corresponding number of specific receptors (see Chapter 1). The occupation of the receptor by a given hormone leads to the activation of the membranal adenyl cyclase. There is a great deal of evidence (7) which suggests that the hormone receptor and the adenyl cyclase are distinct molecules (see, however, ref. 8). The nature of the coupling between the membranal receptor-hormone complex and the membranal adenyl cyclase is not understood. Guanyl nucleotides (9-11), divalent cations (12-19), phosphorylation (20, 21), membrane lipids (22-24), transient changes in membrane structure (25-28), etc. have been invoked as transducers. The reader is referred to Chapter 1 for a review of receptors and the regulation of their activity. In addition to its regulation by hormones, the activity of adenyl cyclase is also controlled by a miscellany of low molecular weight molecules; these include GTP (9-11), adenosine (29, 30), and, in vitro, fluoride. The physiological significance, if any, of this regulation by low molecular weight molecules is not known. Little is known also about the regulation of the activity of guanyl cyclase. The enzyme seems to be predominantly soluble in some, and predominantly particulate in other types of cells (31-34). In a variety of tissues, the neurotransmitter acetylcholine stimulates the accumulation of cyclic GMP (3, 35); an activation of the guanyl cyclase by acetylcholine or by any other hormone has not yet been demonstrated to our knowledge. An activation of the guanylate cyclase of both soluble and particulate preparations from a variety of tissues by gaseous nitric oxide (NO) has been reported (36). The activated state of the enzyme decayed with a half-life of 3-4 hr at 4 °C. The activation of guanyl cyclase by sodium azide, sodium nitrite, hydroxylamine, and sodium nitroprusside described earlier (cf. ref. 36 for references) is probably mediated by NO produced from these compounds. Experiments on the regulation of the activity of bacterial adenyl cyclases and of levels of cyclic AMP have been reviewed (37, 38). It seems that the control of the activity of the adenyl cyclase of the bacterium *Escherichia coli* is related to either the transport or the metabolism of sugars.

Since regulation by cyclic nucleotides depends on changes in their cellular concentration, it is clear that there must exist mechanisms for their destruction (or excretion) as well as their synthesis. There are a variety of soluble and particulate cyclic nucleotide phosphodiesterases which catalyze the hydrolysis of cyclic AMP and cyclic GMP to their respective 5′ derivatives. In most tissues, both a "low K_m-high affinity" and a "high K_m-low affinity" cyclic AMP phosphodiesterase are found. The physiological significance and the relationship between these two forms of the enzyme are not understood at present. The subject of cyclic nucleotide phosphodiesterases and their activators and inhibitors has been reviewed extensively (39–41). It seems that, whereas a number of pharmacologically active compounds, such as, for example, the methylxanthines, isoquinilines, papaverine, and a variety of heterocyclic compounds inhibit the activity of cyclic AMP phosphodiesterases, little is known about the physiological regulation of cyclic nucleotide phosphodiesterase activity. Proteinaceous activators of cyclic AMP phosphodiesterases have been isolated from a variety of tissues (39, 41). These heat-stable proteins bind calcium, and it may well be that the activator protein-calcium complex plays a regulatory role. The specificity of cyclic AMP and cyclic GMP phosphodiesterases with respect to the two substrates appears to vary considerably from tissue to tissue.

As stated earlier, it is generally accepted that perhaps a majority, if not all, of the known effects of cyclic AMP are mediated by protein kinases (42, 43). This assumption is based in part on the demonstrated role of protein kinases in the activation by cyclic AMP of, for example, glycogen phosphorylase kinase (44–51), triglyceride lipase (52–60), or the inactivation by cyclic AMP of glycogen synthetase (61–69). It is also based on genetic evidence which shows that in S49 lymphosarcoma cells a functional protein kinase is required for the induction by cyclic AMP of the synthesis of several enzymes, for the arrest of sensitive cells in the G_1 phase of the replication cycle, and for cytolysis of these cells (70). Since the effects of the activity of the protein kinase are generally reversible upon the removal of cyclic AMP (or, ultimately, of the hormone which stimulates the adenyl cyclase), it follows that generally the phosphorylation of the proteinaceous substrates of the cyclic AMP-activated kinases must also be reversible. (An exception to this generalization may be the hypothetical case in which a phosphoprotein plays a role, for example, in development and where maintenance of the phosphorylated state is required for the function of the protein.) A number of phosphoprotein phosphatases have been described (71–87); these enzymes reverse the effects of protein kinase activity in catalyzing the hydrolysis of the serine and threonine phosphoester linkages previously established as a result of protein kinase activity. Little is known about the regulation of phosphoprotein phosphatases.

It is clear that, if indeed the effects of cyclic nucleotides (or a majority of these effects) are mediated by cyclic nucleotide-dependent protein kinases, then the identification of the substrates of the protein kinases is essential to an understanding of the mechanism of cyclic nucleotide action. It seems likely that there is only a limited number of protein kinases in the different tissues of a given organism; it is known that cyclic nucleotides exert different effects in different tissues. It is probable, therefore, that protein kinases have fairly wide substrate specificity and that the ultimate specificity of the hormone effect resides in the differential distribution of the substrates of the protein kinases in different tissues.

HORMONES THAT ACT VIA CYCLIC NUCLEOTIDES

Sutherland and Robinson (88) listed several criteria which, ideally, should be met before a hormone can be considered to act via cyclic AMP: 1) the concentration of cyclic AMP in intact cells should change when the hormone is added; this change in cyclic nucleotide concentration should precede or accompany the observed physiological effect; 2) it should be possible to mimic the physiological effects of the hormone by the administration of cyclic AMP or of one of its derivatives; 3) the effects of the hormone, cyclic AMP, or its derivatives should be potentiated by inhibitors of cyclic AMP phosphodiesterase; 4) the adenyl cyclase of a broken cell preparation should respond to the hormone in a manner analogous to that of the intact tissue of origin; 5) both structural analogs of the hormone and inhibitors of its activity should have the same order of potency as they do in the intact tissue.

Based on these and related criteria, two generalizations may be made: first, cyclic nucleotides mediate the effects of hormones which act at the cell surface and which need not enter the cell to exert their effects and, secondly, hormones which are derivatives of amino acids, including polypeptides and catecholamines, act via cyclic nucleotides. Insulin and thyroxine may be exceptions to this generalization. (The mechanism of action of insulin is still not understood. It was claimed that insulin reduced the level of cyclic AMP (89–91) or increased that of cyclic GMP (3); these claims are weakened, however, by the lack of a clear-cut correlation between the changes in cyclic nucleotide levels and the physiological effect of the hormone.)

It is generally assumed that changes in the cellular concentration of cyclic AMP, subsequent to hormonal stimulation, and the observed physiological effects of the hormone are causally related. In certain instances, this assumption is borne out by conclusive experimental evidence; in other cases, it remains conjectural. For illustrative purposes, the relationship between two β-adrenergic functions of catecholamines, cellular levels of cyclic AMP, and the physiological effects of the catecholamines are examined here in some detail. One case deals with the glycogenolytic effect of epinephrine in the liver and the involvement of cyclic AMP in the phenomenon. This was the first demonstration of a "second messenger" role of the cyclic nucleotide (92–94) and, in fact, led to the discovery of cyclic AMP.

The other case deals with the relaxation of vascular smooth muscle, induced by epinephrine. Here, also, there is significant evidence which suggests an involvement of cyclic AMP; however, a number of questions regarding the precise function of cyclic AMP in the relaxation of smooth muscle remain unanswered.

Findings on the glycogenolytic effect of epinephrine clearly meet the criteria enumerated earlier and provide conclusive evidence for a role of cyclic AMP. Thus, the early work of Sutherland and co-workers (92, 93) demonstrated an increase in the cyclic AMP content of particulate preparations of liver subsequent to exposure to epinephrine. Epinephrine also increased the cyclic AMP content of rabbit liver slices (94) and of perfused liver (95). Furthermore, the effect of epinephrine on glycogenolysis and on the activation of glycogen phosphorylase could be mimicked by cyclic AMP and its derivatives (96–101). The effects could be blocked by the β-adrenergic blocking agents, dichloroisoproterenol and dihydroergotamine (98), and were potentiated by theophylline (102). Finally, the increase in the concentration of cyclic AMP seemed to parallel the increase in the formation of glucose, as stimulated by epinephrine (94). In summary, it appears that cyclic AMP indeed mediates the glycogenolytic response of liver to epinephrine. A recent report (103), which throws some doubt on the obligatory relationship between the increase in cyclic AMP and the onset of glycogenolysis, should be mentioned, however. Thus it has been claimed that in perfused rat liver low doses of epinephrine (1 or 10 mg/min) led to a breakdown of glycogen which was not associated with a detectable increase in cyclic AMP. At higher doses of epinephrine (100 mg/min), parallel increases in the levels of cyclic AMP and rates of glycolysis were indeed observed. The significance of these observations is not clear. Conceivably, the cyclic AMP is compartmented, or changes in the concentration of cyclic AMP too small to be detected by the methodology employed sufficed to stimulate glycogenolysis. The alternative possibility is that cyclic AMP plays no role in the stimulation of glycogenolysis induced by very low doses of epinephrine; yet other explanations of the rather surprising observation are not excluded. However that may be, at present the evidence for cyclic AMP as the mediator of glycogenolysis in liver, stimulated by epinephrine, still seems convincing.

A role of cyclic nucleotides in the maintenance of tone in vascular smooth muscle is frequently assumed, but the evidence is not entirely conclusive. Vascular smooth muscle can either contract or relax on the addition of catecholamines. The two responses are classified as α (contraction) and β (relaxation) (104). Sutherland and Rall (1) first suggested that the β-induced relaxation of smooth muscle might be related to an elevation of cyclic AMP levels; considerable information concerning such a role of cyclic AMP in smooth muscle is now at hand. The occurrence of the enzymes of cyclic nucleotide metabolism in vascular tissue responsive to β-adrenergic agents, as well as of cyclic AMP (and of cyclic GMP), also supports the hypothesis that cyclic AMP mediates β-induced vasodilation (105–111). However, there remain unexplained data (112, 113) which are seemingly in conflict with the hypothesis. The difficulty in demonstrating conclusively a role of cyclic nucleotides in the maintenance of vascular smooth muscle tone is in part methodological. No entirely satisfactory in vitro test system, relevant to the physiological situation, has been developed. For example, in order to demonstrate relaxation of smooth muscle, it is necessary to start with a preparation that is at least partially contracted. The process by which contraction is achieved may itself influence the cellular content of cyclic AMP. For these and other reasons, it has been difficult to demonstrate a good correlation between the cyclic AMP content of smooth muscle and the degree of relaxation. For example, it was concluded (112) that an increase in cyclic AMP content was neither necessary nor sufficient for the relaxation of rat pulmonary artery previously contracted with serotonin. A number of reports describe the relaxation of vascular smooth muscle by either cyclic AMP or its derivatives (114–117), as well as by hormones and other agents known to stimulate the formation (110, 117) or to inhibit the degradation (118) of cyclic AMP. In certain cases, it is difficult to distinguish between a direct effect of cyclic AMP on the tone of vascular smooth muscle and the effects of 5′-AMP because in at least some types of muscle 5′-AMP itself brings about relaxation (118).

For it to be proven conclusively that cyclic AMP is the causal agent of relaxation, it must be shown that the increase in the level of the cyclic nucleotide *precedes* relaxation. In experiments in which isoproterenol was employed to relax mesenteric artery, previously contracted with histamine, a peak in the concentration of cyclic AMP occurred 2 min after the addition of isoproterenol and *when relaxation was already complete* (119). In other experiments in which different concentrations of isoproterenol were used for the relaxation of the rabbit anterior mesenteric portal vein, a positive correlation between cyclic AMP levels and relaxation, with respect to both dose and time, was observed (120). The correlation, however, did not hold when relaxation was produced by inhibitors of cyclic AMP phosphodiesterase activity such as papaverine, RA-233, or S 22964 (121).

It is conceivable that the lack of a clear-cut correlation between levels of cyclic AMP and relaxation will be explained by a role of cyclic GMP. Vascular smooth muscle contains cyclic GMP (5, 121, 122); the cyclic GMP content of bovine dorsal digital veins and canine lateral saphenous veins increased with increasing tension (122). When vasoconstriction was induced with prostaglandin $F_{1\alpha}$, little change was seen in the level of cyclic AMP, but the ratio of cyclic GMP to cyclic AMP within the tissue increased, suggesting that the two cyclic nucleotides may have opposing roles in the maintenance of vasomotor tone (122). However, the concept that the contraction of vascular smooth muscle is mediated by cyclic GMP, and its relaxation by cyclic AMP, seems to be incompatible with other recent findings which indicate that in canine femoral arteries there is no correlation between cyclic nucleotide levels and isometric tension when a variety of drugs was employed to manipulate tension (113).

It appears then that, although in certain cases, such as that of the stimulation of glycogenolysis by epinephrine, the mediation of the physiological effect of the hormone by cyclic AMP is clearly established, such a relationship has not been proved in many other cases in spite of the general assumption that a cyclic nucleotide mediates the hormonal effect.

CYCLIC AMP-DEPENDENT PROTEIN KINASES AND MECHANISM OF ACTION OF CYCLIC AMP

It was pointed out in the introductory section that there is considerable evidence which suggests that in eukaryotes, at least, a majority, if not all, of the effects of cyclic AMP are mediated by protein kinases (42, 43). That cyclic AMP activates protein kinases was first shown in 1968 in the case of glycogen phosphorylase kinase kinase (46). This enzyme activates glycogen phosphorylase kinase by the transfer of the terminal phosphate of ATP to specific serine and threonine residues of the phosphorylase kinase. Many proteins can act as phosphate acceptors in the reaction catalyzed by the phosphorylase kinase kinase, hence the term "protein kinase" was accepted. In 1969, it was shown that protein kinase inactivated glycogen synthetase by a mechanism similar to the activation of glycogen phosphorylase kinase (123). It seemed then that at least in vitro protein kinase, activated by cyclic AMP, mimicked two important physiological functions of epinephrine, i.e., the activation of phosphorylase kinase and the inactivation of glycogen synthetase.

The biochemical characteristics of all cyclic AMP-dependent protein kinases are similar (protein kinases not activated by cyclic nucleotides are not discussed in this chapter). The γ-phosphate of ATP serves as phosphate donor; other nucleoside triphosphates are ineffective. The phosphate is transferred to a serine or threonine residue of the receptor protein; the phosphoserine and phosphothreonine bonds are acid stable and alkali labile. The sites of phosphorylation of a number of proteins, including β-casein B_1 (124), the α and β subunits of glycogen phosphorylase kinase (125), pyruvate kinase (126), glycogen synthetase (127), histones F_1 and F_{2b} (128, 129), troponin (130), and myelin basic protein (131, 132) have been identified. It seems that the primary structure of the substrate protein in the vicinity of the phosphorylated residue plays an important role with respect to the specificity of the protein kinase. The use of synthetic hexapeptides (133) and octapeptides (134) as substrates for cyclic AMP-activated protein kinases led to the demonstration of a requirement for an arginine residue three residues distal from the phosphorylated serine. In other experiments, it was demonstrated that the shortest peptide phosphorylated was Arg-Arg-Ala-Ser-Val (135).

The mechanism by which cyclic AMP activates protein kinases has been studied extensively, and the reader is referred to recent reviews of the subject (136, 137). Briefly, cyclic AMP-activated protein kinases are composed of two dissimilar subunits, the R (regulatory) subunit with a molecular weight of 90,000–100,000 and a C (catalytic) subunit of 50,000. The R subunit binds cyclic AMP, whereas the C subunit catalyzes the transfer of phosphate from ATP to the proteinaceous substrate. When combined, the R-C unit is inactive. In the presence of cyclic AMP, the R subunit binds the cyclic nucleotide and dissociates from the C subunit; the latter is then enzymically active. Removal of cyclic AMP permits the reassociation of the subunits and returns the protein kinase to the relatively inactive state. The precise conformation of the holoenzyme is not known with certainty and may vary from tissue to tissue. Some evidence suggests the occurrence of protein kinases composed of one R and one C subunit (137); other findings indicate that the protein kinases of tracheal smooth muscle (138), bovine brain (139), bovine heart (140, 141), and skeletal muscle (142) are tetramers rather than dimers.

$$\underset{\text{(inactive form)}}{R_2C_2} + 2\text{ cyclic AMP} \rightleftharpoons 2R - \text{cyclic AMP} + \underset{\text{(active form)}}{2C}$$

Several mechanisms have been proposed whereby the binding of cyclic AMP dissociates the inactive protein kinase into its regulatory and catalytic subunits (143). There is also evidence that the proteinaceous substrates, as well as cyclic AMP, can cause the partial dissociation of protein kinases into their subunits (139, 144, 145). Finally, it seems that autophosphorylation of the R subunit by the cognate catalytic subunit may promote dissociation. Autophosphorylation of the R subunits of the protein kinases of bovine brain synaptic membranes (146) and bovine cardiac muscle (147) has been reported. To what extent autophosphorylation of protein kinases plays a regulatory role is open to question (148).

The use of column chromatography has led to the demonstration of the occurrence of only one form of protein kinase in certain tissues and of multiple forms in others. Thus, a single form of the enzyme has been reported in studies of bovine endometrium (149), amphibian ovary (150) and oocytes (150), rat and guinea pig islets of Langerhans (151), bovine pineal gland (152), toad bladder epithelium (153), rat adipose tissue (154), and bovine tracheal smooth muscle (138). Two forms of the protein kinase were found in bovine thyroid (155), rabbit skeletal muscle (156, 157), dog prostate (158), bovine epididymal spermatozoa (159), rat liver (160, 161) and kidney (162), and monkey skin (163). The occurrence of four or more forms was reported for rabbit reticulocytes (164), bovine mammary gland from lactating animals (165), and rat skeletal muscle (166). The functional significance of, and the relationships between, the multiple forms of protein kinase are not known. In a study of protein kinases from a variety of sources, two main types of the enzyme were distinguished on the basis of their behavior upon DEAE-cellulose chromatography, and a classification of protein kinases on that basis has been suggested (167). Type I is eluted from the DEAE column with 0.05–0.1 M NaCl and Type II protein kinase with 0.15–0.25 M NaCl. The Type I enzyme is more readily dissociated by either salt or histone at 30 °C than is the Type II enzyme; the dissociation can be prevented by MgATP (167). It has been reported that Type I enzyme is not subject to autophosphorylation (148).

It appears that the K_m of most protein kinases examined for cyclic AMP activation is lower than the cellular concentration of cyclic AMP (168), indicating, at first sight, that the protein kinases are at all times in the active state and negating their regulatory role. Presumably, a cellular compartmentation of the protein kinases and cyclic AMP, respectively, would resolve the apparent paradox. However, it has been pointed out (168) that it may not be necessary to invoke compartmentation since it seems that the phosphorylation of physiologically relevant substrates such as, for example, glycogen synthetase, requires higher concentrations of cyclic AMP than does the phosphorylation of commercially available test substrates such as histones or protamine employed under nonphysiological conditions. A possible regulatory role of modulators of cyclic AMP-dependent protein kinases is discussed later.

The following criteria (paraphrased here) have been proposed by Krebs (169) as constituting evidence for the mediation of a given effect of cyclic AMP by a protein kinase: 1) the presence in a cell of a cyclic AMP-dependent protein kinase; 2) the occurrence in the cell of a proteinaceous substrate (of the protein kinase) with a function related to the cyclic AMP-mediated process; 3) alteration of the activity of this substrate in vitro upon phosphorylation; 4) modification of the substrate in response to cyclic AMP in vivo; and 5) the occurrence, in the same cell, of a phosphoprotein phosphatase capable of catalyzing the reverse reaction, i.e., of the dephosphorylation of the phosphoprotein. One might add as a further criterion the requirement that agents which increase the cellular level of cyclic AMP in vivo also activate the protein kinase in vivo. For obvious reasons, it may not be possible to meet all these criteria in any one situation. They come closest to being satisfied in three cases which involve the regulatory action of epinephrine on energy metabolism in liver, skeletal muscle, and adipose tissue. Specifically, they deal with the acceleration of glycogenolysis by the activation of the glycogen phosphorylase kinase (44–51), the reduction of the synthesis of glycogen by the inhibition of the glycogen synthetase (61–69), and the enhancement of lipolysis by the activation of triglyceride lipase (51–60).

To prove the correctness of the hypothesis that cyclic AMP exerts a given effect by the activation of a protein kinase, it should be demonstrated that an increase in the cellular concentration of cyclic AMP is paralleled by an activation of the protein kinase in vivo. A method by which the in vivo state of activation of a protein kinase may be determined has been reported (170). The procedure consists of the assay of the protein kinase, extracted under conditions controlled with respect to salt composition and concentration, in the absence and presence of cyclic AMP. The activity ratio of the protein kinase is then defined as activity in the absence of cyclic AMP versus activity in the presence of cyclic AMP; i.e., a fully activated (in vivo) protein kinase would display a ratio of 1.0. Activation is indicated by an increase in the ratio, e.g., from 0.1 to 1.0. Control of the salt concentration is important because under certain conditions a manipulation of the ionic composition of the buffer, in which extraction occurs, may itself lead to the dissociation of the protein kinase and hence its activation (171). β-Adrenergic agents have been shown to increase the in vivo activity ratios of protein kinases in adipose tissue (170, 172, 173), the thyroid (174), renal cortex (175), and heart (176), whereas glucagon exerted a similar effect also in adipose tissue (170), renal cortex (173), heart (176), and liver (177–179). Insulin had no effect on the heart kinase (176). Other agents reported to increase the activity ratios of protein kinases in vivo include ACTH in adipose tissue (170), parathyroid hormone in renal cortex (170, 175), thyroid-stimulating hormone in the thyroid (180, 181), follicle-stimulating hormone in the testes (182, 183), luteinizing hormone in the corpus luteum (184, 185), and human chorionic gonadotropin in ovarian cells (186). Activation of a protein kinase, however, could not be demonstrated to accompany the induction of steroidogenesis by ACTH (187).

In the examples just mentioned, protein kinase activity was measured with histones as substrates since the true physiological substrates of the enzyme are known with certainty only in the case of the glycogenolytic, the antiglyconeogenic, and the lipolytic effects of the hormones. The literature is replete with reports of proteins serving as substrates of protein kinases; few, if any, of these proteins rigorously fulfill the criteria which, if met, would demonstrate the role of the proteins in the physiological effects of the hormones. In addition to the ones already mentioned, several other enzymes have been observed to act as substrates for protein kinases. These include pyruvate kinase (188–191), which is inactivated by phosphorylation and tyrosine 3-mono-oxygenase (192), RNA polymerase (193–196), reverse transcriptase (197), carbonic anhydrase (198), cholesterol esterase (199), and the cholesterol side chain cleavage enzyme system (200), which are all activated by phosphorylation. Incubation of tyrosine hydroxylase with a protein kinase (201) reportedly decreased the K_m of the hydroxylase for the cofactor 6-methyltetrahydrofolate; however, incorporation of phosphate into the enzyme could not be demonstrated (201).

There are numerous reports (202–229) describing the phosphorylation of membranal proteins by cyclic AMP-dependent protein kinases; however, in only a few cases has phosphorylation been correlated with a change in function. Thus, the phosphorylation of erythrocyte membranes, catalyzed by cyclic AMP-dependent protein kinases, has been described (210). An increase in the phosphorylation of intact turkey erythrocyte membrane, in response to several agents, was paralleled by an increase in cellular levels of cyclic AMP and enhanced uptake of sodium (210). There have been several reports (211–219) of the phosphorylation of both plasma and synaptic membranes derived from nervous tissue; in synaptosomes, phosphorylation lowered the rate of both the uptake and the efflux of calcium, but had no effect on the equilibrium binding of the ion (219). These and similar observations are suggestive, but do not prove an involvement of cyclic AMP and protein kinases in synaptic transmission.

The increase in the permeability of renal medulla and toad bladder to water, brought about by vasopressin (antidiuretic hormone), is thought to be mediated by cyclic AMP (230). Phosphorylation of renal medullary plasma membranes has been demonstrated (220), and there is evidence that phosphorylation occurs on the luminal surface of the medullary cells (221). In apparent conflict with these findings is the observation (222) that cyclic AMP brought about a decrease in the phosphorylation of a specific protein of the plasma membrane of the toad bladder. Experiments performed in the same laboratory (223) indicated a correlation between the phosphorylation of a membrane protein of 49,000 molecular weight and altered rates of sodium transport; there was no consistent relationship between phosphorylation and changes in permeability to water. Other membranes, reported to be phosphorylated by cyclic AMP-dependent protein kinases, include the plasma membranes of liver (224), Ehrlich ascites cells (225), glial and glioma cells (226), fat cells (227), adenohypophyseal cells (228), platelet membranes (229), the membranes of anterior pituitary secretory granules (231), and the microsomal membranes of adrenal cells (232).

β-Adrenergic agents exert positive inotropic (increase in strength of contraction) effects on heart and slow skeletal muscle; they exert positive chronotropic (increase in frequency of contraction) effects on heart muscle and have little, if any, effect on fast skeletal muscle. The positive inotropic and chronotropic effects of β-adrenergic agents have been attributed to their ability to stimulate the formation of cyclic AMP, and attempts have been made to link the effects with the activation of protein kinases and the phosphorylation of proteins. Thus, changes in cardiac contractility induced by isoproterenol and glucagon in perfused heart muscle have been associated with increased ratios of protein kinase activity (233). In other studies on isolated sarcoplasmic reticulum, the uptake of calcium by the sarcoplasmic reticulum of cardiac and slow skeletal muscle was measured at the same time as was the phosphorylation catalyzed by a cyclic AMP-dependent protein kinase. The phosphorylation of a 22,000-dalton subunit of cardiac and slow skeletal muscle sarcoplasmic reticulum paralleled an increase in the uptake of calcium (234–241). Data, obtained in different laboratories, on the behavior of fast skeletal muscle seem to be in conflict. One report (243) describes the phosphorylation of a protein of 95,000 molecular weight and an increase in the uptake of calcium, whereas in a second report it was claimed that under similar conditions neither phosphorylation nor an increase in the uptake of calcium occurred (237). These contradictory reports, as well as the observation that exogenous glycogen phosphorylase kinase can catalyze the phosphorylation of a 95,000-dalton subunit and stimulate the uptake of calcium by the sarcoplasmic reticulum of all three types of muscle (242), suggest caution in the interpretation of the role of phosphorylation in the regulation of the uptake of calcium and muscle contraction.

The phosphorylation, catalyzed by cyclic AMP-dependent protein kinases, of the contractile proteins of both cardiac and skeletal muscle has also been studied. Since the first report on the phosphorylation of troponin by a protein kinase (244), there have appeared several somewhat contradictory accounts concerning the identity of the component acting as phosphate acceptor. The in vitro phosphorylation of both the troponin inhibitory (TN-I) subunit of troponin, catalyzed by glycogen phosphorylase kinase (245), and of the troponin-tropomyosin-binding (TN-T) subunit, catalyzed by exogenous protein kinase (246–248), has been claimed. Recent findings (245) indicate that the extent of phosphorylation of TN-I and TN-T is modulated by the presence of troponin-Ca^{2+}-binding subunit (TN-C), which itself alters the activity of glycogen phosphorylase kinase. However, in extracts of skeletal muscle obtained from rats genetically defective in glycogen phosphorylase kinase, phosphorylation of troponin still occurred (249), pointing to a role for protein kinase. However that may be, there seems to be evidence for a correlation between the degree of phosphorylation of the TN-I subunit of the contractile apparatus and the positive inotropic effects of β-adrenergic agents (250, 251). The validity or relevance of the apparent correlation, however, may have to be re-examined in the light of recent reports describing in vivo and in vitro experiments in which a relationship between phosphorylation and function could not be established (252–255). In brief, at the time of writing, the physiological relevance of the cyclic AMP-mediated phosphorylation of contractile proteins is not firmly established.

Cellular components other than those discussed above serve as substrates of cyclic AMP-dependent protein kinases. The phosphorylation of ribosomes of liver (256–261), adrenal cortex (262, 263), and reticulocytes (264) has been reported. The phosphorylation of liver ribosomes is stimulated by treatment of the animal with glucagon (260, 261); phosphorylation in this case seems to be limited to the S6 protein of the 40 S subunit of the ribosome (261). The physiological significance of the phosphorylation of the ribosomal protein is unknown; ribosomal function, as tested in vitro, was not affected appreciably by the phosphorylation (259). A similar situation obtains with respect to the physiological effect(s) of the phosphorylation of histones. Phosphorylation of histones is stimulated by the administration of either glucagon or of cyclic AMP to perfused liver (265), of glucagon to the whole animal (266), or of cyclic AMP to liver extracts (267). Other nonenzymic proteins phosphorylated by cyclic AMP-dependent protein kinases include unfractionated nuclear proteins (268), nuclear acidic proteins (269), nucleolar proteins (270), myelin basic proteins (271–273), microtubular proteins and tubulin (274–278), and the active A_1 component of cholera toxin (279). The physiological significance of the phosphorylation of these proteins is not known at present.

Mechanism of Action of Cyclic GMP-dependent Protein Kinases

As pointed out elsewhere in this chapter, the precise physiological role of cyclic GMP is not well understood. Even less is known about the mechanism by which cyclic GMP exerts its effect within the cell. The occurrence of cyclic GMP-dependent protein kinases was first observed in arthropods, and the enzyme from that source has been partially purified (280–283). The mechanism by which the arthropod cyclic GMP-dependent protein kinase is activated seems to be analogous to the activation of the cyclic AMP-dependent protein kinases. Furthermore, it seems that the regulatory subunit of mammalian (bovine brain) cyclic AMP-dependent protein kinase inhibits the catalytic subunit of the cyclic GMP-dependent protein kinase obtained from lobster tail (283). The occurrence of cyclic GMP-dependent protein kinases in mammalian tissues has now also been reported (284–296).

The mechanism by which cyclic GMP activates mammalian cyclic GMP-dependent protein kinases is not clear. It seems that the holoenzyme (molecular weight 150,000–160,000) of fetal calf heart (292), fetal calf lung (293), and bovine aorta (294) is dissociated by cyclic GMP into a 50,000–60,000-dalton catalytic subunit and a presumptive cyclic GMP-binding regulatory subunit. However, other experiments (291) on the cyclic GMP-dependent protein kinase of bovine lung indicated that the enzyme was composed of two subunits of identical molecular weight, i.e., 74,000. Regulatory or catalytic activities could not be assigned to either single subunit (291). The enzyme, like the Type II, cyclic AMP-dependent protein kinases, was subject to autophosphorylation (295). Recent work (296) led to the purification and characterization of the lung enzyme; it was concluded that the cyclic GMP-dependent protein kinase was indeed composed of two identical subunits of a molecular weight of 81,000 which bound both cyclic GMP and catalyzed the phosphotransfer reaction. Much as in the case of the cyclic AMP-dependent protein kinases, the physiologically relevant substrates have not been identified definitively. It has been reported (297) that the phosphorylation of two membranal proteins of 100,000 and 130,000 molecular weight, derived from smooth muscle, is catalyzed by a cyclic GMP-dependent protein kinase. The brush border of intestinal epithelium contains a protein which is phosphorylated in a reaction stimulated by cyclic GMP (298).

Modulators of the Activities of Cyclic Nucleotide-dependent Protein Kinases

A proteinaceous inhibitor of cyclic AMP-activated protein kinase was isolated from rabbit skeletal muscle and partially purified by Walsh et al. in 1971 (299). It seems to occur in all tissues in which cyclic AMP-dependent protein kinases are found and has now been obtained in pure form after 430,000-fold purification from rabbit skeletal muscle (300). The inhibitor is stable to heating at 96 °C and to precipitation with 5% trichloroacetic acid. It is an acidic protein with a molecular weight of 11,300, contains 98 amino acid residues, and is devoid of sulfur; its NH_2 terminus is blocked. The inhibitor binds tightly (K_i of approximately 2×10^{-9} M) to the catalytic subunit of the protein kinase and acts competitively with respect to the protein substrates. It seems that the inhibitor blocks the activity of all cyclic AMP-dependent protein kinases, but does not inhibit cyclic GMP-dependent protein kinases. It also has no effect on other cyclic AMP-independent protein kinases. The inhibitor may, therefore, be used diagnostically to differentiate between natively cyclic AMP-independent protein kinases and the catalytic subunits of natively cyclic AMP-dependent protein kinases formed by the fortuitous dissociation of the holoenzyme during extraction.

It has been suggested (168) that one function of the protein kinase inhibitor might be to buffer variations in the basal level of cellular cyclic AMP since the inhibitor is able to neutralize 0.1–0.2-mol fractions of the total cyclic AMP-dependent protein kinase. The interesting possibility that the level of protein kinase inhibitor might be controlled by insulin has also been suggested (301). Apparently, the administration of insulin leads to an increase in the amount of inhibitor, and alloxan depresses its level. If confirmed and extended, these observations might provide an explanation for the well-known antagonism between the effects of cyclic AMP (or the hormones which stimulate its formation) and insulin.

A highly heat-stable protein which stimulates the activity of cyclic GMP-dependent protein kinases has been partially purified from dog heart (302) and rat brain (303). It appears, in fact, that purified cyclic GMP-dependent protein kinases are stimulated by cyclic GMP only in the presence of this proteinaceous modulator. The modulator seems to lack tissue-, species-, or phylum-specificity because modulator isolated from rat brain stimulated cyclic GMP-dependent protein kinases from dog atrium, guinea pig cerebellum and lung, and lobster tail muscle, in addition to the homologous cyclic GMP-dependent protein kinase.

The possibility that modulators, both stimulatory and inhibitory, of cyclic nucleotide-dependent protein kinases play a role in determining the preferential activity of the enzymes toward different proteinaceous substrates is evidently not excluded.

PHOSPHOPROTEIN PHOSPHATASES

If indeed the phosphorylation of proteins is the primary mechanism by which those hormones which act via cyclic nucleotides exert their effects, then the existence of enzymes which catalyze the dephosphorylation of the proteins seems likely. (As noted above, the irreversible phosphorylation of a protein may occur if that protein is required in its phosphorylated form for, for example, an irreversible step in development.)

The occurrence of phosphoprotein phosphatases in a variety of tissues has been reported; these include cardiac sarcoplasmic reticulum (73), uterus (74, 75), granulocytes (76), brain (77), islets of Langerhans (78), reticulocytes (79), toad bladder (80), liver (81–83), skeletal muscle (84–86), spleen (81), and kidney (81). It seems that a phosphoprotein phosphatase obtained from rabbit liver and purified to homogeneity had approximately the same activity toward a variety of phosphorylated substrates tested, including glycogen phosphorylase kinase, phosphorylase a, glycogen synthetase, histone, casein, and troponin-I (82). In another study (83), liver phosphoprotein phosphatase activity was separated into two fractions which displayed identical activities against phosphorylase a and against histone. These preliminary findings suggest that the dephosphorylation of a variety of phosphoproteins is catalyzed by the same phosphatase. It has been reported (304) that cyclic AMP activates a membrane-bound phosphoprotein phosphatase in toad bladder. Conversely, a partially purified phosphoprotein phosphatase from skeletal muscle was inactivated when incubated with a protein kinase and cyclic AMP. The possible occurrence of the regulation of phosphoprotein phosphatase activity by steroids is discussed in the section on steroid-cyclic AMP interactions (87).

CYCLIC AMP-DEPENDENT PROTEIN KINASES AS HORMONAL MEDIATORS

The data presented here, as well as similar findings too numerous to be cited, suggest indeed that many of the effects of those hormones which modulate the cellular concentrations of cyclic nucleotides are mediated by protein kinases. The nature of the ultimate effect, as well as of the specificity, of the hormone resides then in the identity and function of the protein phosphorylated. Only very few substrates of cyclic nucleotide-activated protein kinases have been identified with respect to function; they include glycogen phosphorylase kinase, glycogen synthetase, pyruvate kinase, triglyceride lipase, and, typically, they catalyze key reactions of energy metabolism (44–69). It would, therefore, be premature, in the authors' opinion, to conclude that *all* of the effects of cyclic nucleotides are mediated by protein kinases, and, in this context, the reader is once again referred to the quite different mode of action by which cyclic AMP exerts its effects in prokaryotes (37, 38). Perhaps the strongest evidence for a role of protein kinases as mediators of the effects of cyclic AMP in reactions other than the regulation of energy metabolism comes from studies (70) with mutant strains of SV49 lymphoma cells. High concentrations of dibutyryl cyclic AMP cause the cytolysis of cells with a functional cyclic AMP-activated protein kinase, whereas cells defective in the protein kinase are resistant to the cytolytic effect of the cyclic nucleotide.

CYCLIC AMP, CYCLIC GMP, AND CALCIUM

Clearly, any consideration of the mode of action of cyclic AMP has to relate the effects of this nucleotide to those of cyclic GMP, as well as to those of calcium, which interacts with the cyclic nucleotides at several levels (305–308).

Cyclic GMP was first found in 1963 in rat urine (2). Several years later, it was observed that acetylcholine raised the level of cyclic GMP in perfused heart (309). On the basis of this and similar observations (3–5), Goldberg and his collaborators proposed the so-called Yin Yang hypothesis of biological control. In its essentials, the hypothesis suggests that, in those bidirectional biological systems in which one hormonal agent turns the system on and another hormonal agent turns it off, these signals are mediated by the generally opposing actions of cyclic AMP and cyclic GMP. The hypothesis has been invoked to explain the role of cyclic GMP in several tissues and systems, including the contraction and relaxation of smooth muscle, the aggregation of platelets, the control of the proliferation of cells, etc. As an example of the application of the hypothesis, the proposed role of cyclic GMP in the cholinergic stimulation of the heart is discussed here.

In the heart, β-adrenergic stimulation through agents such as epinephrine or isoproterenol leads to both an increase in the force of the contraction (positive inotropic effect) and an increase in the frequency of contractions (positive chronotropic effect). These effects are physiologically reversed through vagal stimulation mediated by the neurotransmitter, acetylcholine. A 2- to 3-fold increase in the cyclic GMP concentration was found in the heart upon perfusion with acetylcholine (309); furthermore, cyclic AMP levels were diminished as cyclic GMP levels were elevated (310). In these experiments, the dose of acetylcholine was sufficient to produce negative inotropic effects. In other experiments, increased levels of cyclic GMP were found in perfused hearts challenged with doses of acetylcholine which did not lead to changes in contractility (311). In isolated frog heart, the changes in the concentration of cyclic AMP and cyclic GMP followed a cyclic pattern: cyclic AMP reached its highest level just before the peak of the contraction was attained, whereas cyclic GMP was then at its lowest level. Both cyclic nucleotides returned to the basal level prior to the initiation of the next beat (312). Effects of derivatives of cyclic GMP on the contraction of the heart have been observed. Thus, pretreatment of heart papillary muscle with monobutyryl cyclic GMP blocked the normal inotropic response to dibutyryl cyclic AMP (313); the inotropic effects of isoproterenol in the perfused heart were also antagonized by dibutyryl cyclic GMP (311). An antagonism between dibutyryl cyclic GMP and norepinephrine or dibutyryl cyclic AMP, respectively, was also observed in cultured mouse myocardial cells (314). These and similar observations suggest that the Yin Yang hypothesis of the opposing physiological effects of the two cyclic nucleotides may apply to certain systems such as, for example, the autonomic regulation of the heart. The general applicability of the hypothesis, however, is open to question. For example, in the case of smooth muscle, no correlation could be established between muscle tone and levels of cyclic AMP and cyclic GMP (113). In other systems, the cellular concentrations of the two cyclic nucleotides seem to change in parallel rather than in opposing directions. For example, aggregation-competent amoebae of the cellular slime mold, *Dictyostelium discoideum* respond to pulses of either exogenous cyclic AMP or folic acid (both of these compounds serve as chemotactic signals) by the biphasic synthesis of cyclic GMP and of cyclic AMP. The peaks in cyclic GMP levels precede by a few seconds the peaks in cyclic AMP levels (315, 316, 317). Earlier observations (318) demonstrated that cyclic AMP levels in *D. discoideum,* at certain stages of development, oscillated with an amplitude of 30–100-fold over the basal level and a periodicity of approximately 7 min at 22 °C. These findings stress the importance of frequent samplings for the determination of cyclic nucleotide concentrations, and, more fundamentally, they raise the possibility that, in other systems as well, cyclic nucleotide concentrations may fluctuate sharply in response to intra- and intercellular signals.

Interactions of cyclic nucleotides and calcium have been reviewed by Rasmussen and co-workers (305, 306) and by Berridge (307, 308). In contrast with the Yin Yang hypothesis, in which generally opposing functions were assigned to cyclic AMP and cyclic GMP, the hypotheses put forward by Rasmussen (306) and by Berridge (308), respectively, deal primarily with the interactions of cyclic AMP and calcium. Systems of hormonal control are considered by Berridge as either mono- or bidirectional; in both mono- or bidirectional systems cyclic AMP and calcium, respectively, are considered to be the most likely intracellular second messengers. In a monodirectional system, a given hormone brings about an increase in cellular cyclic AMP which then (for example, by the activation of a protein kinase) produces the physiological effect. A second hormone leads to an increase in the cellular concentration of calcium and this ion, in turn, exerts a physiological effect. The two hormonal controls may be interrelated, by, for example, a positive effect of cyclic AMP on the cellular concentration of calcium. Instances of hormonal regulation, most readily explained in terms of this model, have been described. Thus, in the β cells of the pancreas, glucagon causes an increase in cyclic AMP whereas glucose enhances the uptake of calcium by the cells. The increase in the cellular concentration of cyclic AMP may then bring about a further enhancement of the level of cellular calcium by the release of intracellular stores of calcium; the calcium in turn brings about the release of insulin (308). In the adrenal cortex, according to the hypothesis (308), ACTH would enhance the cellular levels of both cyclic AMP and available calcium. Cyclic AMP, it is argued, would stimulate the synthesis of a rate-limiting protein required for the catalysis of one of the early steps in steroidogenesis. The enhanced level of available calcium, possibly derived in part from intracellular stores, would trigger the hydroxylation and release of steroids. Similarly, serotonin, in the salivary glands of insects, stimulates the formation of cyclic AMP and possibly also affects the uptake of calcium by the cells. Cyclic AMP then stimulates the pumping of potassium out of the cell at the apical membrane, whereas calcium enhances the permeability of both basal and apical membranes to chloride. At the same time, cyclic AMP causes the release of stored intracellular calcium, which in turn potentiates the effect of the permeability to chloride.

In bidirectional systems, as in monodirectional systems, one hormone stimulates the formation of cyclic AMP, whereas the second hormone leads to an increase in cellular calcium. Unlike in the case of monodirectional systems, however, in bidirectional systems, the increase in the concentration of cyclic AMP leads to a lowering of cellular calcium and thus reverses the physiological effect of the hormone which brought about the increase in calcium concentration. Hormonal effects, best explained in terms of this model, have also been described (308). Thus, in smooth muscle, acetylcholine brings about the release of calcium into the cytoplasm from internal stores and also increases the uptake of external calcium. This increase in free cellular calcium causes contraction of the muscle. Norepinephrine stimulates the formation of cyclic AMP which, in turn, lowers the cellular calcium and causes relaxation. A second example of bidirectional control is that which regulates the aggregation of blood platelets. ADP or thrombin increases the influx of calcium into the platelets; the calcium then causes changes in the cell surface which lead to aggregation. Protaglandin E_1 or adenosine maintain the unaggregated state by their stimulation of the synthesis of cyclic AMP, which may act by causing the release of calcium.

Both the Yin Yang and calcium hypotheses provide a useful framework for the study of the complex interactions between cyclic AMP, cyclic GMP, and calcium. It has already been pointed out that the Yin Yang hypothesis, with its emphasis on the antagonism between the effects of the two cyclic nucleotides, is probably of only limited applicability. An assessment of the general validity of the calcium hypothesis must be postponed until additional experimental data are at hand. The use of calcium ionophores should provide new insights into the role of calcium, although at times it may be difficult to distinguish between a role of calcium as second messenger and its required participation in a specific reaction, such as, for example, in the case of the reaction catalyzed by glycogen phosphorylase *b* kinase.

GLUCOCORTICOID—CYCLIC AMP INTERACTIONS

A great deal of evidence suggests that adrenal glucocorticoids are also implicated in the control of metabolic functions mediated by cyclic nucleotides. Yet, whereas considerable information is now available regarding the mode of action of steroid hormones in general (cf. Chapter 4 this volume), the basis of the occasional synergism between cyclic nucleotides and glucocorticoids is still not understood (319). The interaction between cyclic nucleotides and glucocorticoids is exemplified by the so-called permissive effect. An early observation bearing on this phenomenon was the finding (320) that the injection of epinephrine failed to increase the level of plasma-free fatty acids in adrenalectomized or hypophysectomized dogs; treatment with hydrocortisone restored the level of released free fatty acids to that of control animals. Analogous observations on the effects of glucocorticoids on the stimulation of lipolysis by cyclic AMP were made by other workers (321–323). Glycogenesis from hexose was decreased by adrenalectomy and restored by hydrocortisone (324); glycogen synthetase activity was abolished by adrenalectomy and restored by glucocorticoids (325, 326). The transport of α-aminoisobutyrate induced in cultured rat liver parenchymal cells was stimulated 2-fold by dexamethasone, which in the absence of glucagon (or dibutyryl cyclic AMP) had no effect (327). Of particular interest in the context of the permissive effect is the finding that hydrocortisone or dexamethasone could substitute for serum normally required for the stimulation of DNA synthesis in 3T3 fibroblasts by the polypeptide fibroblast growth factor (328). Apparently, the fibroblast growth factor exerts its effect by bringing about a transient increase in cyclic GMP (329). It seems then that glucocorticoids interact in some manner with hormones which exert their effects by changes in the concentration of either cyclic AMP or cyclic GMP.

As stressed earlier, the basis of the permissive effect is still not understood. It has been suggested that the glucocorticoids may act at the level of the cyclic AMP phosphodiesterase. Thus, in HTC hepatoma cells, cyclic

AMP phosphodiesterase activity decreased by 25–40% after incubation of the cells for 36 hr or longer in the presence of μmolar concentrations of dexamethasone; the basal level of cyclic AMP increased slightly, and there was a pronounced increase in the cyclic AMP concentration when epinephrine and theophylline were administered (330). In a more direct approach (331), the effect of μmolar concentrations of cortisol on rat testicular and beef heart cyclic AMP phosphodiesterases was tested in vitro, and pronounced inhibition of the activities of the two enzymes was found. The significance of these and similar observations, however, is open to question because, in a majority of cases where withdrawal of glucocorticoids, for example, by adrenalectomy, led to loss of cyclic AMP-mediated activity, neither a marked decrease in the level of the cyclic nucleotides was found, nor did addition of the cyclic nucleotide restore the lost activity. The reader is referred to the review by Amer and Kreighbaum (40) for a listing of papers dealing with an effect of hormones on cyclic AMP phosphodiesterases.

Interactions at the cellular level between hormones which act via cyclic nucleotides and steroids other than glucocorticoids have also been noted. There are several reports which suggest that the aldosterone antagonist spironolactone (332), aldosterone (80), progesterone (74), and estradiol (75) either induce or enhance phosphoprotein phosphatase activity. The original claim (75), however, that the protein induced by estradiol in the rat uterus had phosphoprotein phosphatase activity seems to have been in error (333).

A recent observation by Greengard and his collaborators (334) may be relevant to the interaction of cyclic nucleotides and steroid hormones. The investigators isolated from a variety of tissues a protein which they termed SCARP (steroid and cyclic AMP-regulated phosphoprotein); the protein has an apparent molecular weight of approximately 54,000. Its phosphorylation is controlled in a complex and tissue-specific manner by several steroid hormones as well as by hormones which act via cyclic AMP. The function of SCARP is not known at the time of writing.

CONCLUDING REMARKS: CYCLIC AMP IN EVOLUTION

Cyclic nucleotides are found in prokaryotes and in all phyla of animals. Somewhat surprisingly, the occurrence, or nonoccurrence, of cyclic nucleotides in higher plants is still a matter of debate; the evidence is reviewed in a recent article (335). In any case, the quasi-ubiquity of cyclic nucleotides indicates evolutionary antiquity. It may be instructive, therefore, to compare the functions of cyclic AMP (more is known about it than about cyclic GMP) in the evolutionarily primitive and organizationally simple prokaryote, *E. coli*, in the cellular slime mold, *D. discoideum* (336), which has both single-celled and multicellular stages in its life cycle, and in the highly evolved multicellular eukaryotes.

In the bacterium *E. coli*, starvation for a source of carbon stimulates the formation of cyclic AMP, and, conversely, there is an inverse relationship between the effectiveness of a given compound as source of carbon and cellular levels of cyclic AMP (337). The precise mechanism by which the uptake of a sugar (38) or its metabolism controls the cellular concentration of cyclic AMP in *E. coli* is still not understood. In any case, cyclic AMP facilitates the transcription of a number of genes (probably no fewer than 50 and no more than 200) which, typically, code for proteins not required under all conditions of growth. In a majority of cases, the genes so affected are under at least dual control in that their expression also requires a second, operon-specific signal. The case of the *lac* operon of *E. coli* may serve as an example (37, 38). Effective transcription of this cluster of genes requires the exposure of the bacteria to both a β-galactoside which neutralizes the inhibition of transcription exercised by the operon-specific, proteinaceous repressor *and* a threshold level of cyclic AMP; i.e., the operon is transcribed at a maximum rate only, if bacterial growth is limited by the availability of an effective source of energy (or if cyclic AMP is furnished exogenously). The mechanism by which cyclic AMP stimulates the transcription of certain genes in *E. coli* is known (37, 38). Thus, a noncovalent complex between cyclic AMP and a protein, of 45,000 molecular weight, interacts with the deoxyribonucleotide sequences which constitute the sites at which RNA polymerase binds to a given operon prior to its transcription. The interaction between the protein-cyclic AMP complex and the deoxyribonucleotide sequences in some manner facilitates the binding of the RNA polymerase and hence the transcription of the gene.

As stated earlier, cyclic AMP modulates the transcription of genes which code for proteins not essential under all conditions of growth. Certain of these proteins are enzymes which catalyze reactions that make available to the bacteria sources of carbon not attacked prior to the increase in the cellular concentration of cyclic AMP. One of the functions of cyclic AMP in bacteria then is the mobilization of potential sources of energy. Other bacterial proteins controlled in their synthesis by cyclic AMP include flagellar proteins required for motility and enzymes that catalyze reactions leading to the destruction of antibiotics. Mutants of *E. coli*, defective in either adenyl cyclase or in the synthesis of the protein with which cyclic AMP combines, grow in the absence of cyclic AMP if furnished wih a rich medium. Adaptation, however, of the bacteria to more demanding conditions of growth, which require the full expression of the genetic repertory of the bacteria, does not occur in the absence of cyclic AMP or of the cyclic AMP-binding protein. In a manner of speaking, then, cyclic AMP is required for the physiological "reversible differentiation" of the bacteria.

Cyclic AMP seems to play a somewhat analogous role in the cellular slime mold, *D. discoideum*. This organism grows and replicates in the form of single-celled amoebae as long as furnished with nutrients (336). When starved, the amoebae synthesize and release cyclic AMP in a pulsatile manner. Whether or not formation of cyclic AMP in this organism, too, can be related specifically to low levels of sources of carbon and energy, or whether starvation for any required nutrient suffices to induce the synthesis of cyclic AMP is still not clear (338–340); the mechanism by which starvation leads to the synthesis of cyclic AMP is also unknown. The cyclic AMP released by the starving amoebae acts as chemotactic agent and brings about the formation of aggregates consisting of up to 10^5 amoebae per aggregate. The aggregates eventually give rise to fruiting bodies composed of two types of cells—large, vacuolated, dead stalk cells and viable spores, which, under appropriate nutritional conditions, germinate into amoebae. The transition from vegetative growth to the formation of the fruiting body occurs in the absence of exogenous nutrients and at the expense of preexisting macromolecules. It seems that cyclic AMP, in addition to its intercellular, chemotactic function, also plays a role in the intracellular events required for development. Specifically, an increase in cyclic AMP is associated with the appearance of a number of proteins related to aggregation and the formation of the fruiting body. Conversely, the exposure of the amoebae to exogenous cyclic AMP under appropriate conditions leads to changes in the enzymic constitution of the cells and accelerates development. The mechanism of this effect of cyclic AMP is not known at present; the de novo syntheses of RNA and protein, however, seem to be required. It has been claimed that pre-stalk and pre-spore cells differ from one another in their content of cyclic AMP and calcium in the sense that there is a significantly higher concentration of both cyclic AMP and calcium in the pre-stalk than in the pre-spore cells (341). Certain developmentally defective mutants of the organism can be "rescued" by exposure to pulses of added cyclic AMP (342). In brief then, cyclic AMP is formed by *D. discoideum* in response to starvation; it acts as chemotactic agent and seems to be required for cellular differentiation.

The role of cyclic AMP as the cellular mediator of polypeptide and catecholamine hormones in higher eukaryotes has been discussed at length. It might be pointed out, however, that the major physiological function of cyclic AMP in the prokaryote *E. coli*, i.e., the mobilization of potential reserves of carbon and energy when readily available sources become limiting, is still one of the major functions of cyclic AMP in certain tissues of the multicellular eukaryotes. In the single-celled bacteria, endogenous metabolism regulates the level of cyclic AMP; the cyclic nucleotide then exerts its effects on energy metabolism by facilitating the synthesis of enzymes which catalyze growth-limiting reactions. Whether a similar mechanism of energy mobilization, with a pivotal role for cyclic AMP in the modulation of gene transcription, exists also in multicellular eukaryotes is not known with certainty but is suggested by the fact that the administration of exogenous dibutyryl cyclic AMP or hormones which cause the endogenous formation of cyclic AMP stimulates the synthesis of tyrosine aminotransferase in cultured liver cells (343) and of serine dehydratase in rat liver (344). Both enzymes catalyze reactions which make the carbon skeletons of tyrosine and of serine and threonine, respectively, available as sources of energy. In any case, it seems that, at least, the best studied mechanism of regulation or energy metabolism in multicellular organisms is intercellular and hormonal. Thus, for example, glucagon, released in response to a low level of glucose in the circulation, stimulates the synthesis of cyclic AMP in the relevant target tissues; the cyclic AMP, in turn, enhances the breakdown of glycogen and lipids by the activation of preexisting enzymes.

The function of cyclic AMP as chemotactic agent in the cellular slime mold, *D. discoideum*, has been mentioned. One may ask whether cyclic AMP also acts as a morphogen in other organisms. The prokaryotic *Myxobacteria* are of interest in this context. Somewhat like the eukaryotic slime molds, the myxobacteria also give rise to fruiting bodies upon starvation for specific nutrients such as certain amino acids (345). The natural chemotactic substances, responsible for aggregation and fruiting body formation, have not been identified. However, both exogenously added ADP and cyclic AMP greatly enhance fruiting body formation in *Myxococcus xanthus* (346), and methylxanthines stimulate both the synthesis of cyclic AMP and the formation of fruiting bodies in the same organism (347). These observations suggest that cyclic AMP may act as chemotactic agent also in those prokaryotes in which cellular differentiation and morphogenesis occur. An investigation of a possible role of cyclic nucleotides in the morphogenesis, particularly the formation of the nitrogen-fixing heterocysts, of the multicellular, filamentous blue-green bacteria would be of obvious interest.

An effect of cyclic AMP on cell differentiation has been observed in the moss *Fumaria hygrometica,* in which cyclic AMP enhances the formation of chloronema filaments. The effect is not specific for cyclic AMP; other purine nucleotides also stimulate the formation of the differentiated structures (348).

An involvement of cyclic nucleotides in the formation of the gemmule, i.e., the dormant form of fresh water sponges formed by aggregation of cells within the adult tissue, has been suggested (349). The continuous exposure of hatching sponges of *Spongilla lacustris* to either cyclic AMP or cyclic GMP-enhanced gemmule formation (350); theophylline arrested germination of the gemmules. In another species of fresh water sponge, *Ephydatia fluviatilis*, theophylline stimulated the formation of the gemmules, but no effect of either cyclic AMP or cyclic GMP was observed (351).

An interesting series of observations was made by Robertson and Gingle (352) on the effects of exposing 1-day-old chick embryos to pulses of cyclic AMP. The cyclic nucleotide affected the migration of the cells during early stages of morphogenesis and caused the bending of the primitive streak toward the source of the pulses of cyclic AMP. Earlier work from the same laboratory (353) showed that cyclic AMP enhanced the ability of 1-day-old chick embryo cells to form aggregates and suggested that cyclic AMP served as a relay signal (much as it does in the case of the amoebae of *D. discoideum*). It seems that cells obtained from older chick embryos, e.g., 8–10 days of age, are insensitive to the effects of cyclic AMP when tested for their ability to re-form the tissue of origin after treatment of that tissue with trypsin (354).

In summary, then, there is conclusive evidence that in prokaryotes and in simple, as well as in complex, eukaryotes cyclic AMP is formed in response to starvation. In obligatorily single-celled prokaryotes, the primary function of cyclic AMP is to enable the organism to overcome starvation by facilitating the utilization of new potential sources of food. This requires the expression of genes not transcribed under nutritionally favorable conditions and leads to what might be termed "reversible cellular differentiation." In organisms such as the cellular slime molds, which have both single-celled and multicellular stages in their life cycle, starvation also provokes the synthesis of cyclic AMP. Cyclic AMP is released by the starving amoebae and brings about multicellularity. It seems then that at least in this organism multicellularity is a direct consequence of starvation. Furthermore, there is some evidence, cited earlier, (341) that cyclic AMP may play a determinative role in *D. discoideum* since a correlation seems to exist between the cyclic AMP (and calcium!) content of a given cell and the likelihood of its becoming either a stalk cell or a spore. In the higher, multicellular eukaryotes, cyclic AMP mediates the cellular effects of a large variety of hormones; it has retained its primeval role as regulator of energy metabolism. There is suggestive evidence that it may play the role of morphogen at very early stages of development, i.e., prior to the appearance of neurohormones, sometimes assumed to act as morphogens (for discussion see ref. 355). It may well be legitimate then to consider cyclic AMP as a prehormone in both the phylogenetic and the ontogenetic sense. The possibility that starvation was indeed the primum movens for multicellularity, and cyclic AMP the mediator, is an intriguing one.

ACKNOWLEDGMENT

The authors acknowledge the skillful assistance of Ethel Goren in the preparation of this manuscript.

REFERENCES

1. Sutherland, E. W., and Rall, T. W. (1960). Pharmaçol. Rev. 12:265.
2. Ashman, D. F., Lipton, R., Melicow, M. M., and Price, T. D. (1963). Biochem. Biophys. Res. Commun. 11:334.

3. Goldberg, N. D., O'Dea, R. F., and Haddox, M. K. (1973). Adv. Cyclic Nucleotide Res. 3:155.
4. Goldberg, N. D., Haddox, M. K., Dunham, E., Lopez, C., and Hadden, J. W. (1974). *In* B. Clarkson and R. Baserga (eds.), The Cold Spring Harbor Symposium on the Regulation of Proliferation in Animal Cells, pp. 609–625. Cold Spring Harbor Laboratory, New York.
5. Goldberg, N. D., Haddox, M. K., Nicol, S. E., Glass, D. B., Sanford, C. H., Kuehl, F. A., Jr., and Estenen, R. (1975). Adv. Cyclic Nucleotide Res. 5:307.
6. Goldberg, N. D., and Haddox, M. K. (1977). Annu. Rev. Biochem. 46:823.
7. Perkins, J. (1973). Adv. Cyclic Nucleotide Res. 3:1.
8. Stellwagen, E., and Baker, B. (1976). Nature 261:719.
9. Rodbell, M., Kizans, H. M. J., Pohl, S. L., and Birnbaumer, L. (1971). J. Biol. Chem. 246:1872.
10. Rodbell, M., Birnbaumer, L., Pohl, S. L., and Kizans, H. M. J. (1971). J. Biol. Chem. 246:1877.
11. Londos, C., Salomon, Y., Lin, M. C., Harwood, J. P., Schramm, M., Wolff, J., and Rodbell, M. (1974). Proc. Natl. Acad. Sci. USA 71:3087.
12. Rasmussen, H. (1970). Science 170:404.
13. Birnbaumer, L. (1973). Biochim. Biophys. Acta 300:129.
14. Perkins, J. P. (1973). Adv. Cyclic Nucleotide Res. 3:1.
15. Levitzki, A., Sevilla, N., Atlas, D., and Steer, M. L. (1974). J. Mol. Biol. 97:35.
16. Steer, M. L., and Levitzki, A. (1975). J. Biol. Chem. 250:2080.
17. Steer, M. L., and Levitzki, A. (1975). Arch. Biochem. Biophys. 167:371.
18. Drummond, G. I., and Duncan, L. (1970). J. Biol. Chem. 245:976.
19. Rodbell, M., Birnbaumer, L., Pohl, S. L., and Krans, H. M. J. (1971). J. Biol. Chem. 246:1877.
20. Constantopoulos, A., and Najjar, V. (1973). Biochem. Biophys. Res. Commun. 53:794.
21. Najjar, V., and Constantopoulos, A. (1973). Mol. Cell. Biochem. 2:87.
22. Puchwein, G., Pfeuffer, T., and Helmreich, E. J. M. (1974). J. Biol. Chem. 249:3232.
23. Limbird, L., and Lefkowitz, R. J. (1976). Mol. Pharmacol. 12:559.
24. Rubalcava, B., and Rodbell, M. (1973). J. Biol. Chem. 248:3831.
25. Cuatrecasas, P., Hollenberg, M. D., Chang, K.-J., and Bennett, V. (1975). Recent Progr. Horm. Res. 31:37.
26. Cuatrecasas, P., and Hollenberg, M. D. (1976). Adv. Protein Chem. 30:251.
27. Cuatrecasas, P. (1974). Annu. Rev. Biochem. 43:169.
28. DeHaën, C. (1976). J. Theor. Biol. 58:383.
29. Rall, T. W., and Sattin, A. (1970). Adv. Biochem. Psychopharmacol. 3:113.
30. Clark, R. B., and Perkins, J. P. (1971). Proc. Natl. Acad. Sci. USA 68:2757.
31. Frey, W. H., Boman, B. M., Newman, D., and Goldberg, N. D. (1977). J. Biol. Chem. 252:4298.
32. Limbird, L. E., and Lefkowitz, R. J. (1975). Biochim. Biophys. Acta 377:186.
33. Kimura, H., and Murad, F. (1974). J. Biol. Chem. 249:6910.
34. Chrisman, T. D., Garbers, D. L., Parks, M. A., and Hardman, J. G. (1975). J. Biol. Chem. 250:374.
35. George, W. J., Polson, J. B., O'Toole, A. G., and Goldberg, N. D. (1970). Proc. Natl. Acad. Sci. USA 66:398.
36. Arnold, W. P., Mittal, C. K., Katsuki, S., and Murad, F. (1977). Proc. Natl. Acad. Sci. USA 74:3203.

37. Rickenberg, H. V. (1974). Annu. Rev. Microbiol. 28:353.
38. Peterkofsky, A. (1976). Adv. Cyclic Nucleotide Res. 1:48.
39. Appleman, M. M., Thompson, W. J., and Russell, T. R. (1973). Adv. Cyclic Nucleotide Res. 3:65.
40. Amer, M. S., and Kreighbaum, W. E. (1975). J. Pharm. Sci. 64:1.
41. Chasin, M., and Harris, D. N. (1976). Adv. Cyclic Nucleotide Res. 7:225.
42. Kuo, J. F., and Greengard, P. (1969). Proc. Natl. Acad. Sci. USA 64:1349.
43. Kuo, J. F., Krueger, B. K., Sanes, J. R., and Greengard, P. (1970). Biochim. Biophys. Acta 212:79.
44. Posner, J. B., Stein, R., and Krebs, E. G. (1965). J. Biol. Chem. 240:982.
45. Riley, W. D., DeLange, R. J., Bratvold, G. E., and Krebs, E. G. (1968). J. Biol. Chem. 243:2209.
46. Walsh, D. A., Perkins, J. P., and Krebs, E. G. (1968). J. Biol. Chem. 243: 3763.
47. Drummond, G. I., Harwood, J. P., and Powell, C. A. (1969). J. Biol. Chem. 244:4235.
48. Mayer, S. E., and Krebs, E. G. (1970). J. Biol. Chem. 245:3153.
49. Walsh, D. A., Perkins, J. P., Brostrom, C. O., Ho, E. S., and Krebs, E. G. (1971). J. Biol. Chem. 246:1968.
50. Goris, J., and Merlevede, W. (1974). FEBS Lett. 48:184.
51. Cohen, P., Watson, D. C., and Dixon, G. H. (1975). Eur. J. Biochem. 51:79.
52. Knight, B. L. (1975). Biochem. J. 152:577.
53. Soderling, T. R., Corbin, J. D., and Park, C. R. (1973). J. Biol. Chem. 248: 1882.
54. Hollenberg, C. H., Raben, M. S., and Atwood, E. B. (1961). Endocrinology 68:589.
55. Vaughan, M., Berger, J. E., and Steinberg, D. (1964). J. Biol. Chem. 239: 401.
56. Corbin, J. D., and Krebs, E. G. (1969). Biochem. Biophys. Res. Commun. 36:328.
57. Huttunen, J. K., Steinberg, D., and Mayer, S. E. (1970). Proc. Natl. Acad. Sci. USA 67:290.
58. Huttunen, J. K., Steinberg, D., and Mayer, S. F. (1970). Biochem. Biophys. Res. Commun. 41:1350.
59. Corbin, J. D., Reimann, E. M., Walsh, D. A., and Krebs, E. G. (1970). J. Biol. Chem. 245:4849.
60. Huttunen, J. K., and Steinberg, D. (1971). Biochim. Biophys. Acta 239:411.
61. Belocopitow, E. (1961). Arch Biochem. Biophys. 93:457.
62. Friedman, D. L., and Larner, J. (1963). Biochemistry 2:669.
63. Traut, R. R., and Lipmann, F. (1963). J. Biol. Chem. 238:1213.
64. Craig, J. W., and Larner, J. (1964). Nature 202:971.
65. Rosell-Perey, M., and Larner, J. (1964). Biochemistry 3:81.
66. Larner, J., Villar-Palasi, C., and Brown, N. E. (1969). Biochim. Biophys. Acta 178:470.
67. Schlender, K. K., Wei, S. H., and Villar-Palasi, C. (1969). Biochim. Biophys. Acta 191:272.
68. Soderling, T. R., Hickenbottom, J. P., Reimann, E. M., Hunkeler, F. L., Walsh, D. A., and Krebs, E. G. (1970). J. Biol. Chem. 245:6317.
69. Soderling, T. R. (1975). J. Biol. Chem. 250:5407.
70. Insel, P. A., Bourne, H. R., Coffino, P., and Tomkins, G. M. (1975). Science 190:894.
71. Li, H. C. (1975). FEBS Lett. 55:134.

72. Nakai, C., and Thomas, J. A. (1974). J. Biol. Chem. 249:6459.
73. Tada, M., Kirchberger, M. A., and Li, H. C. (1975). J. Cyclic Nucleotide Res. 1:329.
74. Roberts, R. M., and Bazer, F. W. (1976). Biochem. Biophys. Res. Commun. 68:450.
75. Vokaer, A., Iacobelli, S., and Kram, R. (1974). Proc. Natl. Acad. Sci. USA 71:4482.
76. Layne, P., Constantopoulos, A., Judge, J. F. X., Rauner, R., and Najjar, V. (1973). Biochem. Biophys. Res. Commun. 53:800.
77. Tang, F. Y., and Hoskins, D. D. (1975). Biochem. Biophys. Res. Commun. 62:328.
78. Dods, R. F., and Burdowski, A. (1973). Biochem. Biophys. Res. Commun. 51:421.
79. Lightfoot, H. N., Mumby, M., and Traugh, J. A. (1975). Biochem. Biophys. Res. Commun. 66:1141.
80. Liu, A. Y. C., and Greengard, P. (1974). Proc. Natl. Acad. Sci. USA 71:3869.
81. Japundzic, I., Mimic-Oka, J., and Japundzic, M. (1972). Eur. J. Biochem. 28:475.
82. Khandelwal, R. L., Vandenheede, J. R., and Krebs, E. G. (1976). J. Biol. Chem. 251:4850.
83. Kobayashi, M., Kato, K., and Sato, S. (1975). Biochim. Biophys. Acta 377: 343.
84. Kato, K., and Bishop, J. S. (1972). J. Biol. Chem. 247:7420.
85. England, P. J., Stull, J. T., and Krebs, E. G. (1972). J. Biol. Chem. 247:5275.
86. Kato, K., and Sato, S. (1974). Biochim Biophys. Acta 358:299.
87. Huang, F. L., and Glinsmann, W. H. (1975). Proc. Natl. Acad. Sci. USA 72:3004.
88. Sutherland, E. W., Jr., and Robinson, G. A. (1966). Pharmacol. Rev. 18: 145.
89. Butcher, R. W., Baird, C. E., and Sutherland, E. W. (1968). J. Biol. Chem. 243:1705.
90. Menahan, L. A., and Wieland, O. (1967). Eur. J. Biochem. 9:55.
91. Hepp, K. D., Menahan, L. A., Wieland, O., and Williams, R. H. (1969). Biochim. Biophys. Acta 184:554.
92. Sutherland, E. W., and Rall, T. W. (1958). J. Biol. Chem. 232:1077.
93. Rall, T. W., and Sutherland, E. W. (1958). J. Biol. Chem. 232:1065.
94. Sutherland, E. W., Øye, I., and Butcher, R. W. (1965). Rec. Prog. Hormone Res. 21:623.
95. Robinson, G. A., Exton, J. W., Park, C. R., and Sutherland, E. W. (1967). Fed. Proc. 26:257.
96. Reynolds, R. C., and Haugaard, N. (1967). J. Pharmacol. Exp. Ther. 156: 417
97. DeWulf, H., and Hers, H. G. (1968). Eur. J. Biochem. 6:552.
98. Northrop, G., and Parks, R. E., Jr. (1964). J. Pharmacol. Exp. Ther. 154: 87.
99. Northrop, G., and Parks, R. E., Jr. (1964). J. Pharmacol. Exp. Ther. 145: 135.
100. Assimacopoulos-Jeannet, F., Exton, J. H., and Jeanrenaud, B. (1973). Am. J. Physiol. 225:25.
101. Henion, W. F., Sutherland, E. W., and Pasternak, T. (1967). Biochim. Biophys. Acta 148:106.
102. Triner, L., and Nahas, G. G. (1966). J. Pharmacol. Exp. Ther. 153:569.
103. Okajima, F., and Ui, M. (1976). Arch. Biochem. Biophys. 175:549.

104. Ahlquist, R. P. (1948). Am. J. Physiol. 153:586.
105. Triner, L., Vulliemay, Y., Verosky, M., Habif, D. V., and Nahas, G. G. (1972). Life Sci. 11:817.
106. Ramanathan, S., and Shibata, S. (1974). Blood Vessels 11:312.
107. Amer, M. S. (1973). Science 179:807.
108. Amer, M. S., Gomoll, A. W., Perhach, J. L., Jr., Ferguson, H. C., and McKinney, G. R. (1974). Proc. Natl. Acad. Sci. USA 71:4930.
109. Klenerova, V., Albrecht, I., and Hynie, S. (1975). Pharmacol. Res. Commun. 7:453.
110. Volicer, L., Polgar, P., Rao, S. L. N., and Rutenburg, A. M. (1973). Pharmacology 9:317.
111. Sands, H., Sinclair, D., and Mascali, J. (1976). Blood Vessels 13:361.
112. Daniel, E. E., and Crankshaw, J. (1974). Blood Vessels 11:295.
113. Diamond, J., and Blisard, K. S. (1976). Mol. Pharmacol. 12:688.
114. Hamilton, T. C. (1972). Br. J. Pharmacol. 46:386.
115. Berti, F., Sirtori, C., and Usardi, M. M. (1970). Arch. Int. Pharmacodyn. Ther. 184:328.
116. Berti, F., Bernareggi, V., and Mandelli, V. (1971). Arch. Int. Pharmacodyn. Ther. 192:247.
117. Shepherd, A. P., Mao, C. C., Jacobson, E. D., and Shanboar, L. L. (1973). Microvasc. Res. 6:332.
118. Lugwier, C., Bertrano, Y., and Stoclet, J. C. (1972). Eur. J. Pharmacol. 19:132.
119. Andersson, R. (1973). Acta Physiol. Scand. 87:84.
120. Collins, G. A., and Sutter, M. C. (1975). Can. J. Physiol. Pharmacol. 53: 989.
121. Amer, M. S., Doba, N., and Reis, D. J. (1975). Proc. Natl. Acad. Sci. USA 72:2135.
122. Dunham, E. W., Haddox, M. K., and Goldberg, W. D. (1974). Proc. Natl. Acad. Sci. USA 71:815.
123. Schlender, K. K., Wei, S. H., and Villar-Palasi, C. (1969). Biochim. Biophys. Acta 191:272.
124. Kemp, B. E., Bylund, D. B., Huang, T.-S., and Krebs, E. G. (1975). Proc. Natl. Acad. Sci. USA 72:3448.
125. Cohen, P., Watson, D. C., and Dixon, G. H. (1975). Eur. J. Biochem. 51: 79.
126. Hjelmquist, G., Anderrson, J., Edlund. B., and Engstrom, L. (1974). Biochem. Biophys. Res. Commun. 61:559.
127. Larner, J., and Sanger, F. (1965). J. Mol. Biol. 11:491.
128. Langan, T. A. (1973). Adv. Cyclic Nucleotide Res. 3:99.
129. Farago, A., Romhanyi, T., Antoni, F., Takats, A., and Fabion, F. (1975). Nature 254:88.
130. Huang, T. S., Bylund, D. S., Stull, J. T., and Krebs, E. G. (1974). FEBS Lett. 42:249.
131. Carnegie, P. R., Kemp, B. E., Dunkley, P. R., and Murray, A. W. (1973). Biochem. J. 135:569.
132. Carnegie, P. R., Dunkley, P. R., Kemp, B. E., and Murray, A. W. (1974). Nature 249:147.
133. Kemp, B. E., Benjamini, E., and Krebs, E. G. (1976). Proc. Natl. Acad. Sci. USA 73:1038.
134. Daile, P., Carnegie, P. R., and Young, J. D. (1975). Nature 257:416.
135. Zetterquist, Ö., Regnarsson, U., Humble, E., Berglund, L., and Engström, L. (1976). Biochem. Biophys. Res. Commun. 70:696.

136. Langan, T. (1973). Adv. Cyclic Nucleotide Res. 3:99.
137. Rubin, C. S., and Rosen, O. M. (1975). Annu. Rev. Biochem. 44:831.
138. Sands, H., Rickenberg, H. V., and Meyer, T. A. (1973). Biochim. Biophys. Acta 302:267.
139. Miyamoto, E., Petzold, G. L., Harris, J. S., and Greengard, P. (1971). Biochem. Biophys. Res. Commun. 44:305.
140. Rosen, O. M., and Erlichman, J. (1975). J. Biol. Chem. 250:7788.
141. Hofmann, F., Beavo, J. A., Bechtel, P. J., and Krebs, E. G. (1975). J. Biol. Chem. 250:7795.
142. Beavo, J. A., Bechtel, P. J., and Krebs, E. G. (1975). Adv. Cyclic Nucleotide Res. 5:241.
143. Ogez, J. R., and Segel, I. H. (1976). J. Biol. Chem. 25:4551.
144. Miyamoto, E., Petzold, G. L., Kuo, J. F., and Greengard, P. (1973). J. Biol. Chem. 248:179.
145. Tao, M. (1972). Biochem. Biophys. Res. Commun. 46:56.
146. Maeno, H., Reyes, P. L., Ueda, T., Rudolph, S. A., and Greengard, P. (1974). Arch. Biochem. Biophys. 164:551.
147. Erlichman, J., Rosenfeld, R., and Rosen, O. M. (1974). J. Biol. Chem. 249: 5000.
148. Rangel-Aldao, R., and Rosen, O. M. (1976). J. Biol. Chem. 251:3375.
149. Sanborn, B. M., Bhalla, R. C., and Korenman, S. G. (1973). J. Biol. Chem. 248:3593.
150. Tenner, A. J., and Wallace, R. A. (1972). Biochim. Biophys. Acta 276:416.
151. Montague, W., and Howell, S. G. (1972). Biochem. J. 129:557.
152. Fontana, J. A., and Lovenberg, W. (1971). Proc. Natl. Acad. Sci. USA 68: 2787.
153. Kirchberger, M. A., Schwartz, I. L., and Walter, R. (1972). Proc. Soc. Exp. Biol. Med. 140:657.
154. Corbin, J. D., Soderling, T. R., and Park, C. R. (1973). J. Biol. Chem. 248:1813.
155. Wilson, M. B., and Malkin, A. (1974). Can. J. Biochem. 52:319.
156. Reimann, E. M., Walsh, D. A., and Krebs, E. G. (1971). J. Biol. Chem. 246:1986.
157. Huang, L. C. (1974). Biochim. Biophys. Acta 358:281.
158. Tsang, B. K., and Singhal, R. L. (1973). Can. J. Physiol. Pharmacol. 51: 942.
159. Hoskins, D. D., Casillas, E. R., and Stephens, D. T. (1972). Biochem. Biophys. Res. Commun. 48:1331.
160. Kumon, A., Nishiyama, K., and Nishizaka, Y. (1972). J. Biol. Chem. 247: 3726.
161. Yamamura, H., Kumon, A., Nishiyama, K., Takeda, M., and Nishizuka, Y. (1971). Biochem. Biophys. Res. Commun. 45:1560.
162. Wombacher, H., Reuter-Smerdka, M., and Körber, F. (1973). FEBS Lett. 30:313.
163. Kumor, R., Tao, M., Piotrowski, R., and Solomon, L. (1973). Biochim. Biophys. Acta 315:66.
164. Traugh, J. A., and Traut, R. R. (1974). J. Biol. Chem. 249:1207.
165. Waddy, C. T., and Mackinlay, A. G. (1971). Biochim. Biophys. Acta 250: 491.
166. Zapf, J., and Froesch, E. R. (1972). FEBS Lett. 20:141.
167. Corbin, J. D., and Keely, S. L. (1977). J. Biol. Chem. 252:910.
168. Beavo, J. A., Bechtel, P. J., and Krebs, E. G. (1974). Proc. Natl. Acad. Sci. USA 71:3580.

169. Krebs, E. G. (1973). *In* Endocrinology, Proceedings of the Fourth International Congress. Excerpta Medica, Amsterdam.
170. Soderling, T. R., Corbin, J. D., and Park, C. R. (1973). J. Biol. Chem. 248:1822.
171. Clark, M. R., Azhar, S., and Menom, K. M. J. (1976). Biochem. J. 158: 175.
172. Correze, C., Auclair, D., and Nunez, J. (1976). Mol. Cell. Endocrinol. 5: 339.
173. Catalan, R. E., Castillon, M. P., Corces, V. G., and Avila, C. (1977). Biochem. Biophys. Res. Commun. 74:279.
174. Spaulding, S. W., and Burrow, G. N. (1975). Nature 254:347.
175. DeRubertis, F. R., and Craven, P. A. (1976). J. Clin. Invest. 57:1442.
176. Keely, S. L., Corbin, J. D., and Park, C. R. (1975). J. Biol. Chem. 250: 4832.
177. Sudilovsky, O. (1974). Biochem. Biophys. Res. Commun. 58:85.
178. Higashino, H., and Takeda, M. (1974). J. Biochem. 75:189.
179. Birnbaum, M. J., and Fain, J. N. (1977). J. Biol. Chem. 252:528.
180. Field, J. B., Bloom, G., Kerins, M. F., Chayoth, R., and Zor, U. (1975). J. Biol. Chem. 250:4903.
181. Spaulding, S. W., and Burrow, G. N. (1975). Endocrinology 96:1018.
182. Means, A. R., MacDougall, E., Soderling, T. R., and Corbin, J. D. (1974). J. Biol. Chem. 249:1231.
183. Cooke, B. A., Van Der Kemp, A. J. W. C. M. (1976). Biochem. J. 154:371.
184. Darbon, J. M., Ursely, J., and Leymaire, P. (1976). FEBS Lett. 63:159.
185. Ling, W. Y., and Marsh, J. M. (1977). Endocrinology 100:1571.
186. Azhar, S., Clark, M. R., and Menon, K. M. J. (1976). Endocrine Res. Commun. 3:93.
187. Moyle, W. R., MacDonald, G. J., and Garfink, J. E. (1976). Biochem. J. 160:1.
188. Ljungström, O., Hjelmquist, G., and Engström, L. (1974). Biochim. Biophys. Acta 358:289.
189. Ekman, P., Dahlquist, U., Humble, E., and Engström, L. (1976). Biochim. Biophys. Acta 429:374.
190. Humble, E., Bergland, L., Titanji, V., and Ljungström, O. (1975). Biochem. Biophys. Res. Commun. 66:614.
191. Ljunström, O., Berglund, L., and Engström, L. (1976). Eur. J. Biochem. 68:497.
192. Morgenroth, V. H., III, Hegstrand, L. R., Roth, R. H., and Greengard, P. (1975). J. Biol. Chem. 250:1946.
193. Jungmann, R. A., Hiestand, P. C., and Schweppe, J. S. (1974). J. Biol. Chem. 249:5449.
194. Martelo, O. J., Woo, S. L. C., Reimann, E. M., and Davie, E. W. (1970). Biochemistry 9:4807.
195. Martelo, O. J., and Hirsch, J. (1974). Biochem. Biophys. Res. Commun. 58:1008.
196. Hirsch, J., and Martelo, O. J. (1976). J. Biol. Chem. 251:5408.
197. Lee, S. G., Miceli, M. V., Jungmann, R. A., and Hung, P. P. (1975). Proc. Natl. Acad. Sci. USA 72:2945.
198. Narumi, S., and Miyamoto, E. (1974). Biochim. Biophys. Acta 350:215.
199. Dittman, R. C., Khoo, J. C., and Steinberg, D. (1975). J. Biol. Chem. 250: 4505.
200. Caron, M. G., Goldstein, S., Savard, K., and Marsh, J. M. (1975). J. Biol. Chem. 250:5137.

201. Lovenberg, W., Bruck, E. A., and Hanbauer, I. (1975). Proc. Natl. Acad. Sci. USA 72:2955.
202. Rubin, C. S., and Rosen, O. M. (1973). Biochem. Biophys. Res. Commun. 50:421.
203. Shimomura, R., Matsumura, S., and Nishizaka, Y. (1974). J. Biochem. 75:1.
204. Greenquist, A. C., and Shohet, S. B. (1974). FEBS Lett. 48:133.
205. Avruch, J., and Fairbanks, G. (1974). Biochemistry 13:5507.
206. Fairbanks, G., and Avruch, J. (1974). Biochemistry 13:5514.
207. Guthrow, C. E., Jr., Allen, J. E., and Rasmussen, H. (1972). J. Biol. Chem. 247:8145.
208. Roses, A. D., and Appel, S. H. (1973). J. Biol. Chem. 248:1408.
209. Rubin, C. S. (1975). J. Biol. Chem. 250:9044.
210. Rudolph, S. A., and Greengard, P. (1974). J. Biol. Chem. 249:5684.
211. Weller, M., and Morgan, I. G. (1976). Biochim. Biophys. Acta 433:223.
212. Weller, M., and Morgan, I. (1976). Biochim. Biophys. Acta 436:675.
213. Ueda, T., Rudolph, S. A., and Greengard, P. (1975). Arch. Biochem. Biophys. 170:492.
214. Johnson, E. M., Ueda, T., Maeno, H., and Greengard, P. (1972). J. Biol. Chem. 247:5650.
215. Ueda, T., Maeno, H., and Greengard, P. (1973). J. Biol. Chem. 248:8295.
216. Johnson, E. M., Maeno, H., and Greengard, P. (1971). J. Biol. Chem. 246: 7731.
217. Ehrlich, Y. H., and Routtenberg, A. (1974). FEBS Lett. 45:237.
218. Dunkley, P. R., Holmes, H., and Rodnight, R. (1976). Biochem. J. 157: 661.
219. Weller, M., and Morgan, I. G. (1977). Biochim. Biophys. Acta 465:527.
220. Dousa, T. P., Sands, H., and Hechter, O. (1972). Endocrinology 91:757.
221. Schwartz, I. L., Shlatz, L. J., Kinne-Saffran, E., and Kinne, R. (1974). Proc. Natl. Acad. Sci. USA 71:2595.
222. DeLorenzo, R. J., Walton, K. G., Curran, P. F., and Greengard, P. (1973). Proc. Natl. Acad. Sci. USA 70:880.
223. Walton, K. G., DeLorenzo, R. J., Curran, P. F., and Greengard, P. (1975). J. Gen. Physiol. 65:153.
224. Shlatz, L., and Marinetti, G. V. (1971). Biochem. Biophys. Res. Commun. 45:51.
225. Ronquist, G., and Ågren, G. (1974). Upsala J. Med. Sci. 79:138.
226. Ågren, G., and Ronquist, G. (1974). Acta Physiol. Scand. 92:430.
227. Chang, K.-J., Marcus, N. A., and Cuatrecasas, P. (1974). J. Biol. Chem. 249:6854.
228. Lamay, A., Deschenes, M., Lemaire, S., Poirier, G., Poulin, L., and Labrie, F. (1974). J. Biol. Chem. 249:323.
229. Steiner, M. (1975). Arch. Biochem. Biophys. 171:245.
230. Orloff, J., and Handler, J. S. (1967). Am. J. Medicine 42:757.
231. Labrie, F., Lemaire, S., Poirier, G., Pelletier, G., and Boucher, R. (1971). J. Biol. Chem. 246:7311.
232. Ichii, S., Murakami, N., and Ikeda, A. (1974). Acta Endocrinol. 75:325.
233. Jesmok, G. J., Calvert, D. N., and Lech, J. J. (1976). J. Pharmacol. Exp. Ther. 200:187.
234. Kirchberger, M. A., Tada, M., and Katz, A. M. (1974). J. Biol. Chem. 249:6166.
235. Tada, M., Kirchberger, M. A., Repke, D. I., and Katz, A. M. (1974). J. Biol. Chem. 249:6174.
236. Kirchberger, M. A., and Chu, G. (1976). Biochim. Biophys. Acta 419:559.

237. Tada, M., Kirchberger, M. A., and Katz, A. M. (1975). J. Biol. Chem. 250:2640.
238. Kirchberger, M. A., and Tada, M. (1976). J. Biol. Chem. 251:725.
239. Kirchberger, M. A., Tada, M., Repke, D. I., and Katz, A. M. (1972). J. Mol. Cell. Cardiol. 4:673.
240. LaRaia, P., and Morkin, E. (1974). Circ. Res. 35:298.
241. Nayler, W. G., Dunnett, J., and Berry, D. (1975). J. Mol. Cell. Cardiol. 7:387.
242. Schwartz, A., Entman, M. L., Kaniike, K., Lane, L. K., Van Winkle, W. B., and Bornet, E. P. (1976). Biochim. Biophys. Acta 426:57.
243. Bornet, E. P., Entman, M. L., Van Winkle, W. B., Schwartz, A., Lehotay, D. C., and Levey, G. S. (1977). Biochim. Biophys. Acta 468:188.
244. Bailey, C., and Villar-Palasi, C. (1971). Fed. Proc. 30:1147.
245. Perry. S. V., and Cole, H. A. (1974). Biochem. J. 141:733.
246. Perry, S. V., and Cole, H. A. (1973). Biochem. J. 131:425.
247. Pratje, E., and Heilmeyer, L. M. G., Jr. (1972). FEBS Lett. 27:89.
248. Stull, J. T., Brostrom, C. O., and Krebs, E. G. (1972). J. Biol. Chem. 247: 5272.
249. Gross, S. R., and Mayer, S. E. (1973). Biochem. Biophys. Res. Commun. 54:823.
250. England, P. (1975). FEBS Lett. 50:57.
251. Solaro, R. J., Moir, A. J. G., and Perry, S. V. (1976). Nature 262:615.
252. England, P. J. (1976). Biochem. J. 160:295.
253. Ribolow, H., Barany, K., Steinscheider, A., and Barany, M. (1977). Arch. Biochem. Biophys. 179:81.
254. Lallemont, C., Seraydarian, K., Mommaerts, W. F. H. M., and Sah, M. (1975). Arch. Biochem. Biophys. 169:369.
255. Buss, J. E., and Stull, J. T. (1977). FEBS Lett. 73:101.
256. Loeb, J. E., and Blat, C. (1970). FEBS Lett. 10:105.
257. Eil, C., and Wool, I. G. (1971). Biochem. Biophys. Res. Commun. 43:1001.
258. Eil, C., and Wool, I. G. (1973). J. Biol. Chem. 248:5122.
259. Eil, C., and Wool, I. G. (1973). J. Biol. Chem. 248:5130.
260. Blat, C., and Loeb, J. E. (1971). FEBS Lett. 18:124.
261. Gressner, A. M., and Wool, I. G. (1976). J. Biol. Chem. 251:1500.
262. Walton, G. M., Gill, G. N., Abrass, I. B., and Garren, L. O. (1971). Proc. Natl. Acad. Sci. USA 68:880.
263. Walton, G. M., and Gill, G. N. (1973). Biochemistry 12:2604.
264. Cawthon, M. L., Bitte, L. F., Krystosek, A., and Kabat, D. (1974). J. Biol. Chem. 249:275.
265. Mallette, L. E., Neblett, M., Exton, J. H., and Langan, T. A. (1973). J. Biol. Chem. 248:6289.
266. Langan, T. A. (1969). Proc. Natl. Acad. Sci. USA 64:1276.
267. Langan, T. A. (1968). Science 162:579.
268. Man, N. T., Morris, G. E., and Cole, R. J. (1974). FEBS Lett. 42:257.
269. Johnson, E. M., and Hadden, J. W. (1975). Science 187:1198.
270. Grummt, I., and Grummt, F. (1974). FEBS Lett. 39:129.
271. Daile, P., and Carnegie, P. R. (1974). Biochem. Biophys. Res. Commun. 61:852.
272. Carnegie, P. R., Kemp., B. E., Dunkley, P. R., and Murray, A. W. (1973). Biochem. J. 135:569.
273. Miyamoto, E., and Kakiuchi, S. (1974). J. Biol. Chem. 249:2769.
274. Leterrier, J. J., Rappaport, L., and Nunez, J. (1974). Mol. Cell. Endocrinol. 1:65.
275. Leterrier, J. F., Rappaport, L., and Nunez, J. (1974). FEBS Lett. 46:285.

276. Casola, L., DiMatteo, G., and Augusti-Tocco, G. (1974). Exp. Neurol. 44:417.
277. Sloboda, R. D., Rudolph, S. A., Rosenbaum, J. L., and Greengard, P. (1975). Proc. Natl. Acad. Sci. USA 72:177.
278. Murray, A. W., and Froscio, M. (1971). Biochem. Biophys. Res. Commun. 44:1089.
279. Rosen, O. M. (1976). Biochemistry 15:2902.
280. Kuo, J. F., Wyatt, G. R., and Greengard, P. (1971). J. Biol. Chem. 246: 7159.
281. Donnelly, T. E., Jr., Kuo, J. F., Reyes, P. L., Lin, Y.-P., and Greengard, P. (1973). J. Biol. Chem. 248:190.
282. Donnelly, T. E., Jr., Kuo, J. F., Miyamoto, E., and Greengard, P. (1973). J. Biol. Chem. 248:199.
283. Miyamoto, E., Petzold, G. L., Kuo, J. F., and Greengard, P. (1973). J. Biol. Chem. 248:179.
284. Kuo, J. F. (1974). Proc. Natl. Acad. Sci. USA 71:4037.
285. Nakazawa, K., and Sano, M. (1975). J. Biol. Chem. 250:7415.
286. Takai, Y., Nishiyama, K., Yamamura, H., and Nishizuka, Y. (1975). J. Biol. Chem. 250:4690.
287. Kuo, J. F., Kuo, W.-N., Shoji, M., Davis. C. W., Seerly, V. L., and Donnelly, T. E., Jr. (1976). J. Biol. Chem. 251:1759.
288. Kobayashi, R., and Fang, V. S. (1976). Biochem. Biophys. Res. Commun. 69:1080.
289. Kuo, J. F. (1975). Proc. Natl. Acad. Sci. USA 72:2256.
290. Kuo, J. F. (1975). J. Cyclic Nucleotide Res. 1:151.
291. Gill, G. N., Holdy, K. E., Walton, G. M., and Kanstein, C. B. (1976). Proc. Natl. Acad. Sci. USA 73:3918.
292. Shoji, M., Patrick, J. G., Davis, C. W., and Kuo, J. F. (1977). Biochem. J. 161:213.
293. Kuo, J. F., Patrick, J. G., and Seery, V. L. (1976). Biochem. Biophys. Res. Commun. 72:996.
294. Shoji, M., Patrick, J. G., Tse, J., and Kuo, J. F. (1977). J. Biol. Chem. 252:4347.
295. deJonge, H. R., and Rosen, O. M. (1977). J. Biol. Chem. 252:2780.
296. Lincoln, T. A., Dills, W. L., Jr., and Corbin, J. D. (1977). J. Biol. Chem. 252:4269.
297. Casnellie, J. E., and Greengard, P. (1974). Proc. Natl. Acad. Sci. USA 71: 1891.
298. deJonge, H. R. (1976). Nature 262:590.
299. Walsh, D. A., Ashby, C. D., Gonzalez, C., Cullkins, D., Fischer, E. H., and Krebs, E. G. (1971). J. Biol. Chem. 246:1977.
300. Demaille, J. G., Peters, K. A., and Fischer, E. H. (1977). Biochemistry 16: 3080.
301. Walsh, D. A., and Ashby, C. D. (1973). Rec. Prog. Hormone Res. 29:329.
302. Kuo, W.-N., and Kuo, J. F. (1976). J. Biol. Chem. 251:4283.
303. Kuo, W.-N., Shoji, M., and Kuo, J. F. (1976). Biochem. Biophys. Res. Commun. 70:280.
304. DeLorenzo, R. J., and Greengard, P. (1973). Proc. Natl. Acad. Sci. USA 70:1831.
305. Rasmussen, H., Jensen, P., Lake, W., Friedmann, N., and Goodman, D. B. P. (1975). Adv. Cyclic Nucleotide Res. 5:375.
306. Rasmussen, H. (1970). Science 170:404.
307. Berridge, M. J. (1975). J. Cyclic Nucleotide Res. 1:305.

308. Berridge, M. J. (1975). Adv. Cyclic Nucleotide Res. 6:1.
309. George, W. J., Polson, J. B., O'Toole, A. G., and Goldberg, H. D. (1970). Proc. Natl. Acad. Sci. USA 66:398.
310. George, W. J., Wilkerson, R. D., and Kadowitz, P. J. (1973). J. Pharmacol. Exp. Ther. 184:228.
311. Watanabe, A. M., and Besch, H. R., Jr. (1975). Circ. Res. 37:309.
312. Wollenberger, A., Babskii, E. B., Krause, E. G., Genz, S., Blohm, D., and Bogdanova, E. V. (1973). Biochem. Biophys. Res. Commun. 55:446.
313. Wilkerson, R. D., Paddock, R. J., and George, W. J. (1976). Eur. J. Pharmacol. 36:247.
314. Goshima, K. (1976). J. Mol. Cell. Cardiol. 8:713.
315. Wurster, B., Schubiger, K., Wick, V., and Gerisch, G. (1977). FEBS Lett. 76:141.
316. Mato, J. M., Krens, F. A., Van Haastert, P. J. M., and Konijn, T. M. (1977). Proc. Natl. Acad. Sci. USA 74:2348.
317. Mato, J. M., Van Haastert, J. M., Krens, F. A., Rhijnsburger, E. H., Dobbe, F. C. P. M., and Konijn, T. M. (1977). FEBS Lett. 79:331.
318. Gerisch, G., and Wick, U. (1975). Biochem. Biophys. Res. Commun. 65:364.
319. Jost, J.-P., and Rickenberg, H. V. (1971). Annu. Rev. Biochem. 40:741.
320. Shafrir, E., and Steinberg, D. (1960). J. Clin. Invest. 39:210.
321. Braun, T., and Hechter, O. (1970). Proc. Natl. Acad. Sci. USA 66:995.
322. Goodman, H. M. (1970). Endocrinology 86:1064.
323. Fain, J. N. (1967). Science 157:1062.
324. Friedmann, B., Goodman, E. H., Jr., and Weinhouse, S. (1965). J. Biol. Chem. 240:3729.
325. Mersmann, H. J., and Segal, H. L. (1969). J. Biol. Chem. 244:1701.
326. De Wulf, H., Stalmans, W., and Hers, H. G. (1970). Eur. J. Biochem. 15:1.
327. Kletzien, R. F., Pariza, M. W., Becker, J. E., and Potter, V. R. (1975). Nature 256:46.
328. Gospodarowicz, D. (1974). Nature 249:123.
329. Rudland, D. S., Gospodarowicz, D., and Seifert, W. (1974). Nature 250:741. 741.
330. Manganiello, V., and Vaughan, M. (1972). J. Clin. Invest. 51:2763.
331. Schmiotke, J., Wienker, T., Flügel, M., and Engel, W. (1976). Nature 262: 593.
332. Japundžič, I., Mimič-Oka, J., and Japundžič, M. (1977). J. Biochem. 28: 475.
333. Kaye, A. M., Walker, M. D., and Sömjen, D. (1975). Proc. Natl. Acad. Sci. USA 72:2631.
334. Liv, A. Y.-C., and Greengard, P. (1976). Proc. Natl. Sci. USA 73:568.
335. Amrhein, N. (1977). Annu. Rev. Plant Physiol. 28:123.
336. Loomis, W. F. (1975). *Dictyostelium discoideum*: A Developmental System. Academic Press, New York.
337. Buettner, M. J., Spitz, E., and Rickenberg, H. V. (1973). J. Bacteriol. 114: 1068.
338. Rickenberg, H. V., Rahmsdorf, H. J., Campbell, A., North, M. J., Kwasniak, J., and Ashworth, J. M. (1975). J. Bacteriol. 124:212.
339. Rahmsdorf, H. J., Cailla, H. L., Spitz, E., Moran, M. J., and H. V. Rickenberg (1976). Proc. Natl. Acad. Sci. USA 73:3183.
340. Marin, F. T. (1976). Dev. Biol. 48:110.
341. Maeda, Y., and Maeda, M. (1974). Exp. Cell Res. 84:88.
342. Darmon, M., Brachet, P., and Pereira Da Silva, L. H. (1975). Proc. Natl. Acad. Sci. USA 72:3163.

343. Wicks, W. D. (1974). Adv. Cyclic Nucleotide Res. 4:335.
344. Jost, J.-P., Hsie, A., Hughes, S. D., and Ryan, L. (1970). J. Biol. Chem. 245:351.
345. Wireman, J. W., and Dworkin, M. (1975). Science 189:516.
346. Campos, J. M., and Zusman, D. (1975). Proc. Natl. Acad. Sci. USA 72:518.
347. Coleman, W., personal communication.
348. Handa, A. K., and Johri, M. M. (1976). Nature 259:480.
349. Simpson, T. L., and Rodan, G. A. (1976). Dev. Biol. 49:544.
350. Simpson, T. L., and Rodan, G. A. (1976). *In* F. W. Harrison and R. R. Cowden (eds.), Aspects of Sponge Biology, pp. 83–97. Academic Press, New York.
351. Rasmont, R. (1974). Experientia 30:792.
352. Robertson, A., and Gingle, A. R. (1977). Science 197:1078.
353. Gingle, A. L. (1977). Dev. Biol. 58:394.
354. McGuire, E. J., personal communication.
355. McMahon, D. (1974). Science 185:1012.

International Review of Biochemistry
Biochemistry and Mode of Action of Hormones II, Volume 20
Edited by H. V. Rickenberg
Copyright 1978 University Park Press Baltimore

3
Prostaglandins, Thromboxanes, and Prostacyclin

R. R. GORMAN
Experimental Biology, The Upjohn Company, Kalamazoo, Michigan

The independent discovery by Goldblatt and von Euler that extracts of human seminal plasma and of the vesicular gland of sheep produce a fall in blood pressure and contract smooth muscle established the field of prostaglandin research (1, 2). von Euler named the active agents prostaglandins and established some of the chemical properties of the compounds (3). Almost three decades elapsed between these initial observations and the isolation of crystalline compounds by Bergstrom and Sjovall (4) and the subsequent structure determination by Bergstrom et al. (5, 6) and Samuelsson (7).

In 1964, it was reported that essential fatty acids can serve as precursors of prostaglandins (8–10). Dihomo-γ-linolenic acid (5,8,11-eicosatrienoate) is the precursor for prostaglandins of series 1; arachidonic acid (5,8,11,14-eicosatetraenoate) is the precursor of series 2 prostaglandins, and

5,8,11,14,17-eicosapentaenoate is the precursor of prostaglandins of series 3. The latter is rarely encountered in nature and is probably of little physiological significance because it fails to prevent dermatitis in essential fatty acid-deficient rats (11, 12). Figure 1 gives the representative structure and nomenclature of the cyclopentane ring and side chains that all prostaglandins share in common, as well as the structures of PGE_1 (13,14 double bond), and PGE_2 (13,14 and 5,6 double bond). The parent compound prostanoic acid is also shown and illustrates the carbon skeleton numbering system.

Although it is possible to produce prostaglandins of series 1 and 3 biosynthetically and demonstrate pharmacological activity, very little dihomo-γ-linolenic acid and even less eicosapentaenoic acid are found in vivo (13–16). The levels of arachidonic acid far exceed those of the other precursor molecules; therefore, it is possible that series 2 compounds predominate.

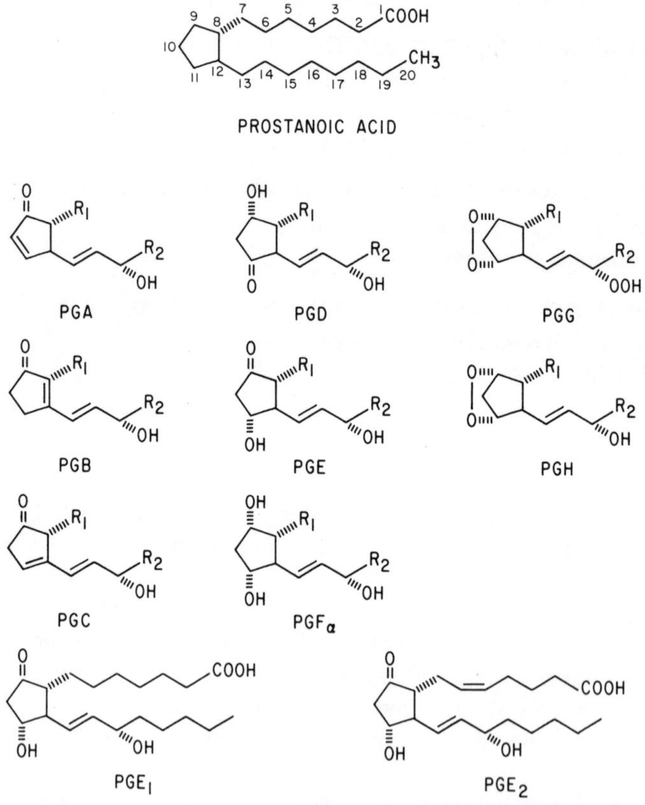

PROSTANOIC ACID

PGA PGD PGG

PGB PGE PGH

PGC PGF$_\alpha$

PGE$_1$ PGE$_2$

Figure 1. Structures of prostaglandins. Prostaglandins of the E, F, or D series are considered primary prostaglandins. The PGAs are produced by the loss of the elements of H_2O from the cyclopentane ring of PGEs. PGBs are products of isomerization of the ring double bond of PGAs. PGCs are produced from PGAs by isomerization of the ring double bond to position 11,12. The unstable endoperoxides PGG and PGH are discussed in more detail later in the chapter.

In the past 10 years, numerous reviews have dealt with the myriad chemical and pharmacological properties of the various prostaglandins (17–21). It is almost impossible to find a tissue or cell type that does not produce or is not affected in some manner by prostaglandins. Table 1 lists some of the cell types and tissues in which prostaglandins have been reported to modulate cellular functions. The table also shows what effect various prostaglandins have on intracellular cyclic nucleotide levels. This item is added because, in general, the regulation of cyclic nucleotide levels seems to be the principal mode of action of prostaglandins. The involvement of the classical prostaglandins in cyclic nucleotide metabolism has been reviewed (38–42); unfortunately, no unifying hypothesis has emerged.

In this chapter, the classical prostaglandin molecules are discussed, but the thrust of the text is on the isolation, characterization, and biological evaluation of newly characterized products of arachidonic acid metabolism, the endoperoxides, thromboxanes, and prostacyclin (PGX, or PGI_2), and the effect of these newly discovered compounds on cyclic nucleotide metabolism. Much of the difficulty associated with previous studies of the mechanism of action of prostaglandins may be explained by the interaction of these until recently unknown compounds with cyclic nucleotide metabolism.

STORAGE AND RELEASE OF PROSTAGLANDINS

Neither the classical prostaglandins, endoperoxides, thromboxanes, nor prostacyclin seem to be stored in any measurable quantities in cells. Instead, the release of these fatty acid molecules from cells or tissues represents neosynthesis and not the release of preformed stored materials (43, 44).

Table 1. Prostaglandin-induced changes in cyclic nucleotide levels

Tissue	Prostaglandin	Change in cyclic nucleotide	Reference
Spleen	E_1	↑cAMP	22
Adipose	E_1, E_2, A_1	↓cAMP	23, 24
Platelets	E_1, E_2	↑cAMP	25–27
Thyroid	E_1, E_2	↑cAMP	28
Diaphragm	E_1	↑cAMP	22
Lung	E_1, E_2	↑cAMP	29
Pancreatic islets	E_1	↑cAMP	30
Neuroblastoma	E_1	↑cAMP	31
Uterus	E_1, $F_{2\alpha}$	↑cAMP	32
Cortical slices	E_1, E_2	↑cAMP	33
Ovary	E_1, E_2	↑cAMP	34
Hybrid cell lines	E_1	↑cAMP	35
Pituitary	E_1, E_2	↑cAMP	36
Peritoneal macrophage	E_1, E_2	↑cAMP	37

The synthesis of prostaglandin-like molecules by the fatty acid cyclo-oxygenase (prostaglandin synthetase) seems to depend on substrate availability because the addition of prostaglandin precursor fatty acids rapidly stimulates prostaglandin biosynthesis (8–10). Negligible amounts of the free acids of polyunsaturated fatty acids are needed for prostaglandin biosynthesis in cells, but considerable levels of precursor acids are found in membrane phospholipids (15, 45, 46). Phospholipids serve as sources of fatty acids for prostaglandin biosynthesis (47, 48), and the rate of synthesis is stimulated by snake venom (48, 49), which contains a high concentration of phospholipase A, which selectively removes the fatty acid from the 2′ position of the phospholipid.

Phospholipids are not the only source of precursor fatty acids for prostaglandin biosynthesis; triglycerides have also been reported to be a source (50). As stated earlier, arachidonic acid is the probable principal prostaglandin precursor fatty acid. Bills and Silver (15) have found that when human platelets are stimulated with the aggregating agent thrombin, arachidonic acid is the only fatty acid released from phosphatidylcholine. Arachidonic acid was also found to be the principal fatty acid released from platelet phosphatidylinositol in response to thrombin, but other fatty acids were released in lesser amounts. These data indicate that the release of arachidonic acid from phospholipid pools is very specific and adds another potential control point in the regulation of prostaglandin biosynthesis.

It has been reported that the mobilization of arachidonic acid is modulated by cyclic AMP (cAMP) in adipose tissue (51), isolated thyroid cells (28), platelets (52), and cultured neuroblastoma, glioma, and fibroblast cell lines (53). In adipose tissue, isolated thyroid cells, and cultured cells, cAMP has been reported to stimulate phospholipase A and to release arachidonic acid from phospholipid pools. The opposite is true in washed platelets. Minkes et al. (52) have found that cAMP selectively inhibits thrombin-induced release of arachidonic acid from phospholipid. That phospholipase A is regulated by cAMP is suggested by the report of a cAMP-dependent phospholipase A in adipose tissue (54).

Recent reports that anti-inflammatory steroids may act through an inhibition of phospholipase A are exciting (55–58) and may provide a tool for biochemical studies of the enzyme. Greaves and McDonald-Gibson (59) were the first to show an effect of anti-inflammatory steroids on prostaglandin biosynthesis. Subsequently, Lewis and Piper (60) found that anti-inflammatory steroids and inhibitors of prostaglandin cyclo-oxygenase blocked the vasodilatation in adipose tissue that accompanies lipolysis. These data were interpreted to imply that anti-inflammatory steroids inhibited the release of prostaglandins from cells because corticosteroids do not directly inhibit the prostaglandin cyclo-oxygenase (61, 62). More recent experiments have shown that anti-inflammatory steroids may control the availability of fatty acid substrate to the cyclo-oxygenase, apparently through an inhibition of phospholipase A (56–58). Outside of these few reports, little is known about the control of this crucial initial step in prostaglandin biosynthesis. There is much evidence that indicates that prostaglandins are synthesized and released coincident with hormonal or mechanical stimulation of cells and that they act as either positive or negative regulators of hormone action (63–66); the prostaglandins are then metabolized at or near their original site of biosynthesis (67–69). Because of these properties, the prostaglandins are known as local hormones.

Figure 2 summarizes the postulated scheme of the initiation of prostaglandin biosynthesis. The precursor fatty acid, in this case arachidonic acid, is stored in the phospholipid pool. Upon either hormonal or mechanical stimulation of the cell, arachidonic acid is cleaved from the phospholipid by phospholipase A and enters into the prostaglandin biosynthetic pathway. The point of possible regulation of the phospholipase A by anti-inflammatory steroids and cAMP is shown.

EVIDENCE FOR PROSTAGLANDIN RECEPTOR

Prostaglandin E Binding Site

As is the case with circulating hormones, the prostaglandins, often called local hormones, have also been reported to have specific binding sites. Evidence for specific binding sites has been presented for prostaglandins of the E, A, and F series. However, because there has been no clear biological response associated directly with either PGA or PGF binding sites, this discussion is limited to specific PGE binding sites. The term "PGE binding site" is used because, in most systems, PGE_1 and PGE_2 have equal binding affinities. Studies of PGE binding have been hampered by the absence of a specific receptor level antagonist. Although 7-oxa-13-prostynoic acid (70, 71), SC-19220 (72, 73), and various polyphloretin phosphate analogues (74, 75) have been purported to be prostaglandin antagonists, they have little if any binding site affinity. In spite of these difficulties, [^{3}H]PGE binding sites have been reported in adipocyte homogenates (76), adipocyte ghosts (77), liver (78), corpus luteum (79), thymocytes (80), uterus (81), thyroid (82), stomach (83), platelets (84), skin (85), and cultured mammalian cells (86). Adequate kinetic and binding specificity measurements have not been done in all of the above studies. From reports of experiments in which some kinetics analysis has been completed, there emerges a remarkably similar set of data, considering that some studies were done with isolated plasma membranes and others with cellular homogenates. Table 2 lists the K_d values for binding of E prostaglandins in a variety of different tissues and species.

Correlation of Prostaglandin Binding with Adenylate Cyclase Activation

Most prostaglandin binding studies have met the four criteria for receptors in general: 1) specificity with regard to ligand binding; 2) saturability;

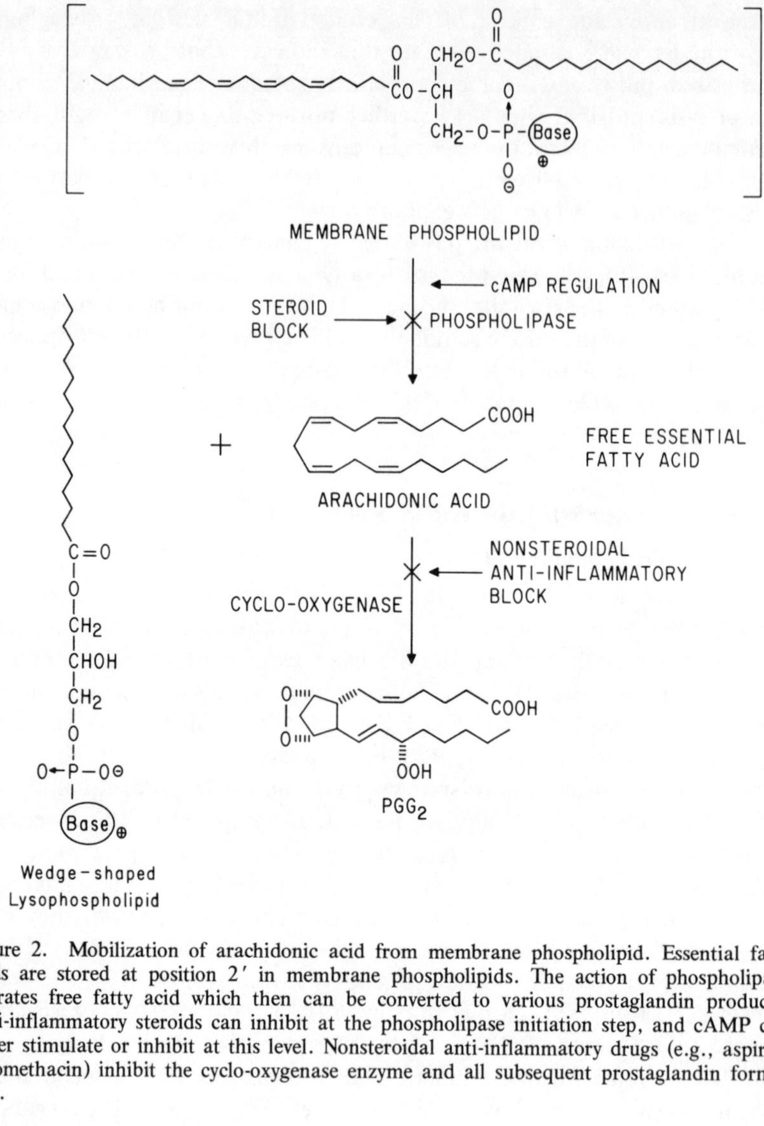

Figure 2. Mobilization of arachidonic acid from membrane phospholipid. Essential fatty acids are stored at position 2′ in membrane phospholipids. The action of phospholipase liberates free fatty acid which then can be converted to various prostaglandin products. Anti-inflammatory steroids can inhibit at the phospholipase initiation step, and cAMP can either stimulate or inhibit at this level. Nonsteroidal anti-inflammatory drugs (e.g., aspirin, indomethacin) inhibit the cyclo-oxygenase enzyme and all subsequent prostaglandin formation.

3) reversibility, and 4) high affinity. Unfortunately, as recently reviewed by Rao (87), there is a relatively weak correlation between PGE binding and any physiological or biochemical parameter, in particular, the activation of adenylate cyclase. This lack of a strong correlation between binding and function is not unique to prostaglandins (88, 89), but it is still disturbing. Recently, Brunton et al. (86) have reported PGE_1 receptors in cultured mammalian cells. These authors found that in clone B82 of murine L cells PGE_1 binds with an affinity of 0.5×10^9 M^{-1}; this binding affinity

Table 2. Comparison of dissociation constants of PGE binding sites

Species and tissue	Preparation	Apparent K_d(M)	Reference
Bovine corpus luteum	plasma membrane	1.3×10^{-9} and 1.0×10^{-8}[a]	62
Beef thyroid	plasma membrane	6.3×10^{-11} and 6.3×10^{-9}	65
Hamster uterus	homogenate	7.7×10^{-10} and 1.3×10^{-8}	64
Rat adipocyte	homogenate	3.0×10^{-9}	59
Rat liver	plasma membrane	1.0×10^{-9} and 2.5×10^{-8}	61
Rat thymocyte	plasma membrane	7.0×10^{-11} and 2.0×10^{-9}	63
Rat adipocyte	plasma membrane sac	2.0×10^{-10} and 2.5×10^{-9}	60
Human platelets	plasma membrane	8.0×10^{-11}	67
Rat skin	plasma membrane	1.5×10^{-7} and 1.9×10^{-7}	68
Cultured mammalian cells	homogenate	2.0×10^{-8}	69

[a]Multiple dissociation constants indicate two classes of binding sites.

was verified by the concentration-dependent activation of adenylate cyclase by PGE_1. This report was the first demonstration in which a prostaglandin receptor showed a direct relationship between PGE binding and adenylate cyclase activation. A lack of correlation between binding and some biological parameter certainly does not invalidate the other PGE binding studies. The similarities of the binding data are compelling, but until correlative data are obtained, one must be careful to differentiate between binding phenomena and true hormone-receptor interactions.

BIOSYNTHESIS AND METABOLISM OF PROSTAGLANDINS

It is not within the scope of this chapter to cover the literature concerning prostaglandin biosynthesis or metabolism. Rather, the discussion attempts to highlight some of the more important aspects of biosynthesis and metabolism that may enable the reader to have an overview of the respective systems.

Biosynthesis of Prostaglandins

A wide variety of tissues and organs contains enzymes that synthesize and metabolize prostaglandins. Prostaglandin production seems to be ubiquitous with respect to both cell type and species (39). For simplicity, the names "prostaglandin cyclo-oxygenase" and "prostaglandin synthetase" are used interchangeably. The term "prostaglandin synthetase" actually encompasses a group of enzymes which converts a precursor essential fatty acid (e.g., arachidonic) into an endoperoxide (e.g., PGG_2) which then serves as a substrate for other enzymic reactions, which eventually yield PGE_2 and other prostaglandins. The cyclo-oxygenase catalyzes a much more limited reaction series, only the formation of the endoperoxide. The postulated mechanism of endoperoxide formation is illustrated in Figure 3 (91). The initial oxygenation occurs at C-eleven. The eleven-peroxy-5,8,12,14-eicosatetraenoic acid formed is converted to an endoperoxide by a concerted reaction involving addition of oxygen at C-15, isomerization of the Δ^{12} double bond, formation of a new carbon-carbon bond between C-8 and C-12, and attack by the oxygen radical at C-9. It is this endoperoxide that serves as critical substrate for all subsequent reactions and which may actually be the important physiological molecule in many cases. Most studies on the mechanism of prostaglandin biosynthesis have used the prostaglandin cyclo-oxygenase from sheep seminal vesicles (90, 91). Extensive studies have also been done

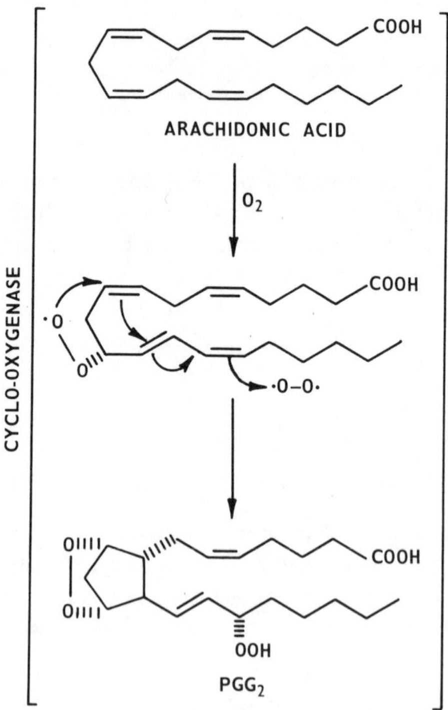

Figure 3. Mechanism of arachidonic acid cyclization to the endoperoxide PGG_2. In the presence of O_2, the cyclo-oxygenase converts essential fatty acids (e.g., arachidonic) to unstable endoperoxides (e.g., PGG_2), which can then serve as substrate for subsequent enzymic transformations. (See text for details.)

on platelets (92, 93), renal medulla (94–96), brain (97), skin (98), rat stomach (99), and bull seminal vesicles (100, 101). In addition, the lipoxygenase from soybeans has proved to be a useful model for studies on mechanisms of fatty acid oxidation (102).

Arachidonic acid is the preferred cyclo-oxygenase substrate. Dihomo-γ-linolenic acid is almost as efficient as arachidonic acid, whereas the conversion of eicosapentaenoic acid to PGE_3 is only 8% as efficient as the conversion of arachidonic to PGE_2 (90).

An important relationship in prostaglandin biosynthesis was found by Anggard and Samuelsson (103) when they incubated arachidonic acid with homogenates of guinea pig lung. Both PGE_2 and $PGE_{2\alpha}$ were formed in the incubation mixture, and these two prostaglandins were not interconvertible (103). These were the first data that indicated that these prostaglandins shared a common precursor molecule. In subsequent work, Samuelsson (104) found that during the biosynthesis of PGE_1 from dihomo-γ-linolenic acid in $^{18}O_2/^{16}O_2$ atmosphere, the two oxygens of the cyclopentane ring contained either 2 atoms of ^{18}O or 2 atoms of ^{16}O. A mixed oxygen structure containing both ^{18}O and ^{16}O was not observed. These data established that the C-9 and C-11 oxygen of the prostaglandin skeleton originated from the same molecule of oxygen and suggested a cyclic peroxide structure that could be isomerized into 1,3-hydroxylketone (PGE series) or reduced to a 1,3-diol PGF series. This was the first experimental evidence for an endoperoxide type of intermediate in prostaglandin biosynthesis, although this type of structure was postulated previously on the basis of theoretical considerations in 1964 (10).

The co-factor requirements for the cyclo-oxygenase are not completely understood. In sheep vesicular gland homogenates, the formulation of PGE_1 from dihomo-γ-linolenic acid can be stimulated by adding boiled or fresh supernatant, glutathione, tetrahydrofolate, and ascorbic acid, but not NADH or NADPH (105, 106). Ascorbate, tetrahydrofolate, and supernatant also stimulate the synthesis of $PGF_{1\alpha}$. One pair of electrons is required for the reduction of 1 atom of oxygen in the biosynthesis of prostaglandins of the E series. The biosynthesis of PGF compounds requires 2 additional electrons for reductive cleavage of the endoperoxide molecule. The source of these electrons is not known, and even definitive proof that PGF compounds are formed enzymically from the endoperoxide is lacking.

The prostaglandin cyclo-oxygenase and the model soybean lipoxygenase have some rather unusual characteristics. When fatty acid substrate is added, there is a short lag period, followed by a period of very rapid (oxygen uptake) synthesis of products; then the enzyme activity is rapidly lost (102, 107). The addition of fresh enzyme to the depleted preparation results in an immediate resumption of synthesis without a lag phase (108). These data indicated that the loss in enzyme activity was not attributable to an accumulation of an inhibitor, but rather to a self-catalyzed destruction of enzyme (109). The disappearance of the lag phase also indicated the presence of an activator apparently produced during the initial lag phase (110). This activator was assumed to be a hydroperoxide when it was shown to be depleted by glutathione peroxidase (107, 110, 111). Lands et al. (112) have developed a complicated kinetic model that embraces the above properties, but the difficulty in purifying the enzyme has complicated its direct study. A recent paper has described a partial purification of the cyclo-oxygenase by affinity chromatography; this should make additional kinetic studies possible (113).

One of the more important discoveries in the study of prostaglandin biosynthesis was made by Vane (114, 115), who was the first to suggest that the mechanism of action of nonsteroidal anti-inflammatory compounds (NOSAC), such as aspirin, was the inhibition of prostaglandin biosynthesis. Since the publication of these initial papers, scores of NOSAC drugs have been used to inhibit the cyclo-oxygenase and to study prostaglandin physiology and biochemistry (116–121). The value of this research tool is obvious. Unfortunately, many studies have been published in which NOSAC were used without adequate analytical data to indicate that the prostaglandin synthesis was actually inhibited in that particular system.

Before leaving the subject of biosynthesis, it should be emphasized that once the endoperoxide is formed, numerous other oxygenated fatty acids can be formed in addition to the classical prostaglandins. Most of the more recent and exciting aspects of prostaglandin research have centered on these other oxygenated products, which are discussed in more detail later in the chapter.

Metabolism of Prostaglandins

The primary prostaglandins have a very short half-life in the circulation. Most of the prostaglandins are almost totally metabolized by a single passage through the lungs (122). Prostaglandins of the A series, dehydration products of PGEs, are not metabolized as rapidly as the E or F prostaglandins (123). For these reasons, it has been suggested that prostaglandins of the A series may act as circulating hormones (124, 125). This hypothesis has found particular support on the basis of experiments in renal physiology, in which PGA compounds are reported to vasodilate and to be natriuretic (126–128). However, careful analysis of PGA_2 production in rabbit and human renal medulla by gas chromatography-mass spectrometry (GC/MS) has failed to detect PGA_2 (129). Apparently, PGA compounds that had been detected previously by radioimmunoassay really represented substances other than PGA_2. PGA compounds may not be produced in measurable amounts in vivo but may be artifacts formed by the dehydration of PGE compounds during extraction and isolation from biological samples (130). There is now a considerable body of literature that supports the following primary sequence for prostaglandin metabolism.

1. The oxidation of the allylic alcohol group at C-15. This oxidation is catalyzed by the 15-hydroxyprostaglandin dehydrogenase, first discovered by Anggard and Samuelsson (131). The distribution of this enzyme is widespread, but the lung, kidney, and placenta have proved to be particularly rich sources (132–136). Jarabak (135) and Schlegel and Greep (137) have purified the enzyme from human placenta. These authors found the enzyme to be noncompetitively inhibited by 15-keto-PGE_2, and 13,14-dihydro-15-keto-PGE_2. The structural requirements for the enzyme substrate have been reported by Sun et al. (138). Bulky group substitution at or near C-15 usually yields an inactive substrate, and the enzyme reacts preferentially with compounds with a keto group at C-9 and a double bond at C-5.
2. The oxidation at C-15 by the 15-hydroxyprostaglandin dehydrogenase is often followed by reduction of the 13,14 double bond. The reduction of substrate by the 15-ketoprostaglandin 13,14-reductase requires a carbonyl group at C-15 (69). The 15-ketoprostaglandin reductase is usually found with the 15-hydroxyprostaglandin dehydrogenase, and the activities of the two enzymes are difficult to separate (137). NADH is the preferred co-factor for the reductase, whereas NAD is a competitive inhibitor and prostaglandin E_2 a noncompetitive inhibitor (137).
3. After oxidation at C-15 and reduction of the Δ^{13} double bond, the 13, 14-dihydro-15-ketoprostaglandin can enter numerous metabolic pathways. The metabolite can be β-oxidized or undergo ω-1 oxidation (69, 139, 140). The major urinary metabolite of PGE_2 in man is a 7α-hydroxy-5,11-diketotetranorprosta-1,16-dienoic acid (141, 142). Figure 4 shows one possible route for the metabolism of PGE_2 from arachidonic acid. Whether or not the metabolism described in this section applies to the endoperoxide-thromboxane system and to prostacyclin (to be discussed later) is not known. However, the identification of a 15-ketothromboxane B_2 in guinea pig lung was described in a recent paper (143); PGX is a substrate for the 15-hydroxyprostaglandin dehydrogenase (144).

ENDOPEROXIDES, THROMBOXANES, AND PROSTACYCLIN

The isolation and characterization of two labile prostaglandin precursor molecules called PGG_2 by Hamberg and Samuelsson and their co-workers (145, 146) and PGR_2 by Nugteren and Hazelhoff (147) have opened a new area of prostaglandin research that may offer a unifying theme to the mode of action of prostaglandins.

Endoperoxides and Thromboxanes

The actual isolation and characterization of the labile molecules, termed endoperoxides, proved difficult because the endoperoxides have a $t_{1/2}$ of 5 min at 37 °C in aqueous buffer. However, their marked stability in anhydrous

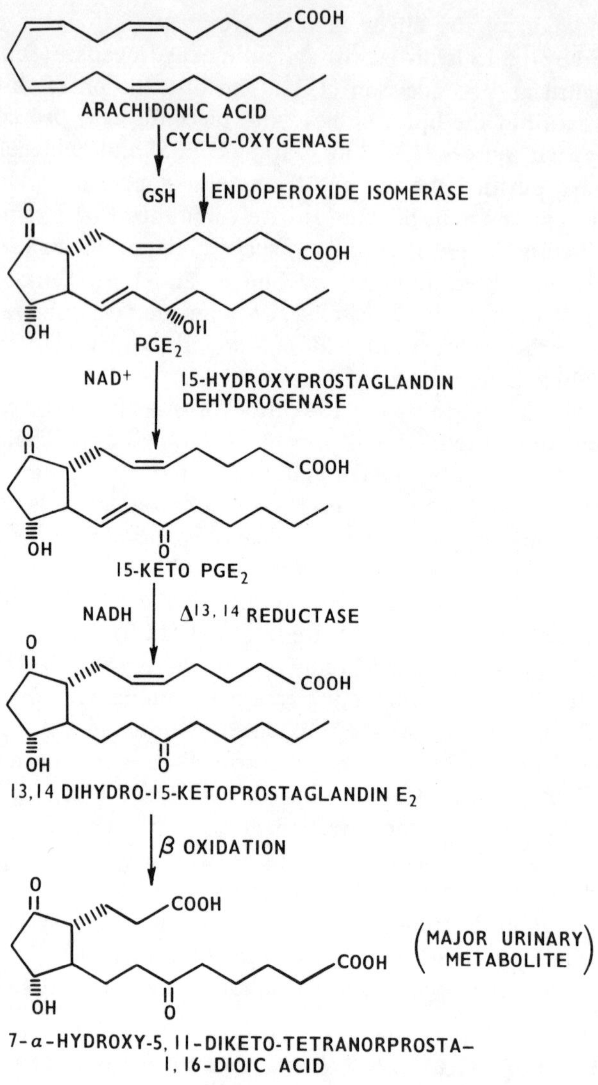

Figure 4. Metabolism of prostaglandin E_2. This is a representative sequence for the metabolism of PGE_2. PGE_2 is dehydrogenated, reduced at the $\Delta^{13,\,14}$ double bond, and then β-oxidized to the primary urinary metabolite 7-α-hydroxy-5,11-diketotetranorprosta-1,16-dioic acid. Many other metabolic sequences are possible. (See text and references for details.)

acetone permits their storage for months at $-70\,°C$ (145). Figure 5 gives the structure of the two endoperoxides, PGG_2 and PGH_2, derived from arachidonic acid. The endoperoxides are formed by the fatty acid cyclo-oxygenase. Originally, it was thought that the endoperoxide PGG_2 could be metabolized in only two directions; either PGG_2 could be converted to PGH_2 through the action of a peroxidase or an endoperoxide isomerase

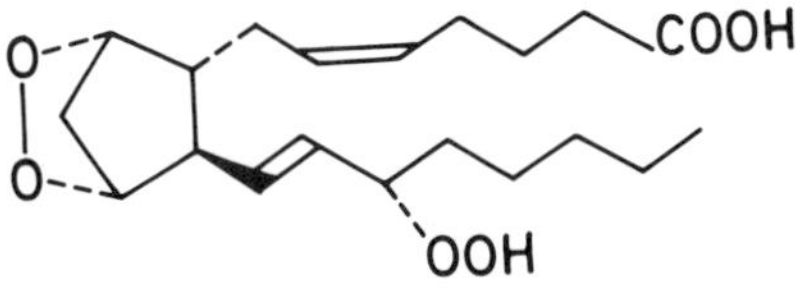

PGG_2

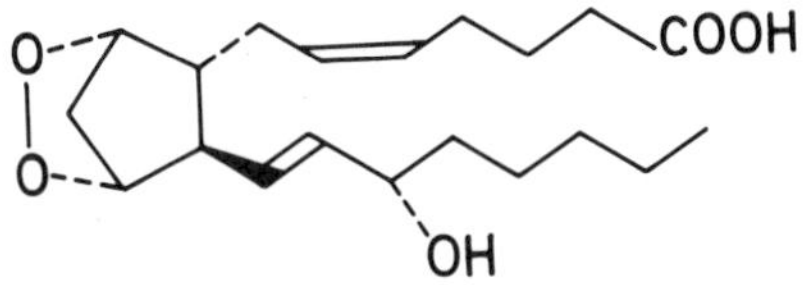

PGH_2

Figure 5. Structural formulas of PGG_2 and PGH_2. PGG_2 (15-hydroperoxy-9α,11α-peroxidoprosta-5,13-dienoic acid) and PGH_2 (15-hydroxy-9α,11α-peroxidoprosta-5,13-dienoic acid) are unstable molecules formed by the cyclization of arachidonic acid by the prostaglandin cyclo-oxygenase. (See text for details.)

would convert PGG_2 to 15-hydroperoxy-PGE_2. Both PGH_2 and 15-hydroperoxy-PGE_2 could be converted to PGE_2. The endoperoxides themselves were found to be potent aggregators of human platelets (146), to contract smooth muscle (148), and to inhibit fat cell ghost adenylate cyclase (149). Based upon these activities, a hypothesis was advanced that endoperoxides were "anti-cAMP" (150); that is, most of the known actions of endoperoxides were opposite to those of agents that are known to increase cAMP. The discovery that endoperoxides are produced in platelets at a rate higher by at least two orders of magnitude than that of the production of prostaglandins of the E or F series, and that most endoperoxide molecules are normally channeled away from E or F synthesis, has necessitated a re-evaluation of the relative significance of E and F prostaglandins in many tissues (151). This imbalance between endoperoxide synthesis and E and F synthesis prevails in the lung and umbilical artery as well as in platelets (152–154). When [1-^{14}C]arachidonic acid was incubated with washed human platelets, three major compounds were found. The first compound was identified by spectrometric analysis to be 12-L-hydroxy-5,8,10,14-eicosatetraenoic acid (HETE). An unstable hydroperoxide precursor of HETE (HPETE) was also isolated. HETE and HPETE were not related to any other products of prostaglandin synthesis, and it was soon recognized that HETE was formed from arachidonic acid by a lipoxygenase unrelated to the prostaglandin cyclo-oxygenase. The second major compound isolated was 12-L-hydroxy-5,8,10-heptadecatrienoic acid (HHT), and the third major compound formed was called PHD or thromboxane B_2 (8-(1-hydroxy-3-oxopropyl)-9,12-L-dihydroxy-5,10-heptadecadienoic acid) (154).

The synthesis of the latter two compounds can be blocked by inhibitors of prostaglandin synthesis such as aspirin or indomethacin, but the lipoxygenase is not inhibited by NOSAC drugs. The finding that both HHT and thromboxane B_2 could be formed from [1-^{14}C]PGG_2 suggested that they were products of endoperoxide metabolism. Experimental evidence showed that thromboxane B_2 was formed from PGG_2 by rearrangement and the incorporation of a water molecule in the endoperoxide structure (151).

Before the isolation of the endoperoxides, another short-lived molecule whose synthesis was known to be sensitive to inhibitors of prostaglandin synthesis had been discovered by Piper and Vane (155). This substance was called rabbit aorta contracting substance (RCS), and after the isolation of endoperoxides, RCS was thought to be an endoperoxide. However, it was soon appreciated that the chemical and biological properties of endoperoxides and RCS were not identical (148).

In a series of elegant experiments, Hamberg et al. (156, 157) determined that an unstable intermediate was formed between PGG_2 and thromboxane B_2 with a $t_{1/2}$ of approximately 32 s. This half-life and the ability of this intermediate to induce platelet aggregation and to contract rabbit aorta led to the suggestion that this intermediate called thromboxane A_2 (TXA_2) and RCS were probably the same molecule. Figure 6 gives the structure and biosynthetic relationship of the various products of arachidonic acid metabolism in human platelets.

On a molar basis, TXA_2 is a more potent inducer of platelet aggregation and constrictor of rabbit aorta than is PGH_2 (157, 158). It is not clear whether or not PGH_2 must be converted to TXA_2 for aggregation to occur (159). Although both PGH_2 and TXA_2 inhibit PGE_1-stimulated cAMP accumulation (TXA_2 is more potent), neither of these compounds decreases basal cAMP levels, (160, 161). This inability of PGH_2 or TXA_2 to decrease basal cAMP levels, while at the same time inducing platelet aggregation, has been difficult to evaluate. PGH_2 and TXA_2 do not induce platelet aggregation in the presence of the Ca^{2+} chelator EGTA (162). These data suggest that the initial event in an aggregation induced by PGH_2 or TXA_2 may involve a movement of Ca^{2+} from the bound to the free state. Once the intracellular Ca^{2+} level increases as a result of release from intracellular Ca^{2+} storing organelles, depolymerization of microtubules occurs, followed by contraction of actomyocin filaments (163–165). The exact site of Ca^{2+} storage is not known, but the dense tubular system (DTS) has been implicated (166, 167). Because PGH_2 and TXA_2 may induce Ca^{2+} mobilization in the DTS, and because Ca^{2+} inhibits platelet adenylate cyclase (168), it is tempting to postulate that PGH_2 and TXA_2 inhibit adenylate cyclase indirectly through Ca^{2+} mobilization. Support for this concept is found in the work of Rodon and Feinstein (169), who recently demonstrated that Ca^{2+} alone can inhibit PGE_1 stimulation of adenylate cyclase in platelet membrane preparations. Agents that increase cAMP inhibit a PGH_2-induced aggregation (Figure 7). The mechanism

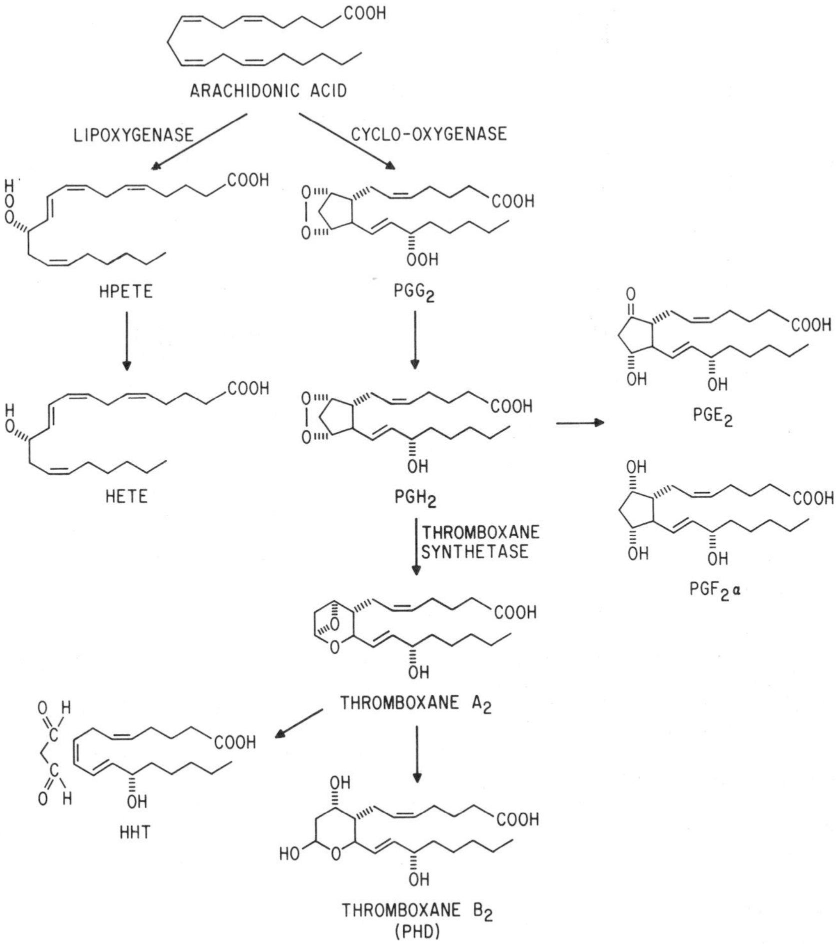

Figure 6. Pathways of arachidonic acid metabolism in human platelets. In the human platelet, arachidonic acid can be metabolized by two separate enzymic pathways. The lipoxygenase pathway forms hydroperoxy-HETE (HPETE) and HETE. This enzyme is inhibited by 5,8,11,14-eicosatetraenoic acid (TYA), but not by nonsteroidal anti-inflammatory compounds (e.g., indomethacin). The cyclo-oxygenase pathways forms the endoperoxide PGH_2, which can then be converted to the various compounds shown. The cyclo-oxygenase pathway is inhibited by both TYA and nonsteroidal anti-inflammatory compounds.

by which cAMP inhibits aggregation induced by endoperoxides is not known, but it may involve an inhibition of Ca^{2+} mobilization, because the addition of the exogenous Ca^{2+} can partially overcome the cAMP block (170). In this context, it is interesting to note that Ca^{2+} ionophore-induced secretion and platelet aggregation are inhibited by dibutyryl-cAMP and PGE_1, which stimulates adenylate cyclase; this inhibition is greatly potentiated by the phosphodiesterase inhibitor theophylline (164, 165).

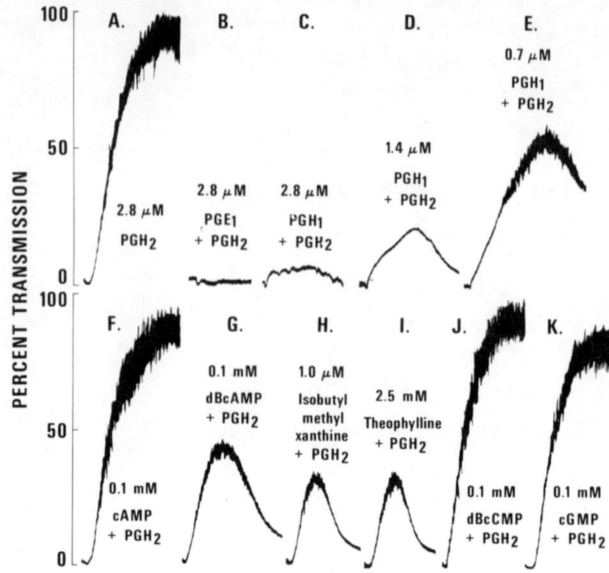

Figure 7. Inhibition of PGH_2-induced human platelet aggregating by agents that increase cAMP levels. *A*, PGH_2-induced platelet aggregation, which is antagonized by a 15-s preincubation with either PGE_1 or the series 1 endoperoxide PGH_1 (*B–E*). Both PGE_1 and PGH_1 are potent stimulators of cAMP accumulation in platelets. A 2-min preincubation with cAMP (*F*) does not inhibit, but a 2-min preincubation with the more lipophilic dibutyryl-cAMP does inhibit a PGH_2-induced aggregation (*G*). The phosphodiesterase inhibitors isobutylmethylxanthine and theophylline are also potent inhibitors of a PGH_2-induced aggregation (*H–I*). Neither dibutyryl-cCMP nor cGMP inhibited the aggregation induced by PGH_2 (*J–K*). Dibutyryl-cGMP does not inhibit or augment the PGH_2-induced aggregation (data not shown). Platelet aggregation is presented as per cent of transmission; 100% represents maximum aggregation.

Prostacyclin

PGH_2 and TXA_2 are powerful inducers of platelet aggregation and inhibitors of cAMP accumulation (157, 160, 161). If these are inducers of platelet aggregation, what are the natural inhibitors of aggregation? PGE_1 is certainly a potent inhibitor of aggregation and stimulator of cAMP accumulation (170–176). However, there is so little homo-γ-linolenic acid in the platelet that it seems unlikely that PGE_1 is the physiological stimulator of cAMP accumulation (15). PGD_2, like PGE_1, stimulates cAMP accumulation, but it is not known whether PGD_2 is produced in amounts sufficient to act as an effective modulator of platelet cAMP (177, 178). PGE_2, which is readily synthesized in platelets, is a poor stimulator of platelet adenylate cyclase and at low concentrations actually enhances an ADP-induced aggregation (155).

The search for a potent endogenously produced prostaglandin that inhibits platelet aggregation and is produced from arachidonic acid has been rewarded. The original discovery came while Moncada et al. were studying arachidonic acid metabolism in isolated pig aorta microsomes (179). Because of the strong vasoconstriction of aortas induced by TXA_2, they were interested to see whether aortas could produce TXA_2 and regulate their own vascular tone. They found that microsomes from aortas did not convert arachidonic acid to any known prostaglandin-like activity that they could identify by their smooth muscle bioassay techniques (179). However, when the endoperoxide [1-^{14}C]PGG_2 was incubated with these same microsomes, a major polar product was formed that did not correspond to known prostaglandin-like activities on thin layer chromatography (180). The muscle bioassay was used in an attempt to identify the unknown microsomal product which they called PGX. This technique did not allow them to identify PGX, but it did tell them what PGX was not. The most useful series of smooth muscles for their bioassay cascade was rabbit aorta, rat colon, and rat stomach. Rabbit aorta is contracted by endoperoxides and TXA_2, but not by PGE_2, $PGF_{2\alpha}$, or PGD_2. Rat colon is contracted by PGE_2 and $PGF_{2\alpha}$, but not by endoperoxides or TXA_2. Rat stomach is contracted to different degrees by the various compounds mentioned. PGX did not contract rabbit aorta, pulmonary artery, or vena cava, so PGX could not be an endoperoxide or TXA_2. Rat colon was not contracted by PGX, thus, by comparison, PGX could not be PGE_2 or $PGF_{2\alpha}$. PGX was also found to be 5–20 times more potent an inhibitor of arachidonic acid-induced platelet aggregation than PGD_2; the antiaggregatory activity of PGX was lost after standing for 20 min at 22 °C or after boiling for 12 s (179). This eliminated the possibility of PGD_2 being PGX. By combining bioassay data and the radio-thin layer chromatography data, as well as from the marked instability of PGX, Gryglewski et al. (180) were able to eliminate PGE_2, $PGF_{2\alpha}$, TXA_2, TXB_2, PGD_2, and HHT as possible candidates for PGX. In addition, none of the above compounds, when incubated with aortic microsomes, could produce PGX.

PGX was also found to be produced in rat stomach fundus (180). Pace-Asciak and co-workers found that a 6(9)-oxy-$PGF_{2\alpha}$ compound, a possible intermediate between PGH_2 and 6-keto-$F_{1\alpha}$, was formed in rat stomach fundus from arachidonic acid and PGG_2 (181, 182). Because of the similar chromatographic properties of 6-keto-$PGF_{1\alpha}$ and the decomposition products of PGX, it was postulated that the unstable PGX was associated with this metabolic pathway. A group of scientists at the Upjohn Company collaborated with the English group and found that the unstable decomposition product of PGX was 6-keto-$PGF_{1\alpha}$ (183). The Upjohn group was able to make 9-deoxy-6,9-epoxy-Δ^5-$PGF_{1\alpha}$ by total organic synthesis (183), and its biological and chemical identity with PGX produced biosynthetically was established. The biosynthetic route of PGX production and its subsequent metabolism to 6-keto-$F_{1\alpha}$ are shown in Figure 8. The trivial name "prostacyclin" has been proposed for PGX (183). It should

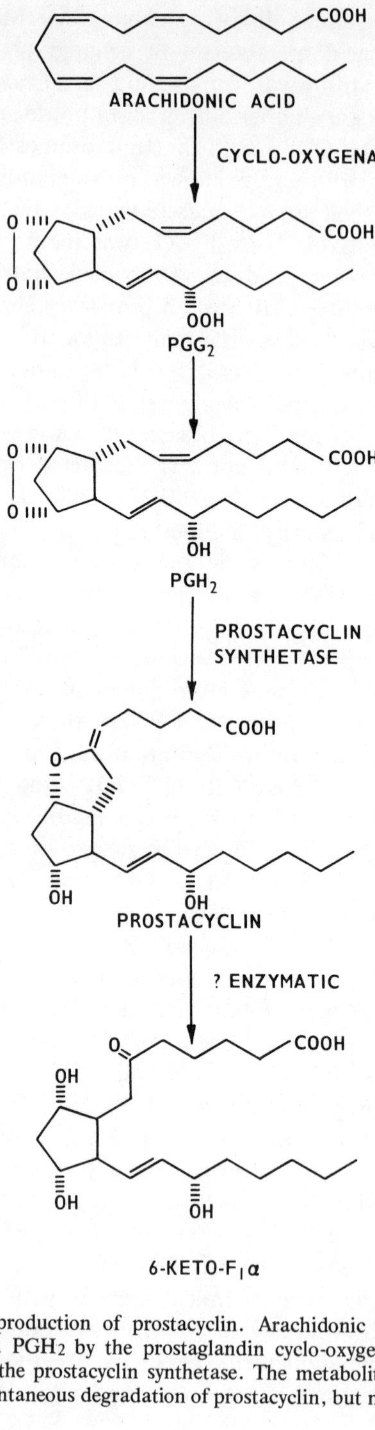

Figure 8. Biosynthetic production of prostacyclin. Arachidonic acid is converted to the endoperoxides PGG_2 and PGH_2 by the prostaglandin cyclo-oxygenase. PGH_2 is then converted to prostacyclin by the prostacyclin synthetase. The metabolite of prostacyclin, 6-keto-$F_{1\alpha}$, can be formed by spontaneous degradation of prostacyclin, but may be formed enzymically in vivo.

be noted that Kulkarni et al. (184) also reported a vasodilating substance from arachidonic acid that undoubtedly is PGX although no chemical characterization was done.

Because prostacyclin is a potent inhibitor of platelet aggregation, it was tested as a stimulator of platelet cAMP accumulation. Gorman et al. (185) and Tatesson et al. (186) found that prostacyclin is a potent stimulator of cAMP accumulation in human platelets.

Prostacyclin was more potent than PGE_1 with respect to both the rate and total amount of cAMP synthesized (185). The prostacyclin directly stimulates adenylate cyclase in platelet microsomal preparations, and on a molar basis, is approximately 10 times more potent than PGE_1 (186).

The studies with endoperoxides and prostacyclin in this report elucidate a novel homeostatic control mechanism that is illustrated in Figure 9. Platelets, when stimulated, produce the endoperoxide PGH_2. The platelet contains an enzyme, thromboxane synthetase, which converts PGH_2 to TXA_2, which brings about platelet aggregation. The vessel wall has the enzyme prostacyclin synthetase, but not thromboxane synthetase (179). Therefore, PGH_2 that migrates from the platelet aggregate is converted to prostacyclin by the vessel (179). Prostacyclin can have two actions: it can dilate the vessel itself or it can migrate to the platelet and increase intracellular cAMP, which in turn reverses the platelet aggregation. The economy of this mechanism is striking. For the first time, it is clear that two compounds, TXA_2 and prostacyclin, which are both produced from PGH_2, have opposite regulatory activities on cAMP production.

It would seem that the balance between the cAMP (adenylate cyclase)-inhibiting activity of the thromboxane system and the cAMP-stimulating activity of prostacyclin controls human platelet aggregation (185).

The interruption of endoperoxide synthesis with aspirin or indomethacin, which was thought to afford some protection against coronary heart disease and strokes, must be re-evaluated as a therapeutic regimen. It is now appreciated that any inhibition of the synthesis of endoperoxide, in addition to blocking TXA_2 synthesis, also compromises the synthesis of prostacyclin; this may be of the upmost importance for the avoidance of vascular disease.

The discovery of the thromboxane and prostacyclin pathways offers a solid basis for studies on the mechanism of action of prostaglandins. It may be that the classical prostaglandins actually mimic the actions of thromboxanes and prostacyclin. Prostaglandins of the E series generally behave like prostacyclin, whereas F series prostaglandins have many biological activities similar to those of thromboxanes. Figure 10 summarizes the biosynthetic relationships discussed in this chapter and emphasizes the pivotal role of the endoperoxide in all prostaglandin and thromboxane synthesis.

It is too early to establish the relative importance of the thromboxane-prostacyclin regulatory systems in the general biochemical control of the

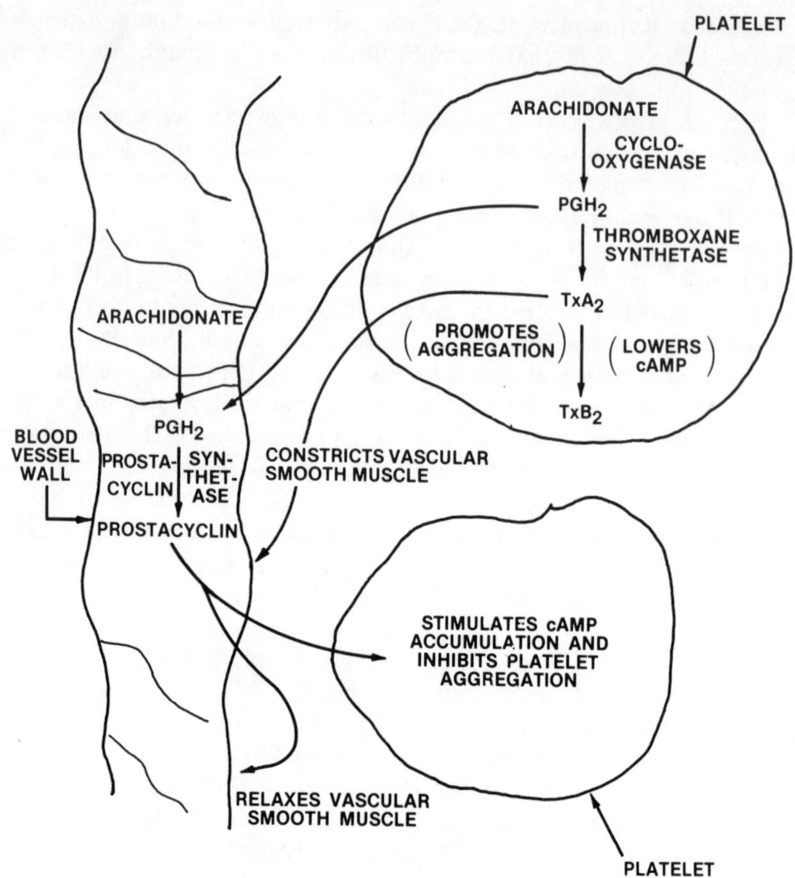

Figure 9. Model of platelet homeostasis. Platelet aggregation and vascular tone are controlled by a balance between the proaggregatory and vasoconstricting actions of TXA_2 and the vasodilating and antiaggregatory actions of prostacyclin. The mechanism responsible for these activities seems to be the reciprocal regulation of cAMP levels. TXA_2 lowers and prostacyclin stimulates cAMP levels.

cell. However, it does seem that in cells containing a thromboxane synthetase the stimulation of the cyclo-oxygenase and subsequent endoperoxide synthesis results in an inhibition of adenylate cyclase (probably through Ca^{2+} mobilization). In cells containing the prostacyclin synthetase, the same stimulation of the cyclo-oxygenase results in an elevation of cAMP (probably through inhibition of Ca^{2+} mobilization). It is now of the utmost importance to characterize cell types on the basis of their ability to synthesize either thromboxanes or prostacyclin. When this is done, we may have a much clearer understanding of both the role and the mechanism of action of the local hormones, the prostaglandins.

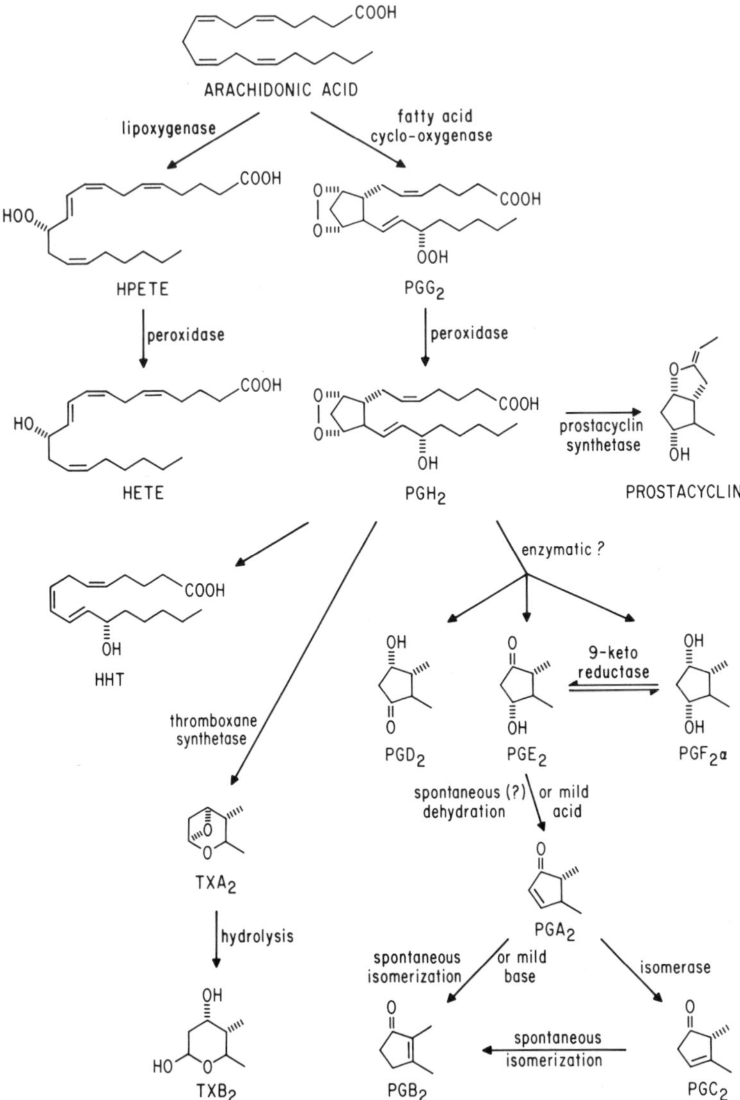

Figure 10. Biosynthetic relationships of prostaglandins and thromboxanes.

REFERENCES

1. Goldblatt, M. W. (1935). J. Physiol. (Lond.) 84: 208.
2. von Euler, U. S. (1934). Arch. Exp. Pathol. Pharmacol. 175:78.
3. von Euler, U. S. (1935). Klin. Wochenschr. 14:1182.
4. Bergstrom, S., and Sjovall, J. (1960). Acta Chem. Scand. 14:1693.
5. Bergstrom, S., Ryhage, R., Samuelsson, B., and Sjovall, J. (1963). J. Biol. Chem. 238:3555.

6. Bergstrom, S., Dressler, C., Ryhage, R., Samuelsson, B., and Sjovall, J. (1962). Arkh. Kemi. 19:563.
7. Samuelsson, B. (1963). J. Am. Chem. Soc. 85:1878.
8. Van Dorp, D. A., Beerthuis, R. K., Nugteren, D. H., and Von Heman, H. (1964). Biochim. Biophys. Acta 90:204.
9. Bergstrom, S., Danielson, H., and Samuelsson, B. (1964). Biochim. Biophys. Acta 90:207.
10. Beal, P. E., Fonken, G. S., and Pike, J. E. (1964). The Upjohn Company, Belgium Patent 659, 884, prior date U.S.A., Feb. 19.
11. Rahm, J. J., and Holman, R. T. (1964). J. Lipid Res. 5:169.
12. Beerthuis, R. K., Nugteren, D. H., Pabon, H. J. J., and Van Dorp, D. A. (1968). Rec. Trau. Chim. Pays Bas. 87:461.
13. Marcus, A. J., Ullman, H. L., and Safier, L. B. (1969). J. Lipid Res. 10:108.
14. Hilditch, T. P. (1956). The Chemical Constitution of Natural Fats, Ed. 3, p. 664. Chapman and Hall, London.
15. Bills, T. K., and Silver, M. J. (1975). Fed. Proc. 34:322.
16. Christ, E. J., and Nugteren, D. H. (1970). Biochem. Biophys. Acta 218:296.
17. Bergstrom, S., Carlson, L. A., and Weeks, J. R. (1968). Pharmacol. Rev. 20:1.
18. Weeks, J. R. (1972). Annu. Rev. Pharmacol. 12:317.
19. Hinman, J. W. (1972). Annu. Rev. Biochem. 41:161.
20. Pike, J. E. (1970). Fortschr. Chem. Org. Naturst. 28:313.
21. Horton, E. W. (1969). Physiol. Rev. 49:122.
22. Butcher, R. W., and Baird, C. E. (1968). J. Biol. Chem. 243:1713.
23. Steinberg, D., Vaughan, M., Nestel, P. J., Strand, O., and Bergstrom, S. (1964). J. Clin. Invest. 43:1553.
24. Butcher, R. W., Pike, J. E., and Sutherland, E. W. (1967). *In* S. Bergstrom and B. Samuelsson (eds.), Prostaglandins, Proceedings of the Second Nobel Symposium, Stockholm, pp. 133. Interscience, New York.
25. Harwood, J. P., Moskowitz, J., and Krishna, G. (1971). Biochim. Biophys. Acta 261:444.
26. Mills, D. C. B., and Smith, J. B. (1971). Biochem. J. 121:185.
27. Salzman, E. W., and Neri, L. L. (1969). Nature 224:610.
28. Burke, G., Chang, L. L., and Szabo, M. (1973). Science 180:872.
29. Stoner, J., Manganiello, V. C., and Vaughan, M. (1974). Proc. Natl. Acad. Sci. USA 70:3830.
30. Johnson, D. G., Thompson, W. J., and Williams, R. H. (1974). Biochemistry 13:1920.
31. Hamprecht, B., and Schultz, J. (1973). FEBS Lett. 34:85.
32. Vesin, M. F., and Harbon, J. (1974). Mol. Pharmacol. 10:457.
33. Berti, F., Trabucchi, M., Bernareggi, V., and Fumagalli, R. (1973). Adv. Biosci. 9:475.
34. Kuehl, F. A., Jr., Humes, J. L., Tarnoff, J., Cirillo, U. J., and Ham, E. A. (1970). Science 169:883.
35. Hamprecht, B., and Schultz, J. (1973). Hoppe Seylers Z. Physiol. Chem. 354: 1633.
36. Borgeat, P., Labrie, F., and Garneau, P. (1975). Can. J. Biochem. 53:455.
37. O'Donnell, E. R. (1974). J. Biol. Chem. 249:3615.
38. Butcher, R. W. (1970). Adv. Biochem. Pharmacol. 3:173.
39. Ramwell, P., and Shaw, J. E. (1970). Rec. Progr. Horm. Res. 26:139.
40. Butcher, R. W., and Baird, C. E. (1970). *In* R. Eigenman (ed.), Proceedings of the Fourth International Congress on Pharmacology, p. 42. Schwabe and Co., Basel.

41. Shaw, J. E., Jessup, S. J., and Ramwell, P. W. (1972). Adv. Cyclic Nucleotide Res. 1:479.
42. Silver, M. J., and Smith, J. B. (1975). Life Sci. 16:1635.
43. Eliasson, R. (1959). Acta Physiol. Scand. 46(suppl. 158):1.
44. Pace-Asciak, C., and Wolfe, L. S. (1968). Biochim. Biophys. Acta 152:184.
45. Hope, W. C., Van Trabert, T. C., and Dalton, C. (1973). American Chemical Society 166th Annual Meeting on Biol. Vol. 45 (abstr.)
46. Haye, B., Champion, S., and Jacquemin, C. (1973). FEBS Lett. 30:253.
47. Lands, W. E. M., and Samuelsson, B. (1968). Biochim, Biophys. Acta 164: 426.
48. Von Keman, H., and Van Dorp, D. A. (1968). Biochim. Biophys. Acta 164: 430.
49. Kunze, H., and Vogt, W. (1971). Ann. N. Y. Acad. Sci. 180:123.
50. Haye, B., Champion, S., and Jacquemin, C. (1976). *In* Advances in Prostaglandin and Thromboxane Research, Vol. 1, Raven Press, New York. B. Samuelsson and R. Paoletti (eds.), pp. 29–34.
51. Dalton, C., and Hope, W. C. (1974). Prostaglandins 6:227.
52. Minkes, M., Stanford, N., Chi, M. M. Y., Roth, G. J., Raz, A., Needleman, P., and Majerus, P. W. (1977). J. Clin. Invest. 59:449.
53. Hamprecht, B., Jaffe, B. M., and Philpott, G. W. (1973). FEBS Lett. 36: 193.
54. Chiappe de Cingolani, G. E., Van der Bosch, H., and Van Deenen, L. L. M. (1972). Biochim. Biophys. Acta 260:387.
55. Cedro, H. K., and Staszewska, J. B. (1974). Abstracts of Second Congress of the Hungarian Pharmacological Society, Budapest, p. 19.
56. Kantrowitz, F., Robinson, D. R., McGuire, M. B., and Levine, L. (1975). Nature (Lond.) 258:737.
57. Tashjian, A. H., Voelkel, E. F., McDonough, J., and Levine, L. (1975). Nature (Lond.) 258:739.
58. Gryglewski, R. J., Panczenko, B., Korbut, R., Grodzinska, L., and Ocetkiewicz, A. (1975). Prostaglandins 10:343.
59. Greaves, M. W., and McDonald-Gibson, W. (1972). Br. Med. J. 2:83.
60. Lewis, G. P., and Piper, P. J. (1975). Nature (Lond.) 254:308.
61. Greaves, M. W., Kingston, W. P., and Pretty, K. (1975). Br. J. Pharmacol. 53:470.
62. Flower, R. J., Gryglewski, R., Cedro, H. K., and Vane, J. R. (1972). Nature (Lond.) 238:104.
63. Kalisker, A., and Dyer, D. C. (1972). Eur. J. Pharmacol. 19:305.
64. Palmer, M. A., Piper, P. J., and Vane, J. R. (1973). Br. J. Pharmacol. 49: 226.
65. Needleman, P., Minkes, M. S., and Douglas, J. R., Jr. (1974). Circ. Res. 34:445.
66. Gimbrone, M. A., Jr., and Alexander, R. W. (1975). Science 189:219.
67. Anggard, E., and Jonsson, C. E. (1971). Ann. N. Y. Acad. Sci. 180:200.
68. Hamberg, M., Israelsson, U., and Samuelsson, B. (1971). Ann. N. Y. Acad. Sci. 180:164.
69. Samuelsson, B., Granstrom, E., Green, K., and Hamberg, M. (1971). Ann. N. Y. Acad. Sci. 180:1126.
70. Fried, J., Santhauakrishnan, T. S., Himizu, J., Lin, C. H., Ford, S. H., Rubin, B., and Grigas, E. O. (1969). Nature (Lond.) 223:208.
71. Fried, J., Lin, C. H., Mehra, M., Kao, W., and Dalven, P. (1971). Ann. N. Y. Acad. Sci. 180:38.
72. Sanner, J. H. (1969). Arch. Int. Pharmacodyn. Ther. 180:46.

73. Sanner, J. H. (1971). Ann. N. Y. Acad. Sci. 180:396.
74. Eakins, K. E., Karim, S. M. M., and Miller, J. D. (1970). Br. J. Pharmacol. 39:556.
75. Eakins, K. E., Miller, J. D., and Karim, S. M. M. (1971). J. Pharmacol. Exp. Ther. 176:441.
76. Kuehl, F. A., Jr., and Humes, J. L. (1972). Proc. Natl. Acad. Sci. USA 69: 480.
77. Gorman, R. R., and Miller, O. V. (1973). Biochim. Biophys. Acta 323:560.
78. Smigel, M., and Fleischer, J. (1974). Biochim. Biophys. Acta 332:358.
79. Rao, C. U. (1974). J. Biol. Chem. 249:7203.
80. Schaumburg, B. P. (1973). Biochem. Biophys. Acta 326:127.
81. Kimball, F. A., and Wyngaarden, L. J. (1975). Prostaglandins 9:413.
82. Moore, W. V., and Wolff, J. (1973). J. Biol. Chem. 248:5705.
83. Miller, O. V., and Magee, W. E. (1973). Adv. Biosci. 19:83.
84. Gorman, R. R. (1974). Prostaglandins 6(abstr.):542.
85. Lord, J. T., Ziboh, V. A., and Warren, S. (1976). *In* B. Samuelsson and R. Paoletti, (eds.), Advances in Prostaglandin and Thromboxane Research, Vol. 1, pp. 291-296 Raven Press, New York.
86. Brunton, L. L., Wiklund, R. A., Van Arsdale, P. M., and Gilman, A. G. (1976). J. Biol. Chem. 251:3037.
87. Rao, C. V. (1975). Prostaglandins 9:579.
88. Lefkowitz, R. J., Roth, J., Pricer, W., and Pastan, I. (1970). Proc. Natl. Acad. Sci. USA 65:745.
89. Rodbell, M., Krans, H. M. J., Pohl, S. L., and Birnbaumer, L. (1971). J. Biol. Chem. 246:1861.
90. Wallach, D. P., and Daniels, E. G. (1971). Biochim. Biophys. Acta 231: 445.
91. Samuelsson, B. (1969). Progr. Biochem. Pharmacol. 5:109.
92. Samuelsson, B., Hamberg, M., Malmsten, C., and Svensson, J. (1976). *In* B. Samuelsson and R. Paoletti (eds.), Advances in Prostaglandin and Thromboxane Research, Raven Press, New York. Vol. 2, pp. 737-746.
93. Malmsten, C., Granstrom, E., and Samuelsson, B. (1975). Biochem. Biophys. Res. Commun. 68:569.
94. Anggard, E., Bohman, S. O., Griffin, J. E., III, Larsson, C., and Maunsbach, A. B. (1972). Acta Physiol. Scand. 84:231.
95. Crowshaw, K. (1971). Nature (Lond.) 231:240.
96. Hamberg, M. (1969). FEBS Lett. 5:127.
97. Wolfe, L. S., Coceani, F., and Pace-Asciak, C. (1967). *In* S. Bergstrom and B. Samuelsson (eds.), Prostaglandins, Nobel Symposium 2, pp. 265-275. Almquist and Wiksell, Stockholm.
98. Jonsson, C. E., and Anggard, E. (1972). Scand. J. Clin. Lab. Invest. 29: 289.
99. Pace-Asciak, C., and Wolfe, L. S. (1970). Biochim. Biophys. Acta 218:539.
100. Foss, P., Takeguchi, C., Tai, H., and Sih, C. (1971). Ann. N. Y. Acad. Sci. 180:126.
101. Yoshimoto, A., Ito, H., and Tomita, K. (1970). J. Biochem. (Tokyo) 68: 487.
102. Lands, W., Lee, R., and Smith, W. (1971). Ann. N. Y. Acad. Sci. 180:107.
103. Anggard, E., and Samuelsson, B. (1965). J. Biol. Chem. 240:3518.
104. Samuelsson, B. (1965). J. Am. Chem. Soc. 87:3011.
105. Van Dorp, D. A. (1967). Progr. Biochem. Pharmacol. 5:71.
106. Samuelsson, B. (1967). Progr. Biochem. Pharmacol. 5:59.
107. Smith, W. L., and Lands, W. E. M. (1972). J. Biol. Chem. 247:1038.

108. Smith, W. L., and Lands, W. E. M. (1972). Biochemistry 11:3276.
109. Lands, W. E. M., LeTellier, P. Z., Rome, L., and Vanderhock, J. Y. (1974). *In* H. J. Robinson and J. R. Vane (eds.), Prostaglandin Synthetase Inhibitors, pp. 1-7. Raven Press, New York.
110. LeTellier, P. R., Smith, W. L. Jr., and Lands, W. E. M. (1973). Prostaglandins 4:837.
111. Cook, H. W., and Lands, W. E. M. (1975). Biochem. Biophys. Res. Commun., 65:464.
112. Lands, W. E. M., Cook, H. W., and Rome, L. H. (1976). *In* B. Samuelsson and R, Paoletti (eds.), Advances in Prostaglandin and Thromboxane Research, Vol. 1, pp. 7-17. Raven Press, New York.
113. Smith, W. L. (1975). Prostaglandins 10:982.
114. Vane, J. R. (1971). Nature (Lond.) 231:232.
115. Vane, J. R. (1972). Hosp. Pract. March:61.
116. Vane, J. R. (1973). Adv. Biosci. 9:395.
117. Vane, J. R. (1974). *In* H. J. Robinson and J. R. Vane (eds.), Prostaglandin Synthetase Inhibitors, pp. 155-163. Raven Press, New York.
118. Ferreira, S. H., and Vane, J. R. (1973). Seminaire "Les Prostaglandin," Paris (INSERM), pp. 345-357.
119. Ferreira, S. H., and Vane, J. R. (1974). Annu. Rev. Pharmacol. 14:57.
120. Flower, R. J. (1974). Pharmacol. Rev. 26:33.
121. Flower, R. J., and Vane, J. R. (1974). Biochem. Pharmacol. 23:1439.
122. Ferreira, S. H., and Vane, J. R. (1967). Nature (Lond.) 216:868.
123. McGiff, J. C., Terragno, N. A., Strand, J. C., Lee, J. C., Lonigro, A. J., and Ng, K. K. F. (1969). Nature (Lond.) 223:743.
124. Lee, J. B. (1973). *In*: P. W. Ramwell (ed.), Prostaglandins, pp. 133-187. Plenum Press, New York.
125. Lee, J. B. (1972). *In* P. W. Ramwell and B. B. Phariss (eds.), Prostaglandins in Cellular Biology, pp. 399-449. Plenum Press, New York.
126. Lee, J. B., Crowshaw, K., Takman, B. H., Attrep, K. A., and Gougoutas, J. Z. (1967). Biochem. J. 105:1251.
127. Gross, J. B., and Bartter, F. C. (1973). Am. J. Physiol. 225:218.
128. Lee, J. B., McGiff, J. C., Kannegiesser, H., AyKent, Y., Mudd, J. G., and Frawley, T. F. (1971). Ann. Int. Med. 74:703.
129. Frolich, J. C., Williams, W. M., Sweetman, B. J., Smigel, M., Carr, K., Hollifield, J. W., Fleischer, S., Nies, A. S., Holmberg, M. F., and Oates, J. A. (1976). *In* B. Samuelsson and R. Paoletti (eds.), Advances in Prostaglandin and Thromboxane Research, Vol. 1, pp. 65-80. Raven Press, New York.
130. Schneider, W. P., Pike, J. E., and Kupiecki, F. P. (1966). Biochim. Biophys. Acta 125:611.
131. Anggard, E., and Samuelsson, B. (1966). Arkh. Kemi. 25:293.
132. Marrazi, M. A., and Matschinsky, F. M. (1972). Prostaglandins 1:373.
133. Tai, H. H., Tai, C. L., and Hollander, C. S. (1974). Biochem. Biophys. Res. Commun. 57:457.
134. Hansen, H. S. (1974). Prostaglandins 8:95.
135. Jarabak, J. (1972). Proc. Natl. Acad. Sci. USA 69:533.
136. Schlegel, W., Demers, L. M., Hildebrandt-Stark, H. E., Behrman, H. R., and Greep, R. O. (1974). Prostaglandins 5:417.
137. Schlegel, W., and Greep, R. O. (1976). *In* B. Samuelsson and R. Paoletti (eds.), Advances in Prostaglandin and Thromboxane Research, Vol. 1, pp. 159-162. Raven Press, New York.
138. Sun, F. F., Armour, S. B., Bockstanz, V. R., and McGuire, J. C. (1976). *In*

B. Samuelsson and R. Paoletti (eds.), Advances in Prostaglandin and Thromboxane Research, Vol. 1, pp. 163–169. Raven Press, New York.
139. Nakano, J., and Morsy, H. H. (1971). Clin. Res. 19:142.
140. Israelsson, U., Hamberg, M., and Samuelsson, B. (1969). Eur. J. Biochem. 11:390.
141. Granstrom, E., and Samuelsson, B. (1971). J. Biol. Chem. 246:5254.
142. Hamberg, M., and Samuelsson, B. (1971). J. Biol. Chem. 246:6713.
143. Dawson, W., Boot, J. R., Cockerill, A. F., Mallen, D. N. B., and Osborne, D. J. (1976). Nature (Lond.) 262:702.
144. Sun, F. F., personal communication.
145. Hamberg, M., and Samuelsson, B. (1973). Proc. Natl. Acad. Sci. USA 70: 899.
146. Hamberg, M., Svensson, J., Wakabayashi, T., and Samuelsson, B. (1974). Proc. Natl. Acad. Sci. USA 71:345.
147. Nugteren, D. H., and Hazelhof, E. (1973). Biochim. Biophys. Acta 326:448.
148. Samuelsson, B., and Hamberg, M. (1974). *In* H. Robinson and J. R. Vane (eds.), Proceedings of the International Symposium on Prostaglandin Synthetase Inhibitors, p. 107. Raven Press, New York.
149. Gorman, R. R., Hamberg, M., and Samuelsson, B. (1975). J. Biol. Chem. 250:6460.
150. Gorman. R. R. (1975). J. Cyclic Nucleotide Res. 1:1.
151. Hamberg, M., and Samuelsson, B. (1974). Proc. Natl. Acad. Sci. USA 71: 3400.
152. Hamberg, M., Svensson, J., Hedquist, P., Strandberg, K., and Samuelsson, B. (1975). *In* Samuelsson and R. Paoletti (eds.), Advances in Prostaglandin and Thromboxane Research, Vol. 1, pp. 495–501. Raven Press, New York.
153. Tuvemo, T., Strandberg, K., Hamberg, M., and Samuelsson, B. (1976). Acta Physiol. Scand. 96:145.
154. Hamberg, M., Svensson, J., and Samuelsson, B. (1974). Proc. Natl. Acad. Sci. USA 71:3824.
155. Piper, P. J., and Vane, J. R. (1969). Nature (Lond.) 223:29.
156. Hamberg, M., Svensson, J., and Samuelsson, B. (1975). Proc. Natl. Acad. Sci. USA 72:2994.
157. Hamberg, M., Svensson, J., and Samuelsson, B. (1976). *In* B. Samuelsson and R. Paoletti (eds.), Advances in Prostaglandin and Thromboxane Research, Vol. 1, pp. 19–27. Raven Press, New York.
158. Needleman, P., Moncada, S., Bunting S., Vane, J. R., Hamberg, M., and Samuelsson, B. (1976). Nature (Lond.) 261:558.
159. Needleman, P., Minkes, M., and Raz, A. (1976). Science 193:163.
160. Miller, O. V., and Gorman, R. R. (1976). J. Cyclic Nucleotide Res. 2:79.
161. Miller, O. V., Johnson, R. A., and Gorman, R. R. In press.
162. Gorman, R. R., Unpublished results.
163. Detwiler, T. C., and Feinman, R. D. (1973). Biochemistry 12:2462.
164. White, J. G., Rao, G. H. R., and Gerrard, J. M. (1974). Am. J. Pathol. 77:135.
165. Feinstein, M. B., and Fraser, C. (1975). J. Gen. Physiol. 66:561.
166. White, J. G. (1972). Fed. Proc. 31:654.
167. White, J. G. (1972). Am. J. Pathol. 66:295.
168. Vigdahl, R. L., Marquis, N. R., and Tavormina, P. (1969). Biochem. Biophys. Res. Commun. 37:409.
169. Rodan, G. A., and Feinstein, M. B. (1976). Proc. Natl. Acad. Sci. USA 73:1829.
170. Gorman, R. R. Unpublished results.

171. Kloeze, J. (1967). Nobel Symposium 2, Prostaglandins, 1966. p. 241. Almquist and Wiksell, Stockholm.
172. Bergstrom, S., Carlson, L. A., and Weeks, J. R. (1968). Pharmacol. Rev. 20:1.
173. Emmons, P. R., Hampton, J. R., Harrison, M. J. G., Honour, A. J., and Mitchell, J. R. A. (1967). Med. J. 2:468.
174. Butcher, R. W., Scott, R. E., and Sutherland, E. W. (1967). Pharmacologist 9:172.
175. Robison, G. A., Arnold, A., and Hartmann, R. C. (1969). Pharmacol. Res. Commun. 1:325.
176. Salzman, E. W., and Levin, L. (1971). J. Clin. Invest. 50:131.
177. Smith, J. B., Silver, M. J., Ingerman, C. M., Kocsis, J. J. (1974). Thrombosis Res. 5:291.
178. Mills, D. C. B., and MacFarlane, D. E. (1974). Thrombosis Res. 5:401.
179. Moncada, S., Gryglewski, R., Bunting, S., and Vane, J. R. (1976). Nature (Lond.) 263:663.
180. Gryglewski, R. J., Bunting, S., Moncada, S., Flower, R. J., and Vane, J. R. (1976). Prostaglandins 12:685.
181. Pace-Asciak, C., and Wolfe, L. S. (1971). Biochemistry 110:3657.
182. Pace-Asciak, C., Nashat, M., and Menon, N. K. (1976). Biochim. Biophys. Acta 424:323.
183. Johnson, R. A., Morton, D. R., Kinner, J. H., Gorman, R. R., McGuire, J. C., Sun, F. F., Whittaker, N., Bunting, S., Salmon, J., Moncada, S., and Vane, J. R. (1976). Prostaglandins 12:915.
184. Kulkarni, P. S., Roberts, R., and Needleman, P. (1976). Prostaglandins 12:337.
185. Gorman, R. R., Bunting, S., and Miller, O. V. (1977). Prostaglandins. 13: 377.
186. Tatesson, J. E., Moncada, S., and Vane, J. R. (1977). Prostaglandins. 13:389.

International Review of Biochemistry
Biochemistry and Mode of Action of Hormones II, Volume 20
Edited by H. V. Rickenberg
Copyright 1978 University Park Press Baltimore

4
Facts and Speculation on the Mode of Action of the Steroid-Binding Protein Complex

J.-P. JOST
Friedrich Miescher-Institut, Basel, Switzerland

Since the discovery (1, 2) of the steroid-receptor complex, a great many laboratories have established its ubiquity in target tissues. The many cellular functions modified by steroids attest to their crucial role in cellular metabolism and differentiation. This chapter examines the current status of research into various aspects of the steroid-receptor complex with special emphasis on the possible role of the steroid-receptor complex in overall and specific gene activation.

CYTOPLASMIC AND NUCLEAR FORMS OF RECEPTOR

Proteins with a high affinity for steroids (K_d 10^{-9}–10^{-12} M) are present in both cytoplasm and nuclei of target cells. Over 10 years ago, Toft and Gorski (3) found, upon homogenization of uterine tissue, that a soluble protein capable of binding labeled estradiol was present in the cytoplasmic fraction. The steroid-binding protein was specific for 17β-estradiol and other biologically active estrogens, and it bound these hormones with high affinity (K_d 10^{-10} M). In general, the steroid-binding molecules were heat labile and nondialyzable, and their protein nature was demonstrated by their sensitivity to proteolytic enzymes.

Over the past decade, sucrose density gradient centrifugation of crude preparations from cytoplasmic or nuclear fractions has been the most widely used method for the separation and determination of the sedimentation coefficient of the steroid receptors. Despite widespread use of this technique, the sedimentation coefficients reported by different laboratories vary considerably (4). Besides differences observed between organs and animal species for a given steroid receptor prepared from a target tissue, the sedimentation coefficient may vary depending on the salt concentration used for the extraction of the receptor and the preparation of sucrose gradients. In addition, polyanions such as heparin produce almost any "desired" sedimentation artifact (4). In the case of the uterine estradiol receptor, upon ultracentrifugation in a sucrose gradient the cytoplasmic estrogen receptor complex sedimented as a discrete band at 8 S (3). Increasing the ionic strength (0.3 M potassium chloride) of these gradients results in the reversible transformation of the 8 S complex to a 4 S form (5). In the same system and using the same techniques, while varying conditions of protein concentration, ionic strength, and incubation time, Stancel et al. (6) concluded that the cytoplasmic species of the estrogen receptor is initially present in uterine cells as a small form with a sedimentation coefficient of 3.8–4.8 S. After homogenization, the small form of the receptor seems to undergo a slow aggregation to yield a more rapidly sedimenting species. From their kinetic studies, they also concluded that the aggregation does not seem to be self-aggregation of estrogen receptor but rather an association with other, not yet defined, uterine proteins present in the crude homogenate. The effect of ionic strength on aggregation of the cytoplasmic receptor has also been studied by several other groups. For example, estradiol receptor, extracted from the cytoplasm of uterine tissues, sediments at about 6 S at physiological ionic strength (0.15 M potassium chloride). This value contrasts with the 4 S or 8 S form of receptor extracted at high or very low salt concentration, respectively (7). As determined by sucrose density gradient centrifugation in the presence of 0.3 M potassium chloride, the cytoplasmic form of progesterone-binding protein isolated from chicken oviduct has a sedimentation coefficient of 3.8 S; in the absence of potassium chloride, it aggregates to 5 S and 8 S (8). Indirect physical measurements suggest that the progesterone receptor exists in the shape of a prolate ellipsoid with a monomeric molecular weight of about 90,000. The estimated axial ratio shows that progesterone receptor has a length 14–18 times greater than its width (8). The findings may explain the apparent discrepancies of molecular weight observed upon analysis by different techniques, e.g., sucrose density gradient centrifugation versus gel filtration. In the presence of calcium chloride, partially purified chicken oviduct progesterone receptor (4–5 S form) dissociates into subunits of 2.6 S with a Stokes radius of 21 Å and an apparent molecular weight of 20,000 (9). The subunit formation occurs in the presence or absence of steroid or diisopropyl fluorophosphate (a protease inhibitor). With the use of the charcoal dextran assay for bound steroids, it was also found that the subunits have the same order of affinity for progesterone as does the intact receptor (K_d 4.10^{-10} M for progesterone) (9). This small form of steroid receptor subunit does not seem to be unique to the chicken oviduct system because for the rat uterus estradiol receptor Rochefort and Baulieu (10) have obtained a 2.4 S form of receptor upon treatment of the crude preparation with high concentrations of mercaptoethanol.

The occurrence of a homogeneous peak of receptor activity, as measured in a sucrose density gradient, does not necessarily indicate a single molecular form of receptor. For example, in their studies on the chick oviduct progesterone receptor Schrader and O'Malley (11) found that the 4 S peak obtained on sucrose gradient could be further separated on DEAE-cellulose into two different progesterone receptors. As in the above experiments, the two proteins had identical hormone binding kinetics and steroid specificity. They did, however, differ in their net negative charges and their physical stability. During the purification on DEAE-cellulose, components A and B were found in equal amounts, and indirect evidence suggests that forms A and B of progesterone receptor are subunits of the 8 S cytoplasmic receptor (11). Similarly, estrogen-binding protein of calf endometrium can be separated into two distinct components on DEAE-cellulose (12). The possible implication of the occurrence of these two forms of cytoplasmic receptor proteins for gene regulation is discussed below.

As measured by sucrose density gradient centrifugation, the nuclear form of steroid receptor differs from the cytoplasmic form by its sedimentation coefficient. For example, the nuclear form of estradiol receptor has a sedimentation coefficient of 4–5 S, whereas the cytoplasmic form of the receptor varies between 4 and 8 S. By contrast, the binding constants for estradiol of the two forms of receptor are practically identical. The question still remains: do the cytoplasmic and nuclear forms of steroid receptor share a common subunit? This important question is discussed in the next section.

It is self-evident that the study of the exact relationship between the cytoplasmic and nuclear forms of receptor as well as their possible role in gene activation will be feasible only when the receptor molecules are purified to homogeneity. This has recently become possible with the introduction of affinity chromatography techniques (13–18). These techniques exploit the unique biological property of the receptor to bind specifically and reversibly to ligands which are covalently bound to an inert insoluble matrix. For example, by using a deoxycorticosterone-bovine serum albumin-Sepharose complex in combination with other standard techniques, O'Malley's group was able to purify the two cytoplasmic forms of progesterone receptor of chick oviduct to homogeneity (17). In the purified form, the two subunits A and B of progesterone 4 S receptor retained the same biological activities as compared with the crude preparation; e.g., they had comparable K_d values for progesterone ($1.1 \cdot 10^{-9}$ M) and were able to bind to nuclei, chromatin, and DNA (17). With the use of a similar approach, human uterine progesterone receptor (18), calf uterus estradiol receptor (14, 15), and glucocorticoid receptor from hepatoma cells (16) were isolated and their biological activities compared with those of the crude preparations.

TRANSLOCATION OF STEROID RECEPTOR FROM CYTOPLASM TO NUCLEUS

The earliest known effect of steroids in their specific target cells is a strong interaction of specific cytoplasmic protein(s) with the entering hormone. This interaction is immediately followed by the appearance of the labeled steroid in the nuclei (1). The pioneering work of Gorski and Gannon (2) and of Jensen et al. (19) has led to the concept that estrogen induces a conformational change in the cytoplasmic receptor protein of the uterine tissue which is followed by the translocation of the steroid-receptor complex to the nucleus. The intracellular transfer to the protein-bound estradiol from the cytoplasm to the nucleus was initially demonstrated in the rat uterus (19). Briefly, Jensen et al. (19) found that the in vitro incubation of labeled 17β-estradiol with uterine tissue resulted in an accumulation of salt-extractable (0.3 M potassium chloride) steroid-receptor complex from the nuclei. The nuclear form of steroid-receptor complex had a sedimentation coefficient of 5 S (sucrose density gradient determination). Such a steroid-receptor complex could not be found in the nuclei of target tissues not previously exposed to the hormone. Concomitant with the appearance of a nuclear form of receptor, the investigators also observed a depletion of total cytoplasmic 8 S receptor. When they used a cell-free system consisting of nonfractionated cytoplasm and nuclei of the target tissue, they also observed that the addition of labeled estradiol led to accumulation of extractable 5 S steroid-receptor complex in the nuclei; no extractable 5 S complex was found when labeled estradiol was incubated with uterine nuclei in the absence of a cytoplasmic fraction.

These observations led Jensen and DeSombre (1) and Jensen et al. (19) to formulate the two-step hypothesis according to which the estradiol-receptor complex of the uterine nucleus is derived from the cytosol by a temperature-dependent process in which the association with the hormone activates the cytosol receptor protein which then migrates to the nucleus (see Figure 1). The observations of Jensen and DeSombre (1) were later confirmed by other groups and with other systems. For example, in the case of glucocorticoid receptor in hepatoma cells, Rousseau et al. (20) found that the binding of dexamethasone and cortisol results in a loss of most of the receptor sites from the cytoplasm and the appearance of an equivalent number of steroid molecules in the nucleus. They also reported that when the steroid is removed from the culture the hormone leaves the nucleus while the level of the cytoplasmic receptor returns to normal. The trans-

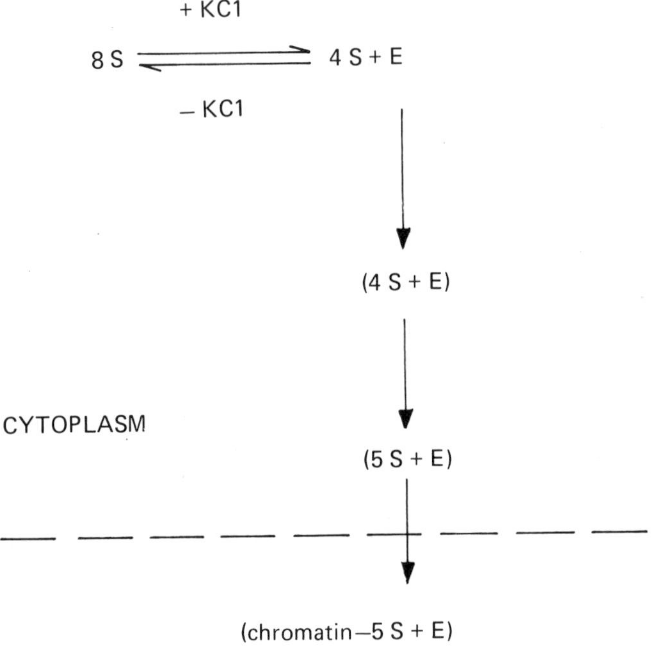

Figure 1. The two-step mechanism of the steroid-receptor translocation. Redrawn from Jensen and DeSombre (1) with permission. E, 17β-estradiol.

location of the receptor from the nucleus back to the cytoplasm does not seem to require either RNA or protein synthesis. However, the transloca-tion of the receptor from the cytoplasm to the nucleus requires specific steroids. The absolute requirement for specific steroids is nicely exempli-fied by the system of induction of tyrosine aminotransferase by dexametha-sone in hepatoma cells in tissue culture. Anti-inducers such as progesterone prevent the translocation of the glucocorticoid receptor from the cyto-plasm to the nucleus and at the same time prevent the induction of the enzyme (20). According to the early work of Jensen and DeSombre (1), the estradiol receptor from the uterus cytoplasm is in a 4–8 S form and upon activation it appears as a 5 S form in the nucleus. According to Jensen and DeSombre (1), the conversion of the 4 S or 8 S to the 5 S form of the receptor depends on a) a cytoplasmic factor(s), b) temperature, and c) the presence of specific steroids. Before these parameters are briefly considered, it should be noted that most studies on the transformation of the receptor have been performed with crude systems consisting mostly of a cytoplasmic fraction containing the endogenous cytoplasmic receptor prepared from target tissues or cells. In 1968, Jensen and associates (19) demonstrated that the conversion of the 4 S cytoplasmic receptor from the calf uterus to the 5 S nuclear receptor was temperature-dependent. The rate of 4 S to 5 S receptor transformation shows a 200-fold increase between zero and 35 °C. The Arrhenius energy of activation is 21.3 kcal mol^{-1} in buffers without KCl, and 19.1 kcal mol^{-1} in buffers containing 0.4 M KCl (21). Only in the case of the estradiol receptor has a "transforming factor" been reported (22) (see Figure 2). This factor is presumably a protein, is present in the cytoplasm of target cells, and has a molecular weight of about 100,000. It specifically requires Ca^{2+} for its receptor transforming activity (22). The role of this putative transforming factor is, however, still obscure. In the light of the results reported by Stancel et al. (23) it is conceivable that the so-called transforming factor is a subunit of the nuclear form of the re-ceptor. Indeed, their results obtained by exposing the nuclear and cyto-plasmic estradiol receptors to 4 M urea and 1 M KCl suggest that both forms of the receptor share a common subunit which binds estrogen; hence the conversion from cytoplasmic to nuclear form may involve the noncovalent addition of a small subunit to the cytoplasmic form of the receptor. Such a possibility would also be consistent with the observation that the cell-free conversion of the cytoplasmic to the nuclear form of the receptor on DNA-cellulose column is a bimolecular process (24). In addition, studies with gel permeation columns and sedimentation rate analysis on sucrose gradient have shown that the 5 S receptor behaves as if it is a larger molecule than the 4 S form, thus ruling out a simple change in shape as the result of the conversion from 4 S to 5 S (24). The conversion of the cytoplasmic to the nuclear form of receptor requires specific steroids. In the case of estrogen receptor, only estradiol and other estrogenic compounds could influence

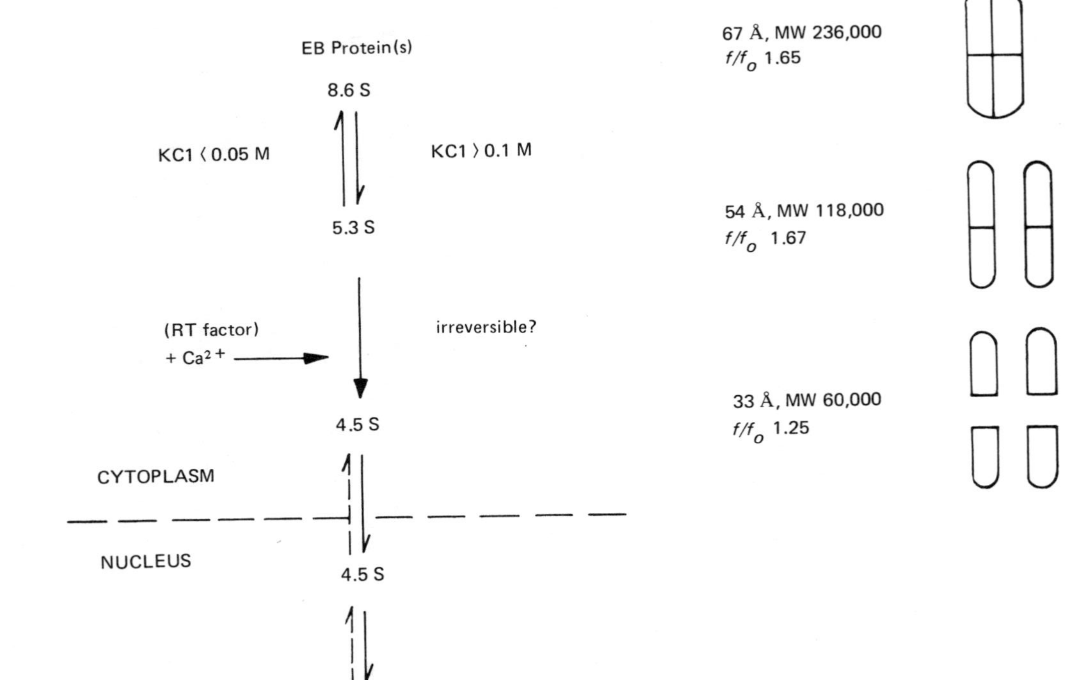

Figure 2. Model of the translocation of the steroid-receptor complex. Redrawn from Puca et al. (22) with permission. RT factor, receptor-transforming factor; EB protein(s), 17β-estradiol-binding protein(s).

the conversion of the 4 S to the 5 S form. In addition, the natural inhibitor, estrone, prevents such a transformation (24). Contrary to the generally accepted dogma that 4 S and 5 S forms of estrogen receptors are cytoplasmic and nuclear forms of the receptor, respectively, and that the 5 S receptor binds to the chromatin, Jackson and Chalkley (25) argued that the binding of cytoplasmic estradiol receptor to chromosomal material is an artifact of isolation and that there is doubt as to the physiological significance of the cytoplasmic estradiol-receptor complex found in association with nuclear material. They found that 4 S estradiol receptor binds cellular membranes, microsomal membranes, and nuclear membranes in a temperature-dependent process requiring the presence of divalent cations, whereas the 5 S receptor binds nucleohistones in a very rapid reaction even at 4 °C with no specificity for target tissues. They further argue that the fact that the 5 S receptor can bind chromatin instantly even at 4 °C raises the possibility that immediately upon tissue disruption at 4 °C the 5 S estradiol-receptor complex could possibly bind to nuclei to give an appearance of having bound to the chromatin before isolation of the nuclei, and thus generate an experimental artifact. They put forward a model based on the following experimental evidence: a) the 4 S receptor binds to membranes; b) the 5 S estradiol receptor has a much reduced affinity for membranes; and c) the 4 S and 5 S estradiol receptor conversion does occur within the cytoplasm. Their model can be summarized as follows: 1) the 4 S receptor binds to plasma membrane; 2) the 4 S estradiol-receptor complex is formed and is then converted to 5 S; 3) the 5 S estradiol-receptor complex dissociates from membranes inside the cytoplasm; 4) estradiol dissociates from the 5 S receptor complex and reassociates with high affinity sites in nuclei; and 5) the 5 S receptor is converted to 4 S receptor, which reassociates with plasma membranes and the cycle starts again. In the model of Jackson and Chalkley, the receptor molecules would play only the role of carrier of steroids from the cell membrane to the nuclei but would not under physiological conditions react with genetic material, e.g., bind to chromatin. The first argument that the steroid receptor binds to cellular membranes is quite plausible; however, the second argument that the steroid-receptor complex does not bind to the genetic material seems to be contradicted by an overwhelming number of reports dealing with the direct interaction of the steroid-receptor complex with nuclear elements. In order to clarify this last point, the following sections review first the effect of estradiol on RNA synthesis and then the interaction of the steroid-receptor complex with RNA polymerase(s), DNA, nuclei, and chromatin.

EFFECT OF STEROIDS ON RNA SYNTHESIS

It is known that in most instances steroid hormones stimulate total protein as well as specific synthesis. This is generally preceded by quantitative and often qualitative changes in the synthesis of cellular RNA. Stimulation of rapidly labeled nuclear heterogeneous RNA, followed by increased production of ribosomal RNA and tRNA, is a frequently observed effect on RNA metabolism (26, 27). Recently, it has been observed that steroid hormones are capable of inducing the synthesis of specific mRNA molecules in target cells and tissues (28–39). For example, in the chicken oviduct, estradiol increases the net amount of ovalbumin mRNA synthesis, whereas progesterone induces avidin mRNA activity (28–31). In chicken and *Xenopus* liver, which is the site of synthesis of egg yolk proteins, estradiol increases the amount of translatable vitellogenin mRNA (32–37). In yet another system, Beato and co-workers (38, 39) have demonstrated that rabbit uteri treated with progesterone had a large increase in uteroglobin mRNA activity. On the basis of the observed effects of steroids on RNA synthesis described above, the following questions can be asked: Are the nuclear steroid-receptor complexes responsible for both the ribosomal RNA and specific mRNA synthesis? And are there, for a given steroid hormone in a given cell, qualitatively different types of receptors responsible for rRNA and for specific mRNA synthesis, respectively, or do different classes of genes have different binding affinities for one and the same receptor? These important questions are discussed in the last section (see also refs. 40 and 41). Among the first experiments bearing on a possible effect of steroid-receptor complex on RNA synthesis are those of Jensen and his associates (42). Using a crude system, they showed that exposure of purified uterus nuclei to unpurified estradiol-receptor complex of uterine cytosol increased the RNA synthetic capacity of the system by 2- to 3-fold. The effect was specific in that the uterine cytosol fraction had no effect on kidney nuclei. This stimulation of RNA synthesis was affected by the transformed cytosol-steroid complex only. They also observed that both types of RNA polymerase activities were enhanced after incubation of their system with estradiol.

Such experiments with crude preparations suggest that the estradiol-receptor complex may influence overall transcription, but such results should not be overinterpreted. Indeed, O'Malley and his associates recently pointed out the dangers in using crude systems (43). With the use of a crude fraction of estrogen complex from chicken oviduct, they showed a stimulation of the rifampicin-resistant RNA synthesis in a cell-free transcription system containing oviduct chromatin as template. Unfortunately, such a crude receptor fraction could also stimulate RNA synthesis by a template-independent process (43). These results rule out the possibility that in their case the increase in RNA synthesis resulted from the steroid-receptor complex and suggest that the receptor preparation was contaminated with polynucleotide phosphorylase activity. Superficially, at least, it seems that the steroid-receptor complex stimulates RNA synthesis in most systems: the thymocytes, however, are an exception. For example, glucocorticoids and androgens added directly to isolated thymus nuclei lead to inhibition of RNA synthesis (44). Sekeris and his associates (45) have shown in thymocyte nuclei the presence of a glucocorticoid receptor with a molecular weight of about 150,000. Addition of this receptor, obtained from DEAE-cellulose, to a thymocyte nuclear preparation inhibited RNA synthesis significantly. It was also shown that cortisol preferentially inhibited ribosomal RNA synthesis. In these experiments, the possible contamination of the receptor preparation with nucleases or proteases was carefully ruled out (45).

POSSIBLE EFFECT OF STEROID-RECEPTOR COMPLEX ON DNA-DEPENDENT RNA POLYMERASE

One of the simplest explanations for the pleiotropic anabolic effects caused by steroids would, of course, be the activation of the DNA-dependent RNA polymerase by a steroid-receptor complex. The first indirect evidence for such a possibility was presented by Yu and Feigelson (46). After cortisone treatment, they found that purified rat liver nuclei had an enhanced ability to incorporate nucleotide triphosphates into RNA in vitro. The hormonal effect was localized exclusively in the nucleolar fraction (RNA polymerase 1 or A). In order to distinguish between the possibilities that cortisone either enhances availability of the DNA template or increases the amount or activity of the RNA polymerase, they carried out the following experiment. The endogenous nucleolar template in isolated nuclei was blocked by actinomycin D, and a synthetic template (polydeoxycytidylate) which does not bind actinomycin D but which codes for polyriboguanylate synthesis was used to evaluate RNA polymerase activity. The results indicate that the increase in RNA synthesis in nucleoli was largely a consequence of elevated activity of RNA polymerase. Later, similar results were obtained by using a reconstituted in vitro system from rat ventral prostate, consisting of crude receptor and polymerase preparations together with calf thymus DNA or prostatic chromatin (47). In this system, it was shown that 5-α dihydrotestosterone-receptor complex stimulated RNA polymerase activity by about 10–30%. Higher levels of stimulation (40–158%) were only observed in the presence of native chromatin. However, from this experiment it is not clear whether the steroid-receptor complex stimulates the polymerase directly or merely increases the availability of the DNA template for the polymerase. As in the case of the rat liver system and cortisone, the androgen preferentially stimulates the nucleolar RNA polymerase. In another study with the same system, Hu et al. (48) purified the androgen 3 S receptor protein to electrophoretic homogeneity. The purified androgen-receptor complex was then incubated with purified calf thymus polymerase B and prostate DNA; the stimulation of RNA polymerase in such a heterologous system was proportional to the amount of 3 S receptor added to the incubation mixture. From these results, it would be tempting to conclude that the steroid-receptor complex can stimulate the RNA polymerase from nontarget tissues. However, because the preparation of 3 S androgen-receptor complex still contained some undefined protein contaminant of high molecular weight, no hard conclusion can be drawn. Moreover, it is still not clear whether in these experiments the steroid-receptor complex stimulates the RNA polymerase directly or creates new binding sites for the polymerase on the chromatin.

The possibility that the steroid-receptor complex may indeed be linked to the RNA polymerase has been studied by Müller et al. (49). They found that both estradiol receptor activity and DNA-dependent RNA polymerase are extracted at the same ionic strength of 0.3 M sodium chloride from oviduct nuclei of diethylstilbestrol-treated quails. Estradiol receptor activity and RNA polymerase 1 (nucleolar RNA polymerase) are combined in a complex which persists through the purification procedure including DEAE-cellulose chromatography, gel filtration, and sucrose density gradient centrifugation. The receptor-polymerase complex has a molecular weight of about 130,000. Omission of only one ribonucleoside triphosphate from the incubation mixture reduces the polymerase activity by less than a third, thus partially excluding nucleotidyl phosphorylase activity. It is surprising, however, that a complex between RNA polymerase and estradiol-receptor complex has such a low molecular weight; possibly, the steroid-receptor complex was binding a subunit of the polymerase which retained some of the activity of the holoenzyme.

An interesting correlation between the sensitivity of RNA polymerase and steroid-receptor complex to specific antibiotics has been presented recently by Lohmar and Toft (50). They show that two polymerase inhibitors, *o*-phenanthrolene and rifamycin AF/013, which interfere with the binding of polymerases to DNA, also inhibit the binding of steroid-receptor complex to nuclei. Conversely, two related compounds which do not interfere with the binding of polymerase to DNA (*m*-phenanthrolene and rifampicin) do not inhibit nuclear binding of the steroid-receptor complex. Because *o*-phenanthrolene is a metal chelator which removes zinc ions bound to the enzyme, the possibility is raised that steroid receptors are metalloproteins. Furthermore, the inhibition by rifamycin AF/013 of the binding of both the RNA polymerase to DNA and the steroid-receptor complex to nuclei suggests that either the two macromolecules share common binding sites on the chromatin or that the binding of the two to the chromatin occurs via a polymerase-steroid receptor complex. Lastly, it is still possible that rifamycin AF/013 reacts with the two proteins independently and that the correlation observed is just fortuitous.

BINDING OF STEROID-RECEPTOR COMPLEX TO DNA

That DNA might be involved in the binding of steroid-receptor complex was suggested by early experiments in which the destruction by DNase of the DNA in nuclei from rat uterus prevented the formation of nuclear estradiol-receptor complex upon subsequent incubation with a cytoplasmic fraction containing 8 S estrogen-receptor complex. Such DNase treatment did not release any component capable of binding either estradiol or the activated cytoplasmic estradiol-receptor complex (51). Studying the interaction between estradiol receptor and DNA-cellulose, André and Rochefort (52) made the following observations. The interaction between the activated estrogen-receptor complex and DNA was reversible and reached an equilibrium more rapidly at 25 °C than at 4 °C. More important, they observed that under equilibrium conditions the interaction between the two macromolecules was not saturable: i.e., when the concentration of the estradiol-receptor complex was increased, the ratio between the estrogen receptor bound to DNA and the unbound estrogen receptor was always constant. An inhibitory effect of the cytosol fraction on the binding of estrogen-receptor complex to DNA was found to be responsible for the pseudosaturation of DNA by estradiol-receptor complex. Thus, André and Rochefort (52) concluded that the interactions observed in vitro between the estrogen receptor and DNA involve a large number of DNA receptor sites with weak affinity for the receptor. If one estradiol binding site per receptor molecule and 6 pg of DNA per cell are assumed, there are over 10^5 acceptor sites per nucleus. This number is far superior to the maximum number of estrogen-receptor complexes found in a target cell nucleus; hence, the quantity of nuclear estradiol-receptor complexes must be determined exclusively by the concentration of the cytosol-receptor-ligand complex and does not seem to be limited by any saturable nuclear process (52). Further studies on the specificity of the binding of estradiol receptor to DNA have shown that the partially purified 5 S estradiol receptor binds to single stranded homopolymer dT or double-stranded homopolymer pairs dG:dT to a higher extent than to calf thymus DNA (53). It was concluded that the high affinity of the steroid-receptor complex for homopolymers in vitro may indicate that such homopolymer stretches on DNA may be the binding sites for the 5 S receptor. In fact deoxyadenylate- and deoxyguanylate-rich regions are present in mammalian DNA in much higher concentration than in bacterial and viral DNA (54).

King and Gordon (55) have proposed that DNA is the acceptor of the steroid-receptor complex and that the nonhistone proteins determine the specificity by controlling which regions of DNA are accessible to the receptor. This means that the nonhistone proteins would play only a passive role. The postulate was that, if DNA is the acceptor site, then the interaction between DNA and the steroid-receptor complex would have characteristics consistent with nuclear binding of the steroids in vivo and in vitro. In the case of estradiol in the uterus, the steroid-receptor complex should be disruptable by 0.3–0.4 M KCl. Moreover, DNA should have limited capacity and high affinity for the estradiol-receptor complex. Lastly, the receptor should require the proper hormone for the attachment of the complex to DNA and should exhibit tissue specificity. King and Gordon found that, indeed, the binding of the estradiol-receptor complex to uterine DNA meets most of these requirements with the exception of the specificity. The lack of specificity of the binding of estradiol-receptor complex, as mentioned above, has also been described by other laboratories. For example, Yamamoto and Alberts (56), using their very sensitive method of sedimentation partition chromatography, have shown that the 5 S form of estradiol receptor binds to DNA at least 15-fold more tightly than does the 4 S form. The 5 S receptor binds equally well to several kinds of double-stranded DNA, bacterial DNA, and polydeoxyadenylate-deoxythymidilate copolymers. By contrast, no interaction was observed with double-stranded reovirus RNA. The fact that the interaction between steroid receptor and DNA seems to be of low affinity and nonspecific with respect to DNA base sequence raises the important question of how such steroid-receptor complexes can regulate the transcription of specific loci on the genome. Yamamoto and Alberts (57) have put forward a model based on an analogy with the DNA-binding properties of the *E. coli lac* operon repressor protein. They believe that estradiol receptor exerts its effect by binding to a small number of high affinity sites on the genome, while also having a low affinity for nonspecific DNA sequences. These nonspecific loci, because of their large number, would completely mask the presence of the high affinity sites. Yamamoto and Alberts (57) estimated that about 10^3 specific binding sites with affinities in the range of 10^{-8}–10^{-10} M could exist without being detected by bulk binding assays currently in use. Not all steroid-receptor complexes have the capacity to bind to DNA. As O'Malley's group has shown (58) in the chick oviduct, the progesterone-binding protein can be separated on DEAE-cellulose into two different components, A and B. Both subunits are able to bind labeled progesterone with high affinity. Furthermore, in in vitro assays, component A binds to DNA from several sources, whereas component B does not bind DNA. Component B binds to purified oviduct chromatin and retains target specificity. In contrast, the DNA-binding component A is unable to react with chromatin (58). The titration experiments of André and Rochefort (52) showing that the ratio between the estrogen receptor bound to DNA and the unbound estrogen receptor was always constant could possibly be explained by the presence of two subunits, A and B. As in the case of the progesterone receptor, one subunit would bind only to DNA, whereas the other one would recognize only chromatin as the binding site, thus giving the ratio 1:1 for bound and unbound steroid-receptor complex in the presence of naked DNA.

BINDING OF STEROID-RECEPTOR COMPLEX TO ISOLATED NUCLEI

As is the case for DNA, the binding sites for estrogen-receptor complex in isolated purified nuclei from several target and nontarget tissues are unsaturable even when the nuclei have bound several times more receptor than found at maximal in vivo stimulation of uterus (59).

By contrast, O'Malley and his associates (60) found that for the chick oviduct and other organs the binding process was a saturable phenomenon in both target and nontarget tissues. However, more nuclear acceptor sites were available in target tissues (9,000 sites per oviduct nucleus) than in nontarget tissues (2,000–3,000 sites per nucleus). The binding constant of the steroid-receptor complex in the nuclei in both target and nontarget tissues was about K_d 10^{-8} M. A second much smaller class of higher affinity binding sites (K_d of about 10^{-11} M) may also exist. Its presence was inferred by Scatchard analysis of the binding kinetics, which showed that the binding was nonlinear. This nonlinearity, detectable only at very low concentrations of added steroid-receptor complex, is an indication for the presence of at least two binding components. With the use of a two-component model, the existence of a very high affinity (K_d 10^{-11} M) binding component was postulated. Because this class occurs probably in extremely low quantity, accurate values for the K_d and number of binding sites could not be determined.

One obvious question is: where does the steroid-receptor complex bind in the purified nuclei? In experiments in which the interaction of the progesterone-binding protein with purified chick oviduct nuclei was studied, O'Malley and associates (61) found that after destroying about 60% of the DNA with DNase prior to incubation of the nuclei with steroid-receptor complex, some 80–85% of the receptor binding capacity was still retained in the nuclei. This means that the progesterone-receptor complex probably binds to DNase-resistant portions of the oviduct genome. In contrast, Baxter et al. (62) have reported that a 33% destruction of the hepatoma nuclear DNA eliminated about 92% of the nuclear binding capacity of the glucocorticoid-receptor complex. Because these experiments are very indirect and hard to evaluate quantitatively, it is difficult to draw any firm conclusions.

INTERACTION BETWEEN STEROID-RECEPTOR COMPLEX AND CHROMATIN

One promising approach to the elucidation of the mode of action of the steroid-receptor complex on the genome has been the use of chromatin. Contrary to the nonspecific binding of the steroid-receptor complex to DNA, it has been shown that the interaction between the hormone-receptor complex and chromatin has a certain degree of specificity. For example, O'Malley's group (63–65) has shown that a crude preparation of chick oviduct progesterone-receptor complex binds specifically to the chromatin of the target organ. Much less binding to spleen, heart, or erythrocyte chromatin was observed. Chromatin from both target and nontarget tissues had a K_d of approximately $3 \cdot 10^{-9}$ M for the progesterone-receptor complex (66). Quantitatively, chick oviduct had about 1,300 binding sites per picogram of DNA, whereas spleen and erythrocyte chromatin had about 840 and 330 binding sites per picogram of DNA, respectively (66). Earlier it was shown that steroid hormones were associated in vivo with histones (67). For example, Tsai and Hnilica (67) found that arginine-rich F3 histone had the highest affinity toward glucocorticoids. However, from their data, the possibility that the binding of the hormone occurred on a minor contaminant of nonhistone protein in the histone preparation cannot be excluded. By using a reconstituted system consisting of pure DNA and histone, it was shown that the binding of labeled progesterone-receptor complex to the DNA-histone was lower than that to native chromatin (64). Furthermore, if histone was exchanged between chromatins of different tissues, no difference in the capacity to bind the progesterone-receptor complex was observed. Because the extent of open template in the complex is greater than in native oviduct chromatin, the presence of open DNA does not seem to be a factor in the extensive binding of the steroid-receptor complex to oviduct chromatin (64). The above results were confirmed by a different experimental approach in which it was shown that if chromatin were treated with antibodies prepared against single-stranded or double-stranded DNA, the extent of binding of the progesterone-receptor complex to the chromatin was not altered (66). As for the nuclei, pretreatment of the chromatin with nuclease specific for single-stranded DNA did not inhibit the capacity of the chromatin to bind the steroid-receptor complex (66). From the above considerations, it is tempting to conclude that the chromatin nonhistone proteins are responsible for the binding of the steroid-receptor complex.

Another method for pinpointing the molecule responsible for the specific binding of progesterone-receptor complex to chick oviduct chromatin has been the reconstitution of chromatin, with the nonhistone proteins from the chromatin of one tissue being exchanged with those from another. Such experiments resulted in a corresponding exchange of the template capacity to accept progesterone-receptor complex so that the hybrid chromatin resembles the chromatin which served as the source of the nonhistone proteins (65, 68). O'Malley's group also isolated a subfraction of the nonhistone proteins of the oviduct chromatin which seems to be responsible for the binding of the progesterone-receptor complex to the chromatin (68). The possible regulatory role, if any, of such nonhistone proteins is not yet known.

An original approach for the study of the interaction of the estrogen-receptor complex with proteins of the chromatin has been introduced by Puca et al. (69). It consists of solubilizing, from the purified nuclei, molecules which, when immobilized on an inert support like agarose, interact specifically with extranuclear steroid-receptor complex in a way that resembles closely the physiological steroid nuclear interaction. A fraction of nuclear proteins with high affinity for the estrogen receptor was isolated from calf uterus by this procedure. When immobilized on CNBr-agarose, this specific protein fraction binds estradiol-receptor complex with high affinity and specificity. Such binding was not observed in the absence of estradiol or with nonestrogen steroid hormones. As demonstrated by Scatchard plots, the binding of estrogen-receptor complex was saturable in a range of protein concentrations which varied from 5 to 50 μg/ml, levels which were from 100 to 1,000 times lower than those found in the crude cytosol (69). With the uses of very low concentrations of this particular fraction of proteins (basic nuclear proteins), the presence of specific acceptor sites was verified by their high affinity for the receptor. A *Kd* of $2 \cdot 10^{-10}$ M was calculated for the receptor-acceptor site interaction. As a control, the interaction of estradiol-receptor complex (from crude cytosol preparation) with casein-agarose and with histone-agarose was studied. The results showed that the binding of the steroid receptor to such a matrix had a different sensitivity to KCl and that it was not modified by the presence or absence of estradiol. Because DNA and RNA were completely eliminated from the preparation, it was possible to exclude an effect of nucleic acids on the interaction of steroid-receptor complex with specific binding protein fractions isolated from the chromatin (69).

During the late ontogeny of primitive vertebrate reproductive tissues, such as in the developing chick oviduct, several cell types develop from the single undifferentiated epithelial cells. The appearance of the mature organ with highly specialized differentiated cells (70, 71) is the concluding stage of specialization of this tissue. The differentiation of the oviduct is accompanied by changes in the chromatin composition such as the accumulation of nonhistone proteins and RNA. During the final stage of the maturation of the oviduct, a decrease in the level of the nonhistone proteins and RNA, as well as in the template capacity of the chromatin, takes place (72). Throughout differentiation, fluctuations in the hormone binding capacity of the oviduct chromatin follow a pattern very similar to that of the level of nonhistone proteins, RNA, and template capacity (72). In addition, at all stages of differentiation the capacity of the chromatin to bind the progesterone-receptor complex is much higher than for nontarget tissues (72). From these observations, it can be concluded that chromatin binding sites in the developing oviduct are established during organogenesis.

If the relationship between the level of nuclear receptor and the number of initiation sites of transcription is now considered more closely, a close positive correlation can be found. Using a method which measures the formation of an RNA chain initiation complex between RNA polymerase and chromatin, O'Malley's (73, 74) group has determined the number of initiation sites available to saturating concentrations of *E. coli* polymerase on chick oviduct chromatin isolated from controls and estradiol-treated chicks.

The number of available initiation sites for RNA synthesis on the chromatin correlates with the endogenous level of nuclear estradiol receptor. Conversely, a decrease in the nuclear concentration of estrogen receptor molecules and the concentration of initiation sites for RNA synthesis occurred during withdrawal of estrogen from chickens previously treated with estradiol. Both the number of receptors and the number of initiation sites declined with a similar half-life. If estradiol were injected again into chickens which had been deprived of estrogen, the number of initiation sites and the level of estradiol receptor increased again in a parallel manner, thus showing that there is a temporal relationship between the two. Simultaneous measurements of RNA chain length and RNA chain elongation rate demonstrated that these parameters remained relatively constant throughout estrogen withdrawal as well as during secondary stimulation (73). These results mean either that the hormone-receptor complex acts directly on chromatin by causing rearrangement of the chromosomal proteins, or that its binding to nonhistone proteins provides recognition sites for the RNA polymerase.

Because most of the above experiments were carried out with the use of RNA polymerase from *E. coli*, the physiological relevance of such experiments could be questioned. That important question was recently examined by O'Malley and his associates. RNA chain initiation in chick oviduct was studied by using both homologous and *E. coli* RNA polymerases (75). The results indicated that chick oviduct RNA polymerase B (nucleoplasmic RNA polymerase) and *E. coli* RNA polymerase competed with each other for the same initiation regions in chromatin. In contrast, when naked DNA was used as template, the majority of the initiation sites on DNA which bind the oviduct RNA polymerase B are different from those used by *E. coli* RNA polymerase. Again, these results strongly suggest that chromatin proteins (most likely nonhistone proteins) are involved in the selection of initiation sites on chromatin for RNA polymerases irrespective of their origin. The fact that prokaryotic polymerase, as well as the homologous eukaryote nucleoplasmic RNA polymerase, recognizes new initiation sites favors the hypothesis that the steroid-receptor complex reacts primarily with the nonhistone proteins of the chromatin rather than with the RNA polymerase itself. It still remains to be shown that the positive correlation between the number of initiation sites for RNA polymerase and the number of hormone receptors on the chromatin reflects any direct cause-effect relationship. This important question can only be answered unequivocally by a genetic approach, or by tedious in vitro reconstruction experiments using highly purified and well defined macromolecules.

STEROID-RECEPTOR COMPLEX AND INDUCTION OF TUMOR VIRUSES

The demonstration of the possible involvement of steroids in the induction of tumor viruses in target cells is a recent development. For example, Parks et al. (76) have demonstrated that treatment of C3H mouse mammary tumor cells with dexamethasone induces an increase in murine mammary tumor RNA sequences, proteins, and even complete virus particles (type B). At concentrations as low as $1 \cdot 10^{-10}$ M, dexamethasone was the most potent inducer of murine mammary virus DNA polymerase. Corticosterone was an effective inducer although concentrations 10–100 times higher than dexamethasone were required for a comparable response. Aldosterone, which is known to have glucocorticoid activity, was also able to induce mammary tumor virus, although less effectively than the other glucocorticoids tested (77). Progesterone, testosterone, 17β-estradiol, and androstendione did not induce the virus (77). In the particular case of murine mammary tumor, the virus could only be induced with corticosteroid in cell lines already making a constitutive level of murine virus RNA sequences. This means that the steroid merely amplified what the cell was already able to do. Another interesting finding was that the cell line KA 31, which contains glucocorticoid receptor and murine tumor sequences in its DNA, was unable to produce mammary tumor virus in the presence of dexamethasone, although KA 31 nuclei had the steroid-receptor complex (77).

In another system, estrogens or compounds having an estrogenic action induced in the uteri of certain strains of mice the synthesis of protein that is serologically identical with the group-specific antigen (gs) of the murine and feline type C RNA tumor virus (78). Similarly, estrogen treatment caused the activation of RNA-directed DNA polymerase. Treatment of ovariectomized mice with estrogen elevated the level of the murine leukemia virus group antigen and the activity of RNA-directed DNA polymerase in the uterus. In this particular case, the level of the viral markers depended on the relative biological potency of the estrogen injected (78).

What evidence exists for the involvement of the steroid-receptor complex in the induction of specific tumor virus particles? To date, the only evidence is indirect and it is the positive correlation between the steroid receptor level in the cells and the synthesis of the virus particles, and the temporal relationship between the appearance of the hormone-receptor complex in the nucleus and the synthesis of the tumor viral RNA. Recently, it was shown that addition of an excess of progesterone, a competitive inhibitor of glucocorticoid receptor (79), completely abolishes the viral RNA accumulation induced by dexamethasone in GR cells (a cell line derived from a spontaneous mammary tumor in the GR strain of mouse). Studying the kinetics of the induction of mouse mammary tumor virus RNA synthesis, Ringold et al. (79) found that after treatment of cells with saturating amounts of dexamethasone (10^{-5} M), the maximum induction of mouse mammary tumor virus RNA was attained within 7 hr, and the apparent lag was quite short (15 min) and might be attributed to the time required for the translocation of the hormone-receptor complex to the nucleus and the transcription and processing of the viral genome. This, together with the dexamethasone dose dependence for the induction of mouse mammary tumor RNA, indicates that there is a positive correlation between the presence of the glucocorticoid-receptor complex and the generation of virus particles. Again, in this particular case, it is not yet proven that there is any direct cause-effect relationship between the two phenomena.

CONCLUDING REMARKS

Before trying to formulate a generalized model for the action of the steroid-receptor complex on gene regulation, it is of vital importance that the sequence of events happening in vivo after the administration of steroid hormone is considered. When the results obtained for hormones such as triiodothyronine, growth hormone, cortisone acting on liver, estradiol acting on uterus, and testosterone acting on prostate are considered, it is always found that stimulation by the hormone of transcription in the target tissues is initially restricted to the nucleolar RNA polymerase (polymerase A or I), and only at a later time is there an effect on the nucleoplasmic RNA polymerase B or II (80). The stimulation of nucleolar RNA polymerase activity by steroids seems to require the synthesis of a specific protein(s). This was inferred from experiments in which puromycin or cycloheximide (81) inhibited the synthesis of a very early protein which was necessary for the stimulation of nucleolar RNA polymerase by estradiol. Similarly, 30 min after an injection of estradiol into spayed or immature rats the synthesis of an early protein was suppressed by actinomycin D (82, 83) or α-amanitin (84). The synthesis of protein(s) was necessary for the stimulation of nucleolar RNA polymerase activity. This means that the very first target of the steroid-receptor complex (see Figure 3) could be a few high affinity binding sites located on the promoter region of specific genes coding for an activating protein(s) or integrator protein (41). This protein(s) could either stimulate the nucleolar RNA polymerase directly or facilitate the binding of the steroid-receptor complex to the promoter region of nucleolar structural genes (with low affinity binding sites). During the same period, the steroid-receptor complex could also bind to the promoter sites of many structural genes located in the nucleoplasm (also with low affinity binding sites). Such binding could facilitate the formation of initiation complexes with nucleoplasmic RNA polymerase, but these initiation complexes would become operational only after a specific gene product of the nucleolus had been synthesized. This latter would be responsible for the processing of pre-mRNA and transport of mRNA from the nucleus to the cytoplasm, thus establishing a feedback control between the expression of nucleolar genes and nucleoplasmic genes. Quantitatively, no more pre-mRNA could be synthesized unless its processing was assured. Such an integrated control could explain why there is, within minutes after steroid administration, rapid binding of a large quantity of steroid-receptor com-

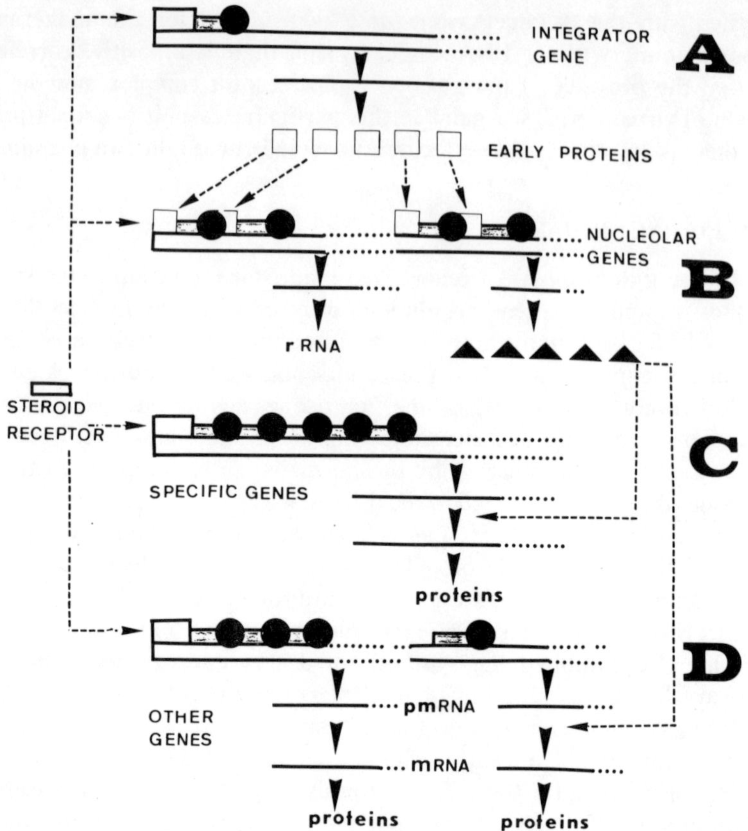

Figure 3. Model integrating the action of the steroid-receptor complex. *A*, the steroid-receptor complex binds to the integrator genes (high affinity binding sites) which are coding for the early proteins or integrator proteins. *B*, the early proteins bind to the nucleolar genes, together with the steroid-receptor complex. (The gene products of the nucleolus are rRNA and some hypothetical proteins responsible for the processing of the pre-mRNAs.) *C*, the steroid-receptor complex binds to few specific genes (high affinity binding sites) coding for specialized proteins such as ovalbumin or vitellogenin in the chicken system. *D*, the steroid-receptor complexes bind to low affinity binding sites on many other genes coding for several proteins. For *C* and *D*, the rate-limiting step of pre-mRNA processing is carried out by the gene product of B.

plex to the chromatin, yet there is still a time lag between the stimulation of the nucleolar RNA polymerase and nucleoplasmic RNA polymerase activity. This model might explain the overall pleiotropic effect of the steroid-receptor complex in many target tissues, but could hardly account for the stimulation of a limited number of specific genes such as ovalbumin and vitellogenin in the chicken oviduct or liver. For this last, but not least, important case, it would be convenient to think that the stimulation of specific genes such as the ones mentioned above could possibly be achieved by the binding of steroid-receptor complex to high affinity binding sites on the regulatory portion of the gene. In addition, the promoter sites of such genes would be able to bind a much larger number of polymerases than the other genes, thus ensuring a steady flow of RNA polymerases transcribing the gene at top capacity. The highly hypothetical element of the model is the specific protein(s), encoded by nucleolar genes and responsible for the post-transcriptional control of genes. For the moment, the existence of such a control mechanism can only be inferred by the general role of the nucleolus in the transfer of genetic information from the nucleus to the cytoplasm (85). The exact way in which the steroid-receptor complex reacts with the genetic material is not yet known; O'Malley recently put forward an interesting hypothesis as to how such an interaction may take place (86). His hypothesis states that in the case of progesterone receptor the subunit B binds to the AP3 fraction of the nonhistone proteins on the chromatin, whereas the subunit A binds to the naked plus strand of DNA in such a way as to enable a molecule of RNA polymerase to occupy an initiation site on DNA.

From the above considerations, it is obvious that we do not yet know how the steroid-receptor complex functions, and substantial new evidence is required before one can propose a definitive model. In the interim, it is obviously necessary to keep an open and critical mind when analyzing new results and interpretations.

ACKNOWLEDGMENT

The author wishes to thank Dr. I. B. Levitan for his critical reading and correcting of the manuscript.

REFERENCES

1. Jensen, E. V., and DeSombre, E. R. (1972). Annu. Rev. Biochem. 41:203.
2. Gorski, J., and Gannon, F. (1976). Annu. Rev. Physiol. 38:425.
3. Toft, D., and Gorski, J. (1966). Proc. Natl. Acad. Sci. USA 55:1574.
4. Chamness, G. C., and McGuire, W. L. (1972). Biochemistry 11:2466.
5. Korenman, S. G., and Rao, B. R. (1968). Proc. Natl. Acad. Sci. USA 61:1028.
6. Stancel, G. M., Leung, K. M. T., and Gorski, J. (1973). Biochemistry 12:2130.
7. Reti, I., and Erdos, T. (1971). Biochimie 53:435.
8. Sherman, M. R., Corval, P. L., and O'Malley, B. W. (1970). J. Biol. Chem. 245:6085.
9. Sherman, M. R., Atienza, S. B. P., Shansky, J. R., and Hoffman, L. M. (1974). J. Biol. Chem. 249:5351.
10. Rochefort, H., and Baulieu, E. E. (1971). Biochimie 53:893.
11. Schrader, W. T., and O'Malley, B. W. (1972). J. Biol. Chem. 247:51.
12. DeSombre, E. R., Puca, G. A., and Jensen, E. V. (1969). Proc. Natl. Acad. Sci. USA 64:148.
13. Ludens, J. H., De Vries, J. R., and Fanestil, D. D. (1972). J. Biol. Chem. 247:7533.
14. Sica, V., Nola, E., Parikh, I., Puca, G. A., and Cuatrecasas, P. (1973). Nature (New Biol.) 244:36.

15. Truong, H., and Baulieu, E. E. (1974). FEBS Lett. 46:321.
16. Failla, D., Tomkins, G. M., and Santi, D. V. (1975). Proc. Natl. Acad. Sci. USA 72:3849.
17. Kuhn, R. W., Schrader, W. T., Smith, R. G., and O'Malley, B. W. (1975). J. Biol. Chem. 250:4220.
18. Smith, R. G., Iramain, C. A., Buttram, V. C., and O'Malley, B. W. (1975). Nature (Lond.) 253:271.
19. Jensen, E. V., Suzuki, T., Kawashima, T., Stumpf, W. E., Jungblut, P. W., and DeSombre, E. R. (1968). Proc. Natl. Acad. Sci. USA 59:632.
20. Rousseau, G. G., Baxter, J. D., Higgins, S. J., and Tomkins, G. M. (1973). J. Mol. Biol. 79:539.
21. Notides, A. C., Hamilton, D. E., and Auer, H. E. (1975). J. Biol. Chem. 250: 3945.
22. Puca, G. A., Nola, E., Sica, V., and Bresciani, F. (1972). Biochemistry 11: 4157.
23. Stancel, G. M., Leung, K. M. T., and Gorski, J. (1973). Biochemistry 12: 2137.
24. Yamamoto, K. R., and Alberts, B. M. (1972). Proc. Natl. Acad. Sci. USA 69:2105.
25. Jackson, V., and Chalkley, R. (1974). J. Biol. Chem. 249:1627.
26. O'Malley, B. W., Aronow, A., Peacock, A. C., and Dingman, C. W. (1968). Science 162:567.
27. Tomkins, G. M., Gelehrter, T. D., Granner, D., Martin, D., Samuels, H. H., and Thompson, E. B. (1969). Science 166:1474.
28. Chan, L., Means, A. R., and O'Malley, B. W. (1973). Proc. Natl. Acad. Sci. USA 70:1870.
29. McKnight, G. S., Pennequin, P., and Schimke, R. T. (1975). J. Biol. Chem. 250:8105.
30. Palmiter, R. D. (1973). J. Biol. Chem. 248:8260.
31. Cox, R. F., Haines, M. E., and Emtage, J. S. (1974). Eur. J. Biochem. 49:225.
32. Berridge, M. V., Farmer, S. R., Green, C. D., Henshaw, E. C., and Tata, J.-R. (1976). Eur. J. Biochem. 62:161.
33. Shapiro, D. S., Baker, H. J., and Stitt, D. T. (1976). J. Biol. Chem. 251: 3105.
34. Wahli, W., Eyler, T., Weber, R., and Ryffel, G. U. (1976). Eur. J. Biochem. 66:457.
35. Mullinix, K. P., Wetekam, W., Deeley, R. G., Gordon, J. I., Meyers, M., Kent, K. A., and Goldberger, R. F. (1976). Proc. Natl. Acad. Sci. USA 73:1442.
36. Jost, J. P., Pehling, G., and Baca, O. G. (1975). Biochem. Biophys. Res. Commun. 62:957.
37. Jost, J. P., and Pehling, G. (1976). Eur. J. Biochem. 66:339.
38. Beato, M., and Arnemann, J. (1975). FEBS Lett. 58:126.
39. Beato, M., and Nieto, A. (1976). Eur. J. Biochem. 64:15.
40. Britten, R. J., and Davidson, E. H. (1969). Science 165:349.
41. Jost, J. P., and Averner, M. (1975). J. Theor. Biol. 49:337.
42. Mohla, S., DeSombre, E. R., and Jensen, E. V. (1972). Biochem. Biophys. Res. Commun. 46:661.
43. Buller, R. E., Schwartz, R. J., and O'Malley, B. W. (1976). Biochem. Biophys. Res. Commun. 69:106.
44. Abraham, A. D., and Sekeris, C. E. (1971). Biochim. Biophys. Acta 247: 562.
45. van der Meulen, N., Abraham, A. D., and Sekeris, C. E. (1972). FEBS Lett. 25:116.

46. Yu, F. L., and Feigelson, P. (1971). Proc. Natl. Acad. Sci. USA 68:2177.
47. Davies, P., and Griffiths, K. (1973). Biochem. Biophys. Res. Commun. 53:373.
48. Hu, A. L., Loor, R. M., and Wang, T. Y. (1975). Biochem. Biophys. Res. Commun. 65:1327.
49. Müller, W. E. G., Totsuka, A., and Zahn, R. K. (1974). Biochim. Biophys. Acta 366:224.
50. Lohmar, P. H., and Toft, D. O. (1975). Biochem. Biophys. Res. Commun. 67:8.
51. Musliner, T. A., and Chader, G. J. (1971). Biochem. Biophys. Res. Commun. 45:998.
52. André, J., and Rochefort, H. (1975). FEBS Lett. 50:319.
53. Sluyser, M., Evers, S. G., and Nijssen, T. (1974). Biochem. Biophys. Res. Commun. 61:380.
54. Shenkin, A., and Burden, R. H. (1974). J. Mol. Biol. 85:19.
55. King, R. J. B., and Gordon, J. (1972). Nature (New Biol.) 240:185.
56. Yamamato, K. R., and Alberts, B. (1974). J. Biol. Chem. 249:7076.
57. Yamamato, K. R., and Alberts, B. (1975). Cell 4:301.
58. Schrader, W. T., Toft, D. O., and O'Malley, B. W. (1972). J. Biol. Chem. 247:2401.
59. Chamness, G. C., Jennings, A. W., and McGuire, W. L. (1974). Biochemistry 13:327.
60. Buller, R. E., Schrader, W. T., and O'Malley, B. W. (1975). J. Biol. Chem. 250:809.
61. Buller, R. E., Toft, D. O., Schrader, W. T., and O'Malley, B. W. (1975). J. Biol. Chem. 250:801.
62. Baxter, J. D., Rousseau, G. G., Benson, M. C., Garcea, R. L., Ito, J., and Tomkins, G. M. (1972). Proc. Natl. Acad. Sci. USA 69:1892.
63. Steggles, A. W., Spelsberg, T. C., Glasser, S. R., and O'Malley, B. W. (1971). Proc. Natl. Acad. Sci. USA 68:1479.
64. Steggles, A. W., Spelsberg, T. C., and O'Malley, B. W. (1971). Biochem. Biophys. Res. Commun. 43:20.
65. Spelsberg, T. C., Steggles, A. W., and O'Malley, B. W. (1971). J. Biol. Chem. 246:4188.
66. Jaffe, R. C., Socher, S. H., and O'Malley, B. W. (1975). Biochim. Biophys. Acta 399:403.
67. Tsai, Y. H., and Hnilica, L. S. (1971). Biochim. Biophys. Acta 238:277.
68. Spelsberg, T. C., Steggles, A. W., Chytil, F., and O'Malley, B. W. (1972). J. Biol. Chem. 247:1368.
69. Puca, G. A., Nola, E., Hibner, U., Cicala, G., and Sica, V. (1975). J. Biol. Chem. 250:6452.
70. Oka, T., and Schimke, R. T. (1969). J. Cell Biol. 41:816.
71. Oka, T., and Schimke, R. T. (1969). J. Cell Biol. 43:123.
72. Spelsberg, T. C., Steggles, A. W., and O'Malley, B. W. (1971). Biochim. Biophys. Acta 254:129.
73. Tsai, S. Y., Tsai, M. J., Schwartz, R., Kalimi, M., Clark, J. H., and O'Malley, B. W. (1975). Proc. Natl. Acad. Sci. USA 72:4228.
74. Kalini, M., Tsai, S. Y., Tsai, M. J., Clark, J. H., and O'Malley, B. W. (1976). J. Biol. Chem. 251:516.
75. Tsai, M. J., Towle, H. C., Harris, S. E., and O'Malley, B. W. (1976). J. Biol. Chem. 251:1960.
76. Parks, W. P., Ransom, J. C., Young, H. A., and Scolnick, E. M. (1975). J. Biol. Chem. 250:3330.

77. Young, H. A., Scolnick, E. M., and Parks, W. P. (1975). J. Biol. Chem. 250:3337.
78. Fowler, A. K., Kouttab, N. M., Kind, P. D., Strickland, J. E., and Hellman, A. (1975). Proc. Soc. Exp. Biol. Med. 148:14.
79. Ringold, G. M., Yamamato, K. R., Tomkins, G. M., Bishop, J. M., and Varmus, H. E. (1975). Cell 6:299.
80. Tata, J. R. (1966). Progr. in Nucleic Acid Res. Mol. Biol. 5:191.
81. Mueller, G. C., Gorski, J., and Aizawa, Y. (1961). Proc. Natl. Acad. Sci. USA 47:164.
82. Mayol, R. F., and Thayer, S. A. (1970). Biochemistry 9:2484.
83. De Angelo, A. B., and Gorski, J. (1970). Proc. Natl. Acad. Sci. USA 66:693.
84. Baulieu, E. E., Alberga, A., Raynaud-Jammet, C., and Wira, C. R. (1972). Nature (New Biol.) 236:236.
85. Harris, H. (1968). Nucleus and Cytoplasm, pp. 142. Clarendon Press, Oxford.
86. O'Malley, B. W., and Schrader, W. F. (1976). Sci. Am. 234:32.

International Review of Biochemistry
Biochemistry and Mode of Action of Hormones II, Volume 20
Edited by H. V. Rickenberg

5 Biochemical Endocrinology of Insect Growth and Development

S. SRIDHARA,[1] J. NOWOCK, and L. I. GILBERT
Northwestern University, Evanston, Illinois

Research was supported by Grants AM 02818 and HD 08450 from the National Institutes of Health, Bethesda, Maryland, and PCM 76-03620 from the National Science Foundation.

[1] An established Investigator of the American Heart Association.

Although the study of insect endocrinology is surely warranted on the basis of the economic and medical importance of insects to man, this chapter demonstrates that these animals, which undergo dramatic changes in shape and form during their life history, are also excellent model systems for research in biochemical endocrinology. Indeed, genetic analysis of the fruit fly ***Drosophila melanogaster*** at the endocrinological level, as well as of cell lines derived from this insect, adds a new dimension to the study of mechanisms of hormone action at the molecular level. For example, temperature-sensitive mutants can be readily visualized which at the restrictive temperature cannot complete the biosynthesis of a hormone, cannot deplete membrane, cytosol, or nuclear receptors, cannot properly phosphorylate a critical nonhistone protein, etc.

As with higher vertebrates, a number of peptide hormones control aspects of insect lipid and carbohydrate metabolism (1–3), and, to make the analogy complete, both glucagon-like and insulin-like molecules have been detected in insect endocrine glands (4). However, the uniqueness of insect metamorphosis reflects primarily the activities of steroid and sesquiterpenoid hormones, ecdysones and juvenile hormones, respectively (Figure 1). This chapter is mainly devoted to a consideration of their possible mechanisms of action.

Prior to discussing the general roles of these two growth hormones, it is of importance that the reader understand the terminology used by the insect physiologist. The characteristic most responsible for the evolutionary success of insects is their cuticle. It is a secretory product of the epidermis, is composed mainly of proteins and chitin, and serves as both skin and exoskeleton. Although there is some evidence for cuticle growth by intussusception, the capacity to increase in surface area is generally limited. Therefore, insect growth necessitates the periodic replacement of the old cuticle by a new and larger version; this occurs through molting. This process is initiated by the detachment of the cuticle from the epidermis (apolysis), continues through replication of epidermal cells, secretion of a new cuticle, digestion and resorption of the inner layers of the old one, and terminates in the shedding (ecdysis) of the remnants of the old cuticle. Thus, development proceeds through distinct stages (instars) which are divided by molts. Although the evolutionary advantages for insects to adopt a regimen of discontinuous growth through molts rather than continuous growth through cuticle extension by intussusception are not understood, the evolutionary benefit is clearly seen in the manifestation of metamorphosis, which introduced enormous adaptive potential. With the exception of Ametabola, insects proceed through a sequence of larval-larval molts with only minor alterations in morphological features before they transform into adults. This transformation occurs either in one step with a larval-adult molt (Hemimetabola) or with an intermediate larval-pupal molt before metamorphosis is completed by a molt to the adult (Holometabola).

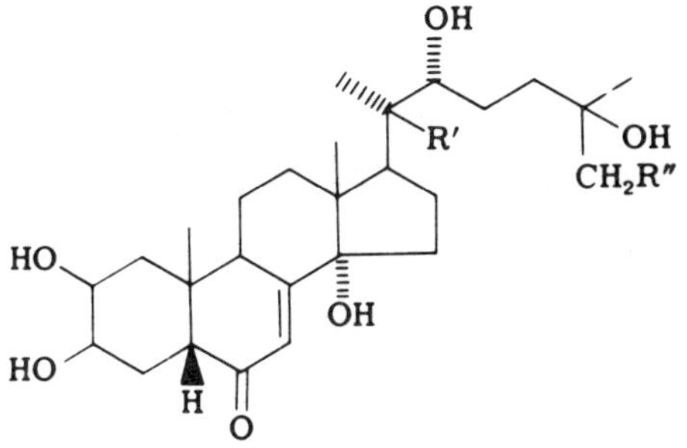

A. α-Ecdysone R′ = R″ = H

B. β-Ecdysone R′ = OH; R″ = H

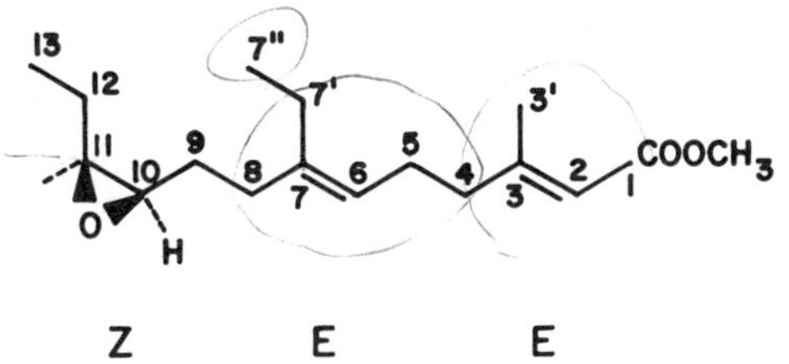

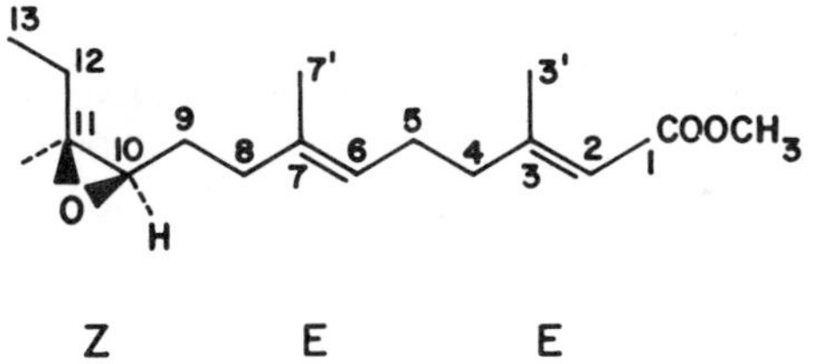

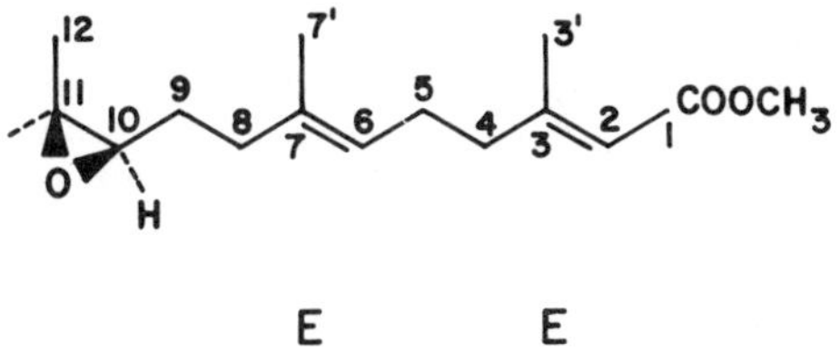

Figure 1. Structures of ecdysones (*top*) and juvenile hormones (*bottom*). For juvenile hormones; JH I (*top*); JH II (*middle*); JH III (*bottom*).

Flies (Diptera), which belong to the latter group, exhibit an even more advanced feature; at the end of the last larval instar, the larva contracts and the larval cuticle hardens and tans to form a protective case (puparium). Within this puparium, the metamorphic molts to pupa and then to adult occur.

The Diptera have been favorite organisms, not only for the study of genetics, but also for the examination of gene activity and its control. This is due in good measure to the presence of polytene chromosomes in the larval salivary glands and in a variety of other tissues, the existence of imaginal discs, the ease in raising these insects, and their short life cycle. Although the phenomenon of puffing at specific gene loci and its correlation with hormone titer and developmental stage are discussed in detail under "Polytene Chromosomes and Puffing Phenomenon," it should be mentioned here that this system allows the visual detection of the activity of specific chromosomal regions when the animals, or even isolated glands, are challenged with the appropriate hormone. The imaginal discs are nests of embryonic cells that grow but do not differentiate during larval life, and eventually give rise to highly complex structures such as eyes, genitalia, wings, antennae, etc., at metamorphosis. These imaginal discs can be extirpated, transplanted, followed with genetic markers, and studied in vitro. In conjunction with the latter statement, it should be mentioned that invertebrates, and particularly insects, are magnificent objects for microsurgery. Their hardiness, the lack of a classical antibody system, and the absence of blood vessels, which are replaced by a slowly circulating pool of hemolymph contained within a hemocoel, allow even the most drastic surgery. This includes parabiosis, extirpation, and transplantation of brains (supraesophageal ganglia), as well as ligation of particular body regions. Indeed, with thread, razor blade, wax, and forceps, one can obtain an amazing amount of information regarding the basis of endocrine control in insects as has been amply demonstrated by the classic studies of Wigglesworth, Fraenkel, Piepho, Pflugfelder, Bodenstein, Williams, etc. (1). (It is somewhat embarrassing that replacement of thread, razor blades, etc. by ultracentrifuges, radiotracers, mass spectrometers, etc. has not added nearly as much information as should be warranted by their cost.)

From the above, the exquisite potential of the insect system for the examination of growth processes and their control by hormones can be readily appreciated. The classic studies have been discussed in some detail in recent reviews (1, 5, 6) and are not reiterated here. Furthermore, the authors' laboratory has recently reviewed the current knowledge of the chemistry and metabolism of the two major growth hormones, ecdysone and juvenile hormone, in another volume of this series (7), as well as the mechanisms used by the insect in maintaining critical titers of these hormones during development (8). Therefore, these subjects are not considered, and the major emphasis is on the action of the molting hormone, ecdysone, and the juvenile hormone at the biochemical and molecular levels.

In a general way, the scheme of endocrine control is as presented in Figure 2 showing the development of the silkmoth *Hyalophora cecropia* as an example (9). A hormone from the brain acts in a tropic manner upon the prothoracic glands, which then release the polyhydroxylated steroid-α-ecdysone. This molecule is quickly converted to β-ecdysone (22-hydroxy-

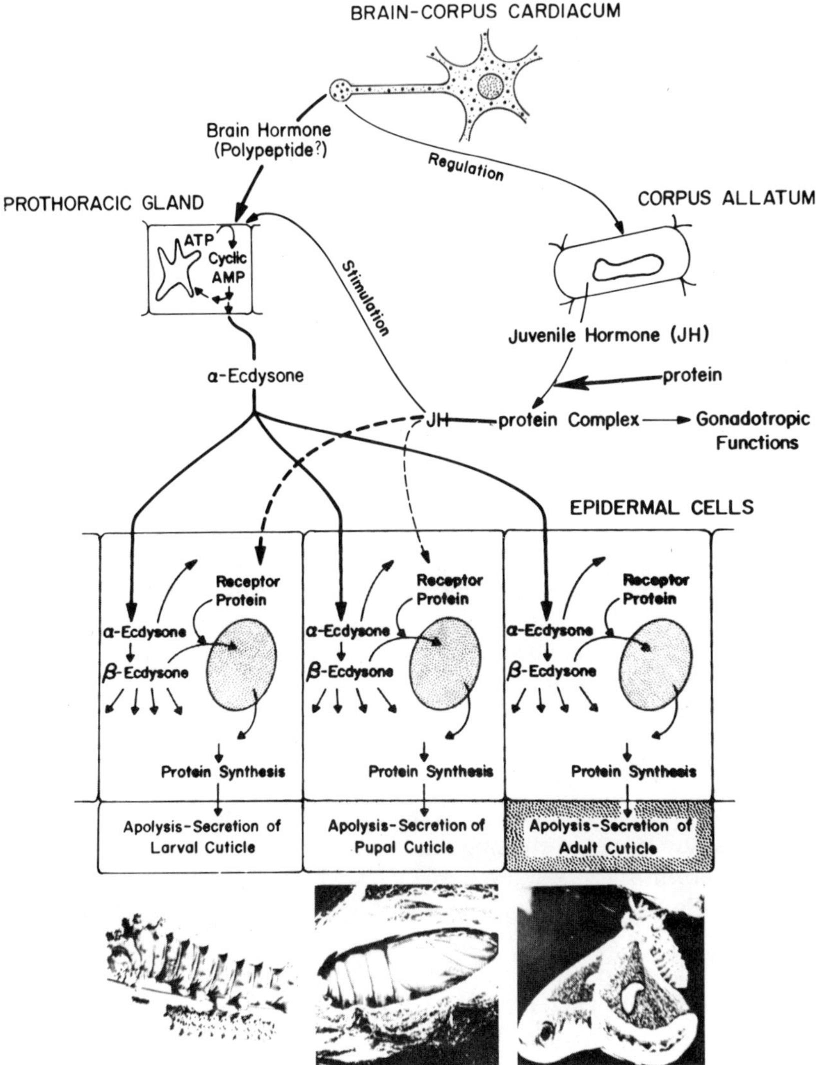

Figure 2. Speculative scheme for the endocrine control of insect development. Reproduced from Gilbert (9) with permission of Academic Press Inc.

α-ecdysone) by peripheral tissues; β-ecdysone seems to be the hormone that elicits the molting cycle beginning with apolysis (1, 7, 9). The primary site of hydroxylation is the insect fat body, which functions analogously to the mammalian liver and adipose tissue. β-Ecdysone, therefore, is the primary stimulus for molting, although the process is indirectly initiated by environmental cues impinging upon the brain, either directly or via other components of the nervous system. The morphological consequence of the molt, i.e., whether it is larval-larval, larval-pupal, or pupal-adult, depends on the titer of juvenile hormone. This juvenile hormone titer results from the synthesis and the release from the corpora allata, the degradation by specific hemolymph esterases, the protection by a specific hemolymph-binding protein from general esterases, and finally from the accumulation of hormone by lipophilic compartments within the insect's tissues (7, 8, 10–12). When in relatively high concentration, juvenile hormone directs a larval-larval molt. The decrease in the juvenile hormone titer to undectable levels in the last larval instar results in a larval-adult (Hemimetabola) or, as in our example, a larval-pupal molt which is followed by a molt to the adult (Holometabola). The action of juvenile hormone can thus be viewed either as favoring larval syntheses or as inhibiting differentiation to the adult. It should be mentioned that, in the preponderance of female adult insects studied, juvenile hormone is again secreted by the corpora allata and is requisite for normal vitellogenesis (Table V of ref. 6 lists effects of corpora allata and juvenile hormone on vitellogenic protein production in various female insects). The juvenile hormone, therefore, is not only a unique hormone in the sense of its structure (the only known animal hormone possessing an expoxide function), but it seems to be one of the few true developmental agents whose structure has been elucidated. For the developmental biologist, it offers the opportunity to drastically alter the course of developmental events both in vivo and in vitro, the result depending on the juvenile hormone concentration at a critical period.

NEUROHORMONES

Because even the most primitive invertebrates are replete with neurosecretory cells, it is surely not surprising that neurohormones play a bewildering array of roles in the highly evolved insects. It is beyond the scope of this chapter to discuss insect neuroendocrinology in detail (see for example ref. 2) although it should be noted that the products of insect neurosecretory cells are probably peptides or small proteins and act to control various aspects of metabolism and physiology [hyperglycemic hormone, adipokinetic hormone, cardioaccelerator hormone, bursicon (cuticle-tanning hormone), eclosion hormone (controls timing and behavior of adult moth ecdysis), diuretic hormone, etc.]. Although the amino acid sequence of the adipokinetic hormone has been elucidated recently (13), the paucity of tissue and laborious bioassays used have hindered success with other insect neurohormones. This problem is exemplified in studies with the so-called diapause hormone of the commercial silkworm *Bombyx mori*. It was more than 25 years ago that it was first demonstrated that a hormone secreted from the subesophageal ganglion of the female causes the eggs laid by the female moth to enter a state of arrested development (diapause) at a specific stage of embryogenesis. Subsequent environmental cues terminate diapause and elicit renewed development. Although the exact structure of this peptide hormone is not yet known, it seems to be composed of two active principals (3,300 and 2,000 daltons) (14). The above should convince the reader of the diversity of neurohormones in insects. However, it has been extremely difficult to elucidate their structure or carry out biochemical studies because of the availability of only minute quantities of neurohormones. For example, to obtain the above information on the diapause hormone, a number of extractions of adult heads had to be conducted, including one of two million heads (15).

The neurohormone of most concern here is that which presumably controls the synthesis of α-ecdysone by the prothoracic glands. This tropic neurohormone is released by specific groups of neurosecretory cells and has been termed brain hormone, activation factor, ecdysiotropic hormone, and prothoracicotropic hormone (PTTH). It is considered here in some detail because of its obviously important role in indirectly regulating growth and development via the prothoracic glands.

The PTTH probably is a polypeptide (perhaps with a nonproteinaceous ligand), but its exact structure is not yet known partially because of the indequacies of the bioassays which have involved injection of extracts into brainless larvae or pupae (1). Not only must one wait days, and in some cases weeks, for the result, but the hemolymph also contains proteases and esterases which may well degrade the hormone. The obvious bioassay is a chemical or biochemical event marking the activation of prothoracic glands maintained in vitro. Conditions for culturing prothoracic glands have only been perfected in the last few years, but their use has already resulted in the elucidation of the product of the prothoracic glands as α-ecdysone (16–18) and the demonstration that the cyclic activity of the prothoracic glands during the instar could be correlated with their ability to secrete α-ecdysone in vitro (19). Inactive prothoracic glands in vitro would, therefore, be ideal for the bioassay of brain hormone because the secretory product could conveniently be quantitatively determined by radioimmunoassay (7). This approach has yet to be used for PTTH purification but has given us some insights into how brain hormone may activate the prothoracic glands.

Previous studies have demonstrated correlations between prothoracic gland activity and RNA synthesis (1, 9) as well as gland-specific alterations in the permeability of individual prothoracic gland cells (2). Despite the fact that these studies really do not explain PTTH action, the observations that the brain hormone is a peptide and that it can affect membrane permeability led the authors' laboratory to study the possible role of cyclic nucleotides as intermediaries between brain hormone and prothoracic gland activation (20–22).

Using the last instar larva of *Manduca sexta*, the authors first identified the exact stages at which the prothoracic glands were either maximally or minimally active (19) and assayed the glands for adenylyl cyclase activity. These first studies revealed that in the presence of aminophylline active glands produced four times as much cAMP as inactive glands. It should be mentioned that of all the tissues tested (fat body, gut, nervous system, etc.) the minute prothoracic gland, composed of only a few hundred cells, exhibited the highest adenylyl cyclase activity. The difference in adenylyl cyclase activity between active and inactive prothoracic gland disappeared in the absence of aminophylline, suggesting that the active glands also possess potent phosphodiesterase activity. Perhaps gland regulation is at the level of the cAMP phosphodiesterase. Although the injection of cAMP or the dibutyryl derivative did not elicit development in bioassay animals (brainless pupae) in vivo, phosphodiesterase inhibitors stimulated inactive prothoracic glands to produce α-ecdysone in vitro. These data as well as other studies which suggested a role for the cyclic nucleotides in insect development (23–26) led the authors to examine the cAMP content of the prothoracic glands of *Manduca* during the critical last larval instar, with the use of a competitive protein binding assay. Figure 3 reveals that the cAMP content of the prothoracic glands increased between days 2 and 3 of the last instar and reached a peak on day 4. This was followed by a rapid decrease to the basal level during days 5–6. As can be seen clearly from Figure 3, the endogenous cAMP concentration increases just prior to an increase in α-ecdysone secretion by day 4 prothoracic glands in vitro. This important small surge in ecdysone titer is presumably responsible for reprogramming the epidermal cells so that they will secrete pupal cuticle several days later when exposed to about eight times as much ecdysone on day 8 (8, 19). These data do not prove that PTTH acts via cAMP, but they are very suggestive of such a mechanism and do not seem to be fortuitous.

How might the absence of another peak of cAMP prior to the second and larger α-ecdysone surge by the prothoracic glands be explained? Our current hypothesis is that gland activation occurs by both cAMP-dependent and cAMP-independent mechanisms and that the cAMP-independent mechanism does not require RNA synthesis, in contrast to the cAMP-dependent event. Alternatively, there may be different brain hormones acting during the two periods of α-ecdysone secretion, and each may use a different mechanism. The latter suggestion is supported by the studies of Gersch (2), who has identified two prothoracic gland activation factors secreted by the cockroach brain.

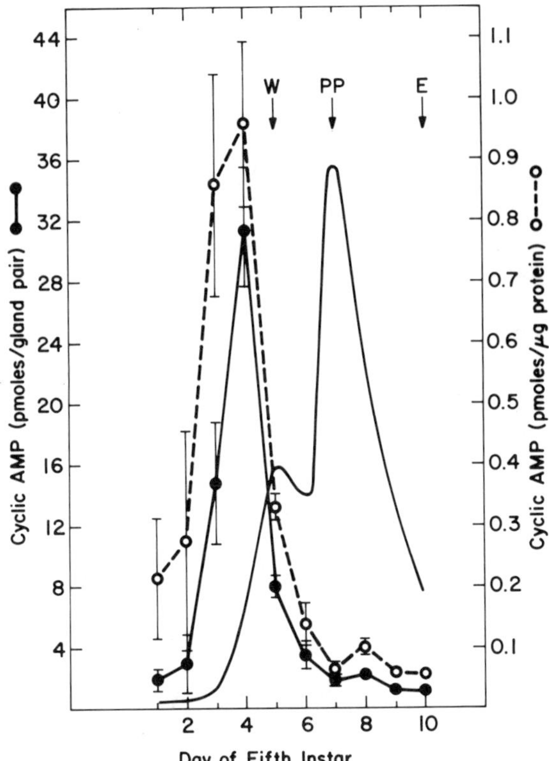

Figure 3. Endogenous prothoracic gland cyclic AMP level during the last larval instar of *Manduca sexta*. W, PP, and E represent the onset of the wandering stage, pharate pupal development, and ecdysis, respectively. *Solid line without points* represents the variations in the ability of the prothoracic gland to secrete α-ecdysone. Reproduced from Vedeckis et al. (22) with permission of Elsevier/North-Holland.

The brain-prothoracic gland axis of *Manduca* seems to be analogous to the vertebrate pituitary-adrenal axis; in the latter case, adrenal cortical steroidogenesis is stimulated by ACTH, possibly by cAMP-dependent and -independent mechanisms (27). If the postulate that cAMP is an intermediary in the stimulation of the prothoracic glands by brain hormone is accepted, then the next question concerns the means by which cAMP accomplishes this task. No data have been published regarding the protein kinases of insect prothoracic glands, but the authors suggest that, as in the adrenal cortex system (28), accumulation of hormone precursor may be the rate-limiting step in the activity of the prothoracic glands (7). This precursor exists as a constituent of the insect hemolymph lipoproteins, and cAMP may change the permeability of the cells comprising the prothoracic glands to allow its entrance and subsequent conversion to α-ecdysone. The exact role of cAMP in brain hormone action still remains to be established, but, as is seen in subsequent sections of this chapter, the recent use of insect tissue and organ culture has stimulated new approaches to the question of insect hormone action. Assuredly, the problem of brain hormone activation of the prothoracic glands will be solved in the near future with the in vitro approach.

Thus, α-ecdysone is secreted by the prothoracic glands as soon as it is synthesized, enters the hemolymph, and is hydroxylated to β-ecdysone by peripheral tissues (e.g., fat body) at a rate dependent upon the developmental state. Juvenile hormone is synthesized and secreted by the corpora allata, also as rapidly as it is synthesized, enters the hemolymph where it is protected from nonspecific esterases by being bound to a specific binding protein (7, 8, 10 12), and then acts upon target tissues. The remainder of this chapter considers the action of molting hormone (ecdysone) and juvenile hormone at the biochemical level. It should be emphasized that this chapter is restricted to only a few systems and is not meant to be comprehensive.

HORMONAL CONTROL OF DNA SYNTHESIS

Replication

Growth is an integral feature of development and takes the form of either cell multiplication or increase in cell size. In insects, the latter is generally a result of polyploidy or polyteny. Indirect evidence that cell proliferation is under hormonal control was provided by the temporal correlations of ecdysone release and a wave of epidermal DNA synthesis (mitosis) at the beginning of the molting cycle, as judged by histological and autoradiographic techniques (29). Because cell division is not required for apolysis and cuticle deposition, the existence of a direct causal connection is conjectural.

Wing discs in *Ephestia kühniella* exhibit two phases of growth (30): a proliferative phase of exponential growth that coincides with the regenerative capacity of the anlage (31) and a second differentiative phase in the pharate pupa that is probably elicited by a high titer of ecdysone in the absence of juvenile hormone. The shift from one phase to the other is characterized by a shortening of the S phase (from 12 to 5.8 h) (32). The postulated ecdysone-mediated mitotic response seems to be asynchronous. Although the number of cells doubles during this period (33, 34), the asymmetric distribution of mitoses in the discs makes it doubtful that all cells undergo a single division. Indeed, clonal analyses in *Drosophila* support the concept of mitotic heterogeneity (35–37). It thus seems that the capacity of the cell to divide as a result of hormonal stimulation is determined by both its developmental history and positional information.

There is no cogent evidence that the proliferative growth phase is hormonally controlled. For example, the imaginal discs of *Drosophila* grow exponentially throughout the larval period, although a detailed analysis of genetic mosaics indicated a depression of mitotic rate prior to ecdysis (36, 38). This suggests either a possible inhibitory role for ecdysone or that regulation is attained by varying the ratio of juvenile hormone (JH) to ecdysone (39).

Ecdysone

In most of the studies conducted on the control of DNA synthesis by ecdysone, isolated imaginal discs have been used (39–44). The essential absence of ecdysone metabolism (42, 45) in the discs makes them excellent objects for the analysis of ecdysone action at the cellular level. Because these studies were generally concerned with morphogenesis and imaginal differentiation, the effect of hormones on DNA synthesis was usually studied only during the differentiative phase. As an indicator of DNA synthesis, [^{3}H]thymidine was generally the precursor of choice, and its incorporation was measured by radioassay of acid-insoluble, base-resistant material or by autoradiography. Despite the potential of the in vitro approach, the results were equivocal, perhaps because of variables such as species, age of donor, type and concentration of ecdysone, and even the experimental regimen. The data on the imaginal discs of Diptera led to the conclusion that concentrations of α- and β-ecdysone in the physiological range stimulated DNA synthesis, at least in a transient manner (39, 42–44). In contrast, in the Lepidoptera the two ecdysones seem to have different effects; incorporation into DNA was promoted by α-ecdysone, whereas β-ecdysone alone had no effect, but inhibited the α-ecdysone-dependent response (40, 41). The lack of a differential effect in Diptera is in accord with the low or undetectable level of α- compared to β-ecdysone in this insect order (18, 46, 47). The latter fact results from a very efficient C_{20} hydroxylase in peripheral tissues that quickly converts α- to β-ecdysone (7). In contrast, however, α-ecdysone stimulated cell multiplication in the diploid cell line Kc derived from *Drosophila* embryos (48). The addition of larger concentrations of α-ecdysone, as well as of β-ecdysone or of noninsect ecdysones to the cell line, led to morphological differentiation and cessation of multiplication (49, 50). When the cells were treated with submorphogenetic doses of hormone, only α-ecdysone stimulated cell growth, whereas β-ecdysone had no effect on the rate of multiplication. If the absence of conversion of α-ecdysone to β-ecdysone is confirmed in these cell lines, then the above results would strongly indicate a differential role for α- and β-ecdysone. That is, β-ecdysone induces differentiation accompanied by a loss of proliferative potential, and α-ecdysone promotes cell multiplication.

Juvenile Hormone

Studies on the effect of JH on DNA synthesis are inconclusive because in these investigations extremely high doses of JH were generally used under in vitro (44, 51) conditions. Concentrations in the range of 10^{-4} M, which are above the solubility limit of JH and several orders of magnitude higher than the in vivo JH titer ($\sim 10^{-8}$ M) (52), were applied to obtain an effect. This regimen precludes any meaningful interpretation.

The data thus far available demonstrate that the hormonal control of DNA synthesis in insects requires further in-depth investigation. There is a need to correlate indicators of DNA synthesis, such as the in vivo and in vitro incorporation of labeled thymidine into DNA, with determinations of DNA polymerase activity. Studies on DNA polymerase in insects have been conducted only on *Drosophila* (53), and a developmental analysis of enzyme activity is available only for embryogenesis (54). Although ecdysones were found in eggs, their role in embryonic development is not clear. Unfortunately, during the larval stages characterized by a high rate of DNA synthesis, determinations of polymerase activity are impaired by a high level of nuclease activity (54, 55) and require elaborate purification procedures.

HORMONAL CONTROL OF TRANSCRIPTION AND TRANSLATION

Ecdysone

Polytene Chromosomes and Puffing Phenomenon Studies of the giant polytene chromosomes, characteristic of certain tissues of Diptera, have contributed greatly to our knowledge of chromosomal structure and function. In particular, the chromosomes of the salivary gland cells of fruit flies (*Drosophila*) and midges (*Chironomus*) attain their large size by repeated replication of the chromosomes without strand separation, resulting in structures large enough to be analyzed by low power microscopy. The appearance of these chromosomes as a series of light and dark bands and the close correlation between the number of bands and genetic functions have led to the suggestion that the study of these chromosomes may answer critical questions relating to the control of gene activation and repression at the molecular level. The appearance of more than 20 reviews on polytene chromosomes during the past 10 years attests to both the importance of the subject and progress in the field. Because this discussion is restricted to recent studies on the role of insect hormones in the puffing phenomenon, the reader can consult the following reviews for the older literature and other aspects of the field (56–63).

For our purposes, the significant observations are that certain chromosome bands seem to loosen and swell (puff) through development (64–66) and that several of these puffs occur specifically in response to ecdysone (67–69). It has been demonstrated that the puffs probably form from single bands (70) and represent active gene sites because there is not only an accumulation of protein and RNA but also active RNA synthesis (71–74). Furthermore, the large puffs synthesize RNA at a higher intensity than the smaller ones. There are instances when puffs or puff-like swellings can be induced in the presence of inhibitors of RNA synthesis; the puffs, however, do not attain the normal size and do not show incorporation of [^{3}H]uridine (73, 75). Based on these and other results, the puffs are considered to be sites of high transcriptional activity (76). Therefore, the following questions can be asked: a) What is the relationship of the puffing pattern to normal development? b) Is there a direct correlation between puffing activity and hormone-stimulated gene products which would explain hormone-dependent biochemical and morphological changes? It is to these questions that attention is now directed.

Both in *Drosophila* and *Chironomus*, among the total number of bands (approximately 5,000 and 2,000, respectively), only a few puff at any one time, e.g., less than 10% in *Chironomus tentans* (71). The puffing pattern varies in a species-specific manner throughout development as existing puffs regress and new ones appear. These changes in puff pattern can be temporally correlated with larval-larval and larval-pupal ecdyses or with intermolt periods, indicating that particular puffs may be required for a defined developmental stage. Maximum puffing activity is usually observed around larval-pupal ecdysis, the pattern following a particular sequence which, particularly in the case of *Drosophila* (68, 69), seems to correlate well with the expected, as well as measured, changes in ecdysone titer. Injection of ecdysone into early third (last) instar *Drosophila* larvae or fourth (last) instar *Chironomus* larvae, or incubation of the respective salivary glands in medium containing ecdysone, elicits the same pattern and sequence of puffing as would be seen normally later in the instar when the ecdysone titer rises, although there may a shift in the time sequence (76–79). These results indicate that ecdysone may be the primary inducer of the normal puffing pattern. Definitive evidence that ecdysone is indeed the inducer of the puffs comes from the study of the *Drosophila* mutant 1(2)gl or lethal giant larvae, in which the defective ring glands (source of α-ecdysone) cannot synthesize and/or release α-ecdysone at the appropriate stage. That it is the ring gland and not the target which is defective was demonstrated by the transplantation of wild type ring glands into mutant larvae which in turn underwent normal development. The salivary gland chromosomes of these mutant larvae do not exhibit the normal puffing activity characteristic of the wild type, although these mutant glands retain the capacity to respond to ecdysone. This was demonstrated by exposing the mutant glands in vitro to β-ecdysone and observing the entire sequence of the normal puffing pattern (80).

Detailed studies have been carried out on the time sequence of puffing, effect of inhibitors of protein and RNA synthesis, and varying concentrations of α- and β-ecdysone (81–84). These investigations reveal that the earliest chromosomal response to ecdysone occurs within 10–15 min and that the appearance of these new puffs can be prevented by blocking RNA synthesis, but not protein synthesis. These facts support the postulate that puffs are sites of transcriptional activity. Simultaneously, a few existing puffs begin to regress. At specific subsequent time intervals, other puffs appear while the early ones regress. Both phenomena are prevented by blocking either RNA or protein synthesis. In *C. tentans*, for example, the first puff to appear is I-18-C, while I-19-A regresses; the second one to appear is IV-2-B, with several others arising later. However, the total number of bands giving rise to large puffs is comparatively small (84). In *D. melanogaster*, however, just before puparium formation more than 100 bands seem to puff in a specific sequence. The evolution of late puffs can be prevented by blocking protein synthesis after the appearance of early puffs. These and other findings have led to the hypothesis of sequential puff induction (sequential gene activation) whereby the translational products of one set of puffs are requisite for the induction of the next series of puffs (72, 83). In both *Drosophila* and *Chironomus*, the size that puffs attain and the time sequence depend on the concentrations of ecdysone (physiological levels 5×10^{-7} to 1×10^{-9} M) used. Not only does ecdysone induce certain puffs, but it is also required for the regression of certain puffs which are active during the pharate pupal stage; these are repressed by the presence of ecdysone and are active only in its absence (85).

It should be noted here that ecdysone is not the only agent that induces puffing, although only ecdysone elicits normal puffing activity. A wide variety of substances such as tryptophan, vitamin B_6, respiratory inhibitors, ions, and conditions and treatments such as recovery from anaerobiosis and heat shock cause the regression of pre-existing puffs and the appearance of 4–6 new puffs. The response is all-or-none and seems to result from the attempt of the cell to maintain homeostasis when challenged by environmental changes (57, 86). These systems, although unphysiological, have aided considerably in the analysis of the role of puffing in gene activity.

Having summarized the evidence showing that ecdysone does elicit the normal puffing sequence and having accepted the postulate that puffing is a consequence of active transcription (see subsequent discussion), the significance of puffing in terms of cell function can be explored; i.e., does transcription at puffed sites lead to the synthesis of specific polypeptide products? The function of the larval salivary gland is to produce certain polypeptide secretions that form a silken cocoon in *Chironomus* and a glue to fix the puparium to the substrate in the case of *Drosophila*. The *Chironomus* secretion consists of six or seven polypeptides as analyzed by SDS-urea gels, whereas that of *Drosophila* yields about five polypeptides. In the midge *A. lucidus*, the secretion contains a protein that is rich in hydroxyproline, and the suppression of its production correlates with the regression of a specific Balbiani ring (87) (BR; an exceptionally large puff). By analyzing the polypeptide pattern of secretion in the midges *C. tentans* and *Chironomus pallidivittatus* and their hybrid progeny, it was concluded that a specific polypeptide (as judged by gel electrophoresis) was coded by BR 4 on chromosome IV (88, 89). Further studies indicated that the activity of BR 1 and 3 on the same chromosome could also be correlated with the appearance of two specific polypeptides (90). As yet, however, it has not been demonstrated definitively that each puff is actually responsible for the synthesis of one polypeptide.

Similar correlative studies were not reported for *Drosophila* until recently, although rapid progress has occurred during the past few years. Cross-breeding and cytogenetic techniques to localize the genes responsible for certain polypeptides in the salivary gland secretion revealed that polypeptide 3 corresponded to band 68C on the third chromosome, whereas polypeptide 4 seemed to be related to band 3C on the X chromosome (91). More importantly, puffing and regression at these bands correlate with the maximum and minimum synthesis of secretory proteins, respectively. Unfortunately, these puffs appear prior to the release of ecdysone and, in fact, regress soon after being exposed to ecdysone. Studies with heat-shocked *Drosophila hydei* have shown a probable correspondence between induction of the mitochrondrial enzymes tyrosine transaminase with puff 2-48BC and NADH dehydrogenase with 4-81B (92, 93). When the proteins synthesized in the salivary glands of normal and heat-shocked larvae of *D. melanogaster* were labeled with [S^{35}]methionine and analyzed by electrophoresis, six new polypeptide bands appeared simultaneously with the regression of certain puffs and the appearance of about nine new puffs (94). A close correlation was noted between the maximum activity of puff 87B and the incorporation of label into a specific protein. This protein (70,000 daltons) is the major protein synthesized during heat shock. These results were confirmed with *D. hydei,* in which six similar protein bands and six puffs appeared after heat shock (95). In fact, treatment with arsenite, 2,4-dinitrophenol, vitamin B_6, or release from anaerobiosis—all treatments that induce puffs similar to heat shock puffs—caused the appearance of the same six strongly labeled polypeptides (96). The recent observation that a *Drosophila* cell line responds to heat shock by altered protein synthesis may be potentially important in this regard. Indeed, the proteins induced by heat shock co-migrated with the proteins resulting from similar treatment of salivary glands (97). In particular, the most prominent band from the heat-shocked cell line corresponded to the 70,000-dalton protein mentioned above. The observation that actinomycin D prevented induction of this protein is further evidence that heat shock acts at the transcriptional level.

Although the above preliminary investigations suggest a correlation between gene activity (puffing) and a cytoplasmic product, they certainly do not prove that the protein was derived from the transcriptional product of the puff. This is particularly true in view of the fact that much of the nuclear RNA in eukaryotes turns over within the nucleus; giant heterologous nuclear RNAs (hnRNAs), probable precursors of mRNA, must be processed before being transported to the cytoplasm (98). It is necessary to show that puff transcripts are, in fact, transported to the cytoplasm and act as mRNAs within the cytoplasm to yield specific products.

Evidence that puff transcripts are either mRNAs, or at least precursors of mRNAs, includes the following.

1. Ribosomal RNA, 4 S, and 5 S RNAs, which are the major classes of RNA present in cells, do not hybridize to puffs, whereas total extracted hnRNA hybridizes to many different chromosome regions, including puffs (99–101).
2. α-Amanitin, an octapeptide toxin which specifically inhibits RNA polymerase II of eukaryotes, an enzyme believed to be involved in the synthesis of mRNA and hnRNA, abolishes the incorporation of precursors into puff RNA (82, 102–104).
3. The puff transcripts, either present in the puff or in the nucleoplasm, contain poly A, which is a characteristic of many eukaryotic mRNAs (104, 105).
4. Cytoplasmic mRNA as well as poly A-containing mRNA present on the polysomes hydridize back to either single or multiple corresponding puffs (97, 105–107).

Finally, one example clearly traces the transcript of a puff from its site of production in the chromosome to the mRNA component of polysomes in the cytoplasm. This success was a consequence of the development of microdissection techniques resulting in the isolation of clean and uncontaminated cellular and nuclear components (108) (Figure 4 A). The work has been carried out on the Balbiani rings of *C. tentans*, especially BR 2 of chromosome IV, which also contains BR 1 and 3. Not only are these BRs rapidly labeled by RNA precursors, but electron microscopic observations also indicate their association with granules that exhibit the size and staining properties of ribonucleoprotein particles (RNP) (71, 109). BR 2 is the largest of the BRs present in chromosome IV, and its RNA exhibits a size distribution characteristic of growing nascent RNA chains, the maximum size of 75 S being characteristic of the final BR 2 product (110). The RNA extracted from the BR 2 region hydridizes only to the BR 2 site when assayed by in situ hybridization, indicating that this RNA is in fact a de novo transcript. It constitutes about 50% of the total chromosomal RNA, is characterized by a specific base composition, and accumulates in the nucleoplasm in a time-dependent manner (111). Much of the nucleoplasmic RNA is 75 S and hybridizes exclusively to BR 2 and BR 1, indicating that the transcripts of the two BRs are approximately the same size and accumulate in the nucleoplasm at the same rate, but are separate and discrete products. Finally, RNA with the above characteristics comprises about 1.5% of the total cytoplasmic RNA and hydridizes exclusively to BR 1 and BR 2 sites as does the nucleoplasmic RNA (107) (Figure 4 B–D).

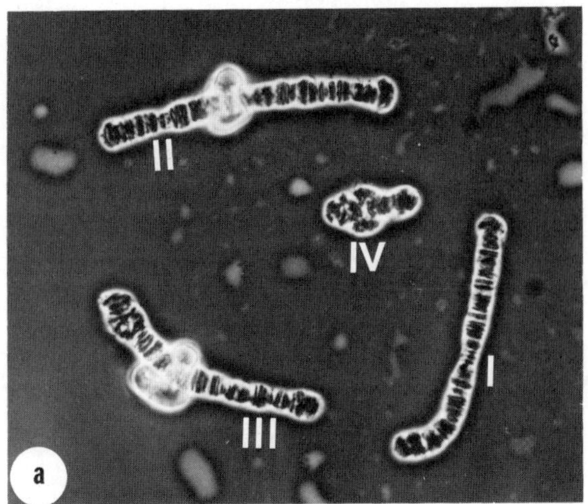

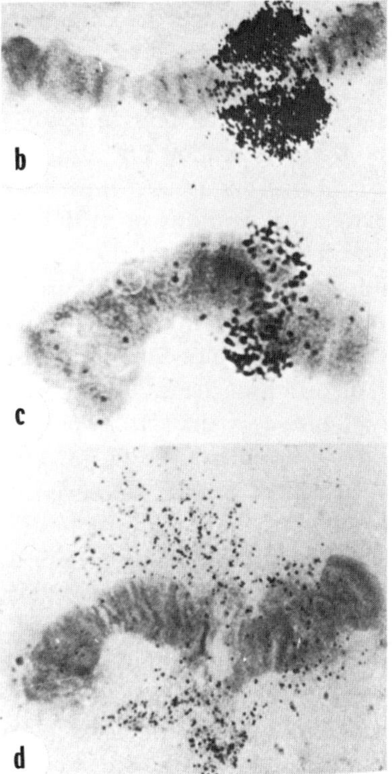

Figure 4. *a*, isolated chromosomes (I-IV) from a salivary gland cell of *Chironomus tentans*. Reproduced from Daneholt et al. (260) with permission of Cold Spring Harbor Laboratory. *b-d*, in situ demonstration of Balbiani ring 2 RNA in nuclear sap and cytoplasm. The radioautographs show chromosome IV after hybridization with BR 2 RNA (*b*), nuclear sap RNA (*c*), and cytoplasmic RNA (*d*). Reproduced from Lambert (106) with permission of Cold Spring Harbor Laboratory.

It was demonstrated recently that more than 40% of the cytoplasmic 75 S RNA was present in heavy polysomes and that the mRNA isolated from these polysomes hydridized only to BR 2. These results strongly indicate the messenger nature of the puff transcripts (63). Translational products of either the heavy polysomes or the mRNA have yet to be characterized, but the possibility exists that this mRNA codes for polypeptides 1, 2, 4, and 5 of the salivary gland secretion. The molecular weight of this RNA (15–35 $\times$ 10^6), if it is polycistronic (90), is great enough to code for all four polypeptides having a combined molecular weight of about 10–12 $\times$ 10^5. If not polycistronic, the mRNA could code for polypeptide 1 (5 $\times$ 10^5), which is the predominant polypeptide of the secretion.

If puffing is an expression of transcriptional activity culminating in the synthesis of polypeptide products as the above suggests, then substances (e.g., ecdysone) that induce puffing can be assumed to act at the transcriptional level. However, not all puffing activity is related to ecdysone surges, and, indeed, the most intensively studied BRs of *C. tentans* are not induced by ecdysone but exist throughout development. Studies of the *Drosophila* mutant 1(2)gl, which does not exhibit the normal puffing sequence, show that the salivary glands synthesize secretion (glue) but store it in the absence of ecdysone (112). Exogenous ecdysone not only elicits the normal puffing sequences in this mutant but also causes the release of the glue into the lumen (80, 112). Therefore, it must be assumed that in both normal and mutant larvae, the synthesis of mRNA for the secretory polypeptides occurs prior to the appearance of the puffs induced by ecdysone. Thus, the ecdysone-induced puffs may be required to produce products required later, e.g., release of secretion (79) (by unknown mechanisms) or the synthesis of hydrolytic enzymes for gland histolysis (113) or both.

It should also be noted that of the 2,000 bands observed in *C. tentans* and the 5,000 in *Drosophila*, less than 10% puff at any time and only about 1–2% exhibit extensive puffing. From numerous studies, including investigations in which hybridization was employed, it is clear that 5,000–10,000 species of mRNA are present in the cytoplasm of *Drosophila* salivary glands at any one time (114) and that RNA synthesis occurs along the entire chromosome (115). It could, therefore, be argued that puff sites are unique and that results obtained from studying puffs cannot be extrapolated to all active genes. However, autoradiographic studies of chromosomes from glands incubated in the presence of [^{3}H]uridine reveal that label accumulates throughout the chromosomes, only the intensity being greater at the puff site. Measurement of endogenous RNA polymerase activity in vitro of fixed polytene chromosomes also demonstrates the presence of RNA polymerase molecules at all sites (Figure 5A and B) (115). This is confirmed by immunofluorescent localization of RNA polymerase throughout the polytene chromosomes (Figure 5C) (116). The data indicate that transcription at puff sites is probably the same as at other chromosomal loci, but it

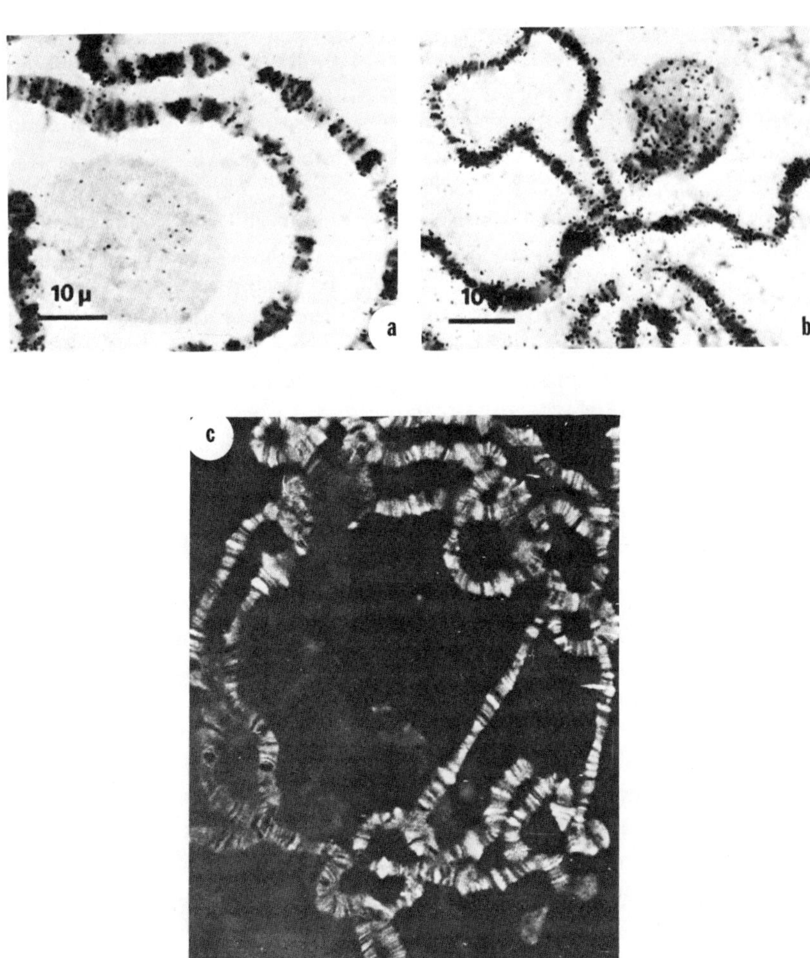

Figure 5. *a* and *b*, autoradiographs of polytene chromosomes from *Drosophila hydei* salivary glands after transcription in situ. The slides were incubated with a reaction mixture containing [^{3}H]UTP, nonradioactive ATP, CTP, GTP, and appropriate buffer at 37°C for 90 min. In *a*, the incubation buffer did not contain Mg^{2+} and in *b* it contained both Mn^{2+} and Mg^{2+}, the nucleolus being labeled in the latter. Reproduced from Pages and Alonso (115) with permission of Academic Press Inc. *c*, polytene chromosomes stained with antibodies against *Drosophila melanogaster* RNA polymerase II and viewed by indirect optics. The fluorescence pattern seems to reflect the specific distribution of RNA polymerase throughout the chromosomes and probably includes molecules of RNA polymerase I. Reproduced from Plagens et al. (116) with permission of Springer-Verlag.

occurs at a more rapid rate in the former case. The mechanisms by which ecdysone might induce these puffs is discussed under "Mechanisms of Action of Insect Hormones."

Cuticular Tanning Because both the deposition of cuticle by the epidermis and molting are vital components of insect growth and development and are influenced by ecdysone, aspects of cuticle maturation are discussed here. The outer cuticular layer, the exocuticle, is hardened and tanned (sclerotization) as a result of the cross-linking of proteins via quinones. The exocuticle is secreted shortly after apolysis and the initiation of epicuticle formation. The epidermis not only secretes phenolic compounds that are responsible at a later stage for hardening and darkening, and the wax necessary for surface hydrophobia, but also produces phenoloxidase and other enzymes required for the tanning process. Consequently, it is not surprising that epidermal cells are the site of RNA and protein synthesis prior to cuticle deposition. At ecdysis, when the insect sheds the partially digested old cuticle, the new cuticle is soft and pale but gradually becomes hard and dark; the color varies from light amber to dark brown or black. Although the sclerotization process has been examined in many insects, the most intensely studied system is puparium formation in flies, where the soft, white larva (maggot) shortens, contracts, and the old larval cuticle becomes dark and hard. It is within the puparium that larval-pupal ecdysis and pupal-adult apolysis occur. Indeed, the most common bioassay for ecdysone involves ligation of fly larvae posterior to the prothoracic gland (ring glands) prior to ecdysone secretion. The abdomen remains larval but undergoes puparium formation (pupariation) when injected with molting hormone (117).

Primarily because of the work of Karlson, Sekeris, and their colleagues a great deal of information is now available on the quinone tanning of the fly (*Calliphora*) puparium and its control by ecdysone (118, 119). They demonstrated initially that *N*-acetyldopamine is the cuticular sclerotizing agent and that it is derived from tyrosine (120). The metabolism of tyrosine follows two different routes, depending on the developmental stage during the last instar (121). Route 1 (Figure 6) is active in early last instar larvae, whereas route 2 is the pathway of choice in larvae shortly before pupariation. The rerouting of tyrosine is a consequence of ecdysone action, probably by hormonal induction of DOPA decarboxylase, which mediates the conversion of 3,4-dihydroxyphenylalanine (DOPA) to dopamine (Figure 6). Evidence for this enzyme induction includes the observations that injection of ecdysone into the abdomens of ligated larvae elicited a significant increase in DOPA decarboxylase activity within a few hours and was repressed by inhibitors of RNA and protein synthesis (122). Assuming that induction resulted from the transcription of new mRNA, the authors studied RNA synthesis and observed that [^{32}P] incorporation into nuclear RNA increased within 1 hr of ecdysone administration and that

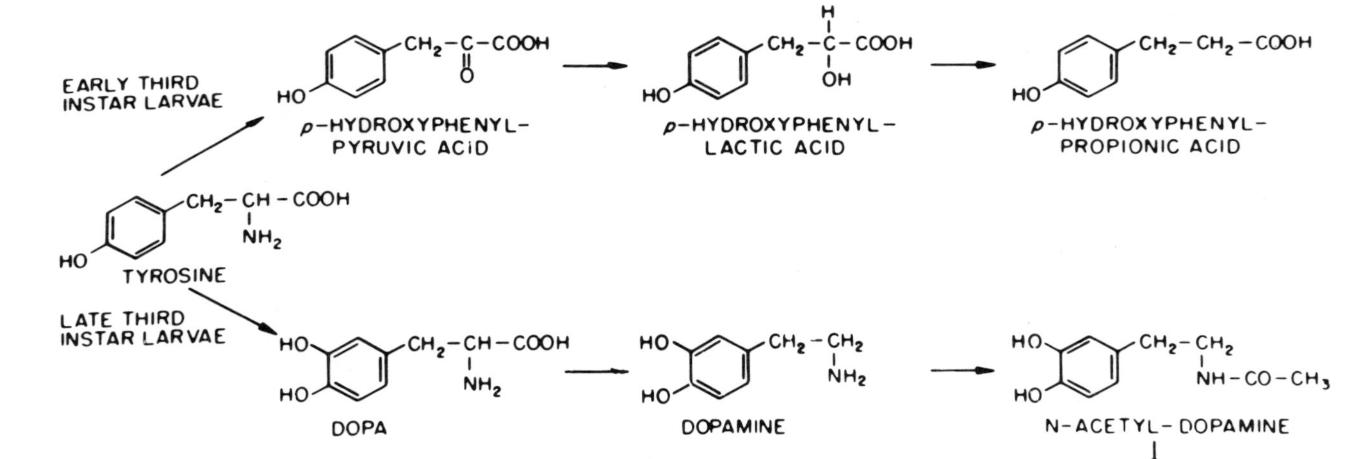

Figure 6. Pathway of tyrosine metabolism in the blowfly *Calliphora erythrocephala*. Reproduced from Karlson and Sekeris (118) with permission of Academic Press Inc.

this increase was reflected in the microsomal RNA 3–4 hr later (123). Further studies showed that DNA-like RNA had increased along with ribosomal precursor and ribosomal RNAs. The measurement of template activity of the isolated RNA with a cell-free rat liver microsomal system showed stimulation of DOPA decarboxylase synthesis and was accepted as conclusive proof for ecdysone induction of mRNA for the enzyme in the epidermis (124). This assumption has been questioned on the basis of experimental regimen and lack of follow-up (6), but a recent series of papers from the same group tends to confirm the earlier conclusion. The results of these latest experiments are noted below.

Injection of α-amanitin prior to ecdysone administration prevented the ecdysone-dependent production of DOPA decarboxylase but did not do so if the inhibitor were injected 3 hr after the ecdysone (125). The enzyme from the integument, purified to homogeneity, was found to consist of two polypeptides (~ 45,000 and 50,000 daltons), and an antibody was obtained (126). By employing double labeling and immunoprecipitation techniques, it was demonstrated that the increase in enzyme activity resulted from stimulation of de novo enzyme synthesis (127). Finally, poly A-containing RNA was isolated from pharate pupae and shown to direct protein synthesis in a cell-free system containing labeled amino acids. A protein produced under these conditions was immunochemically similar to DOPA decarboxylase, and the radioactivity associated with the antigen-antibody complex co-migrated with purified DOPA decarboxylase upon SDS-acrylamide electrophoresis (128). Unfortunately, the product of the cell-free system was not tested for enzymic activity. This is of interest because it is not clear whether the single band of radioactivity observed after SDS-urea gel electrophoresis resulted from the synthesis of only one or both of the subunits. If, indeed, the original reports of the appearance of enzyme activity were confirmed, then it would be important to determine how both subunits are produced stoichiometrically and how they interact to form the active enzyme in vitro.

Although the accumulated evidence supports the concept of a de novo increase in DOPA decarboxylase activity paralleling specific mRNA synthesis at the time of endogenous ecdysone release or after hormone injection, it may not be induction in the strictest sense. That is, enzyme activity and the corresponding mRNA are present prior to hormone stimulation, and the maximum ecdysone-elicited increases observed are in the order of 3- to 4-fold above the basal level (125, 127, 128). Although the possibility of pre-existing mRNA that is either activated or translocated was considered by the investigators (128), they eliminated the hypothesis on the basis of their data. The question can be resolved by preparing cDNA from the purified mRNA preparation and using it as a quantitative probe for measurements of mRNA present before and after induction.

Similar but less extensive studies have been carried out with *Sarcophaga bullata* (flesh fly), in which injection of β-ecdysone (20 μg) into young third instar larvae caused precocious induction of DOPA decarboxylase (129). Again, the observed increase was about 5-fold after about 48 hr although the puparium formed was abnormal. The fact that imaginal wing discs cultured in vitro can be stimulated by ecdysone to undergo morphogenesis and deposition of the cuticle, which then tans, can be put to advantage for the study of the induction of DOPA decarboxylase. β-Ecdysone at the optimum level of 2×10^{-6} M induced all the morphogenetic and other changes observed in vivo in cultured wing discs of *S. bullata*. Whereas cuticle was secreted at about 72 hr after exposure to ecdysone in vitro, DOPA decarboxylase increased from 48 hr on and reached a maximum at 96 hr. As expected, the production of enzyme was sensitive to actinomycin D and cycloheximide. In view of the long delay in the appearance of the enzyme, it was suggested that the hormone triggered a series of events that ultimately elicited synthesis of the enzyme and that the hormone did not induce the formation of the enzyme directly (129).

Another example of ecdysone involvement in the regulation of DOPA decarboxylase activity is derived from work with the ovary of the mosquito *Aedes aegypti* (130). The egg chorions are soft and white at oviposition but darken and harden within about 4 hr. In a variety of experiments, the high level of DOPA decarboxylase characteristic of the mature oocyte was produced by the ovaries in response to ecdysone (131). As in the case of *Sarcophaga* wing discs, appearance of the enzyme was detected 24 hr after injection of ecdysone into the adult female, and the maximum response occurred at 96 hr (about a 4-fold stimulation). In view of both the long time interval and the absence of data regarding the fate of injected ecdysone, it cannot be concluded that ecdysone induces enzyme synthesis. The interpretation becomes more complicated when it is considered that the ovaries themselves produce ecdysone in response to a blood meal (see under "Vitellogenesis"). That DOPA decarboxylase is required for tanning was demonstrated by injection of the enzyme inhibitor (DL)-3-(3,4-dihydroxyphenyl)-2-hydrazino-2-methyl propionic acid, which prevented tanning (132).

Thus, studies on the induction of DOPA decarboxylase, particularly in *Calliphora* have provided the basis for a model to explain the mode of action of ecdysone. Although the model might eventually prove correct, the following suggests that it may not apply to every situation. Sclerotization is not restricted to larval-pupal metamorphosis but occurs after both larval and adult ecdysis. In fact, in *Calliphora*, as well as in other insects, a second peak of DOPA decarboxylase occurs just before adult ecdysis and seems to be an ecdysone-independent event (129, 133, 134). In *Calliphora* and several other insects, an ecdysone peak is observed on day 4 of pharate adult development, but enzyme activity reaches a maximum at day 10, a time interval that casts doubt on the induction of the enzyme by ecdysone. [It should be mentioned that the activity of phenoloxidase, another enzyme critical for the tanning process, parallels the ecdysone (133, 135) titer curve in *Calliphora* and seems to be induced by ecdysone. In this case, enzyme activation is a result of the action of a cuticular activator protein on an inactive proenzyme (136).]

Although the available evidence demonstrates that *N*-acetyldopamine is the critical precursor in the tanning of various insects, little is known of the factors controlling the texture of the cuticle. Why is the larval cuticle of the fly relatively soft and pale, whereas that of the puparium and adult is hard and dark? Because *N*-acetyldopamine is the precursor at all developmental stages, the difference could result from the manner in which this precursor is used within the cuticle, and, indeed, it seems that the difference may be attributed to two enzymes with different specificities within the cuticle. One enzyme activates the β-position of *N*-acetyldopamine to cross-link proteins and leads to a colorless cuticle, whereas the other activates the aromatic ring leading to quinone cross-linking and a dark, hard cuticle. This supposition is based on excellent chemical evidence and preliminary enzymology (137, 138). It would be of interest to solubilize these enzymes and to measure their respective activities so as to determine whether JH and ecdysone control their titer. It might be hypothesized that the proportion of the first enzyme is high in the presence of JH and that its level falls as the JH titer decreases and/or that the second enzyme is elicited by ecdysone in the absence of JH. If the hypothesis is correct, then changes in cuticle texture during metamorphosis could be explained on an endocrine basis.

Finally, ecdysone may control sclerotization indirectly by regulating the provision of precursor substrates. In almost all Diptera and Lepidoptera studied, the concentration of protein in the hemolymph reaches a maximum level during the last larval instar when the larvae cease feeding (139). Subsequently, when the ecdysone titer increases to its maximum before larval-pupal ecdysis, the protein content of the hemolymph decreases dramatically. This is a consequence of tissue sequestration of specific proteins (140) and of the transport of hemolymph protein into the cuticle (141). The former phenomenon may be ecdysone-dependent. For sclerotization to occur, tyrosine must be available in sufficient quantity (Figure 6). In addition to tyrosine, other precursors that provide tyrosine for the tanning process have been identified in considerable quantities in fly hemolymph at the appropriate period, e.g., tyrosine-*o*-phosphate in the fruit fly, tyrosine-*o*-phosphate and the dipeptide β-alanyltyrosine in the flesh fly, γ-glutamyl-L-phenylalanine in the housefly, and the unusually large protein, calliphorin, in the blowfly (142–146). Calliphorin is the major protein in larval and pharate adult *Calliphora* and is unusually rich in the aromatic amino acids tyrosine and phenylalanine. Proteins that are similar or equivalent to calliphorin have been demonstrated also in the hemolymph of a variety of other insects. Of possible importance are the observations that these compounds exist at peak concentrations just before pupariation and decrease soon thereafter; e.g., calliphorin decreases from about 7 mg per larva to 3 mg per pupa (139, 146). Because these quantitative changes are correlated temporally with increased ecdysone titers and cuticular tanning, it is possible that their mobilization is also controlled by ecdysone. The accumulation of the dipeptides noted above during the last stages of larval growth, followed by their abrupt disappearance during puparium formation, suggests control for the latter process by ecdysone (143, 147). In fact, a specific hydrolase for β-alanyl-L-tyrosine has been isolated and characterized from *S. bullata* and is precociously induced by ecdysone very much as is DOPA decarboxylase (148). The induction, even in the presence of ecdysone, is suppressed by inhibitors of translation and transcription. In contrast to the instances of enzyme induction discussed above, ecdysone also seems to inhibit the synthesis of certain proteins, e.g., calliphorin. Synthesis of calliphorin begins during the second half of the second instar and continues up to the cessation of feeding during the last instar when synthesis ceases during a 24-hr period. This cessation of synthesis occurs slightly earlier than in the case of other proteins (139). The addition of β-ecdysone to in vitro cultures of fat body inhibits the production and release of calliphorin within a few hours, a result agreeing with the in vivo data (149). Perhaps the hydrolytic products of calliphorin are used in the tanning and hardening of the fly puparium. The possibility thus exists that ecdysone not only elicits new transcription and translation needed for puparium formation but that it also shuts down larval protein synthesis.

Imaginal Discs As noted in the introduction, many of the prospective adult structures of holometabolous insects exist as primordia during larval life (38). Most of these primordial cells are of ectodermal origin and are arranged in specific structures, the imaginal discs. The discs originate at specific sites from invaginations of the epidermis, and the resulting pocket then subsequently folds back up on itself, forming a double-lined invagination. The inner portion comprises the disc proper, which will give rise to the respective adult structure (wing, leg, eye, antenna, etc.); the outer portion forms an envelope, the function and fate of which is not known. Thus, imaginal discs represent a population of cells segregated from the larval epidermis; however, after development into adult structures, the discs evaginate and become integral components of the epidermis. The latter occurs after apolysis in the last larval instar and marks the beginning of disc metamorphosis. Although the mechanism of evagination is poorly understood, it seems to require alterations in cell surface properties, a contractile system which modifies cell morphology (150, 151), and, as exemplified by the Lepidoptera, the auxiliary action of a special muscle (152).

The recent increase in interest in the biochemistry of the imaginal discs is in part a consequence of the development of mass isolation procedures, particularly for *D. melanogaster* discs, and of the fact that disc morphogenesis can be induced in vitro with ecdysone (153). Thus far, there is only one example of the induction of a particular protein by ecdysone in imaginal discs, i.e., that of DOPA decarboxylase (129) (see under "Cuticular Tanning"). Therefore, studies have been concerned mainly with the effects of ecdysone on total RNA and protein synthesis. As noted above, the first morphogenetic response during larval-pupal development (absence of juvenile hormone) is evagination, which occurs in vivo and in vitro within 20 hr of exposure to ecdysone. Evagination does not proceed entirely normally in vitro; rather, it results from the penetration of the expanding disc proper through the static peripodial membrane. Nevertheless, this morphological response can be conveniently measured and correlated with biochemical parameters; it thus provides an excellent test system.

The in vitro culture of discs with β-ecdysone increased the incorporation of [^{3}H]uridine into RNA; the stimulation was dose-dependent with a concentration of 2×10^{-6} M ecdysone producing the maximum response and doses greater than 2×10^{-5} M being inhibitory. Extrapolation from kinetic data indicates that stimulation may occur within 20 min of hormone application (129). Incorporation of precursor reached a maximum after 4 hr and was followed by a slight decline over the next 16 hr of exposure. On removal of the hormone from the medium, the rate of incorporation declined to the base level within 3–4 hr. Part of the increased incorporation of [^{3}H]uridine resulted from an enhancement of the saturable component (facilitated diffusion) of the uptake system (154, 155). However, an increase in the precursor concentration to a level at which the uptake rate essentially resulted from simple diffusion (hormone independent) still resulted in an elevated rate of incorporation. Furthermore, because the RNA synthesized in the presence of β-ecdysone seemed to be degraded at a slower rate than the RNA recovered from control discs, it was concluded that at least a portion of the hormone-stimulated increase in precursor incorporation resulted from a net increase in RNA. General support for this assumption came from an analysis of the DNA-dependent RNA polymerases in which nuclei isolated from imaginal discs preincubated with β-ecdysone showed higher enzyme activity than control nuclei (156). Polymerase I (identified on the basis of α-amanitin insensitivity and elution profile from DEAE-Sephadex columns) was consistently stimulated, whereas polymerase II (inhibited by α-amanitin) showed only minor fluctuations. To correct for large variations in the data, the alterations in enzyme activity were expressed as the ratio of polymerases (I:I + II), the supposition being that polymerase II activity is not affected by β-ecdysone. However, this assumption is not sufficiently documented and is even contradicted by some of the data. Nevertheless, the overall effect seems to be a pronounced increase in polymerase I activity. The synthesis of at least some new mRNA is required because α-amanitin inhibited ecdysone-mediated evagination (157). This has been elegantly demonstrated by in situ hybridization of [^{3}H]-labeled cytoplasmic RNA from β-ecdysone-stimulated discs to salivary gland polytene chromosomes (158). The polyadenylated RNA fraction contained several species that hybridized to loci on the polytene chromosomes which differed from the loci hybridizing the poly(A) RNA from unstimulated control discs. Other poly(A) RNA species found in the controls were absent from the ecdysone-stimulated disc RNA (analogous to puff regression?). In general, there was an increased capacity for hybridization in the case of the hormone-stimulated poly(A) RNA fraction, but the chromosomal loci to which the poly(A) RNA hybridized did not correspond to sites that normally puff or regress in response to ecdysone (see under "Polytene Chromosomes and Puffing Phenomenon").

An unusual feature emerging from the polymerase studies was the increase in activity observed after ion exchange chromatography of the extracted enzymes, in contrast with other systems in which increases in enzyme activity during short incubation periods with hormones were detectable only in isolated nuclei, but not when solubilized and fractionated (159, 160). Such early changes are thought to result from either a shift from the free to the bound form of the enzymes or the existence of factors that modulate template capacity or enzyme activity. These would either be lost or not observed when the enzymes are solubilized and tested on an exogenous template. Any increases in activity seen at longer intervals result from an actual increase in the number of enzyme molecules. The increase found in the polymerase ratio (I:I + II) because of ecdysone could also be elicited by incubating discs with inhibitors of protein synthesis, and mixing experiments indicated that the lower activity ratio in control discs resulted from an inhibitory factor (156). The data suggested that this inhibitor turns over rapidly and normally acts by depressing polymerase I activity. Ecdysone is, therefore, assumed to repress the gene coding for the inhibitor. Because polymerase I activity remains elevated even after hormone withdrawal, this repression seems to be irreversible and contrasts with observations on the pattern of incorporation of [^{3}H]uridine into whole discs. In the latter case, removal of ecdysone results in a decrease in incorporation, necessitating the postulation of an additional, hypothetical control mechanism. Therefore, the above data obtained with *Drosophila* imaginal disc material after column chromatography should be interpreted cautiously. Although several groups have studied the RNA polymerases of this insect with the use of various tissue sources from diverse developmental stages (e.g., whole third instar larvae, imaginal discs, cell line) (161–165), a common concern was the lability of the enzymes, especially polymerase I, and the significant levels of contamination of tissue homogenates with proteases and nucleases (55, 165, 166). These hydrolytic enzymes tend to co-chromatograph with polymerase I during the initial purification steps and may degrade either the polymerase or the transcriptional products. Other possibilities include modification of the template altering initiation or read-through or modification of the transcript itself.

From this, it is apparent that the change in the ratio of polymerase activities may not be a true reflection of an in situ control mechanism, but a consequence of artifactual alterations during tissue processing.

Ecdysone also stimulated protein synthesis in the imaginal discs and the kinetics followed those of RNA synthesis, at least up to 10 hr (167), as would be expected of a precursor-product relationship. Dose-response studies indicated that the optimum ecdysone concentration for maximum stimulation was the same as that required for evagination, 2×10^{-7} M, an order of magnitude lower in concentration than that required for maximum RNA synthesis. Control experiments demonstrated that the incorporation of [^{3}H]leucine into protein reflected a true measure of net protein synthesis; i.e., addition of hormone resulted neither in a differential change in the specific activity of the precursor pool nor in the rate of degradation of newly synthesized proteins. After removal of the hormone, incorporation of [^{3}H]leucine persisted at the rate achieved at the moment of withdrawal, in contrast with the data on RNA synthesis and evagination in which continuous hormone exposure is required, indicating relatively short half-lives for certain essential species of RNA and protein. The interpretation of the results became more difficult when protein synthesis in the various subcellular fractions was followed by double labeling and SDS-acrylamide gel electrophoresis (168, 169). Although the gel banding patterns exhibited no qualitative differences in samples from ecdysone-treated discs, quantitative changes were evident. The increase in total labeling was most profound in the core ribosomal proteins as opposed to the $17{,}000 \times g$ pellet (nuclei, mitochondria, cell debris) and cytosol ($105{,}000 \times g$ supernatant), in which the incorporation of label was only slightly enhanced. Two observations are of particular interest because they deviate from the data on total protein synthesis. First, all four fractions display increases in protein synthesis within the 2 hr lag period, and the ribosomal proteins and KCl-washed fraction reached a maximum after 3 hr. Secondly, the cytosol and ribosome fraction contained classes of proteins, the synthesis of which depended on the continuous presence of ecdysone. From this evidence, it was proposed that ecdysone-stimulated protein synthesis is regulated by two basic mechanisms. The first is expressed by the specific pattern of hormone-dependent proteins found in the analysis of subcellular fractions. These proteins may exert regulatory functions that are closely involved in transcriptional control, although their presence is not essential for the maintenance of the elevated rate of total protein synthesis once it is attained. The establishment of this enhanced level of protein synthesis is thought to be controlled by the second mechanism operating at the post-transcriptional level and is characterized by an increase in the activity of the translational components. The lag period observed for the stimulation of general protein synthesis may, therefore, be determined by the production of certain rate-limiting elements in the translational machinery such as initiation and elongation factors, specific tRNAs, translatable mRNAs, or structural elements such as the endoplasmic reticulum. However, evidence for these possibilities is lacking at present.

Proof for the existence of these control mechanisms requires the analysis of specific proteins to permit a correlative study of kinetics of synthesis and accumulation of mRNA and protein. Two classes of proteins and their mRNAs should be examined for such an analysis. First, there should be mRNAs which are produced in response to ecdysone at short time intervals, and their production should occur even in the absence of protein synthesis, e.g., the RNAs produced by early puffs in *Drosophila* salivary glands upon ecdysone stimulation (although neither the products of transcription nor translation have been identified; see under "Polytene Chromosomes and Puffing Phenomenon"). Analogous systems in higher organisms include the induced protein in rat uterus (170) and the tyrosine aminotransferase of hepatoma cells (171). Secondly, mRNAs for proteins which are produced in bulk quantities and comprise the major translational output as a consequence of hormonal stimulation should be analyzed, e.g., the cuticular proteins, tanning enzymes, etc. (see under "Vitellogenesis").

Pupal Wings The large pupal wings of saturniid moths represent a potentially important system for the study of hormone action. Diapausing pupae are stimulated to initiate adult development when injected with ecdysone, and the pupal wings, composed of epidermal cells, produce adult cuticle and cuticular structures. Brainless diapausing *H. cecropia* pupae responded to the injection of ecdysone with an increased rate of incorporation of [^{3}H]uridine into RNA during the first 24 hr (172), and this stimulation could also be obtained with ecdysone in vitro (173). Sucrose gradient analysis indicated an enhanced synthesis of all classes of RNA (6). This is in accord with observations that polysomes are formed rapidly during early pupal-adult development (6, 174) and that total wing RNA content increased about 4-fold within 2 days after termination of diapause (175).

Recently, the naturally occurring polyamines (putrescine, spermidine, and spermine) have attracted great interest as possible regulators of growth processes (176, 177). In various organisms, the rate of synthesis and accumulation of RNA increases simultaneously with the concentration of polyamines, and polyamine levels fluctuate with the hormonal status of the cell (178, 179), suggesting an involvement of the polyamines in hormone action. The polyamines are thought to exert their effects by stimulating chain elongation, decreasing rates of RNA degradation, enhancing polymerase activity by counteracting product inhibition, etc. (180, 181). Ornithine decarboxylase catalyzes the initial step in the polyamine biosynthetic pathway, i.e., the decarboxylation of L-ornithine to putrescine; its activity has been studied recently in pupal wing epidermis exposed to ecdysone (182). Injection of β-ecdysone into brainless pupae stimulated the de novo synthesis of ornithine decarboxylase from a nearly undetectable level to a peak of 18 hr; after this, activity declined. [It should be pointed out that there are several examples in which ornithine decarboxylase activity increases in hormone-stimulated target tissues along with RNA polymerase activity, specifically that of type I (183).] The authors have also found that RNA polymerase I of wing epidermis starts increasing at about 8 hr after administration of ecdysone to pupae of another silkmoth, *Antheraea polyphemus*, and reaches very high values at 24 hr. The simultaneous increases in ornithine decarboxylase and RNA polymerase I activity assume importance with the recent demonstration that ornithine decarboxylase modulates RNA polymerase activity by increasing the rate of initiation (184). Determination of the polyamine content revealed a concomitant increase in the spermidine level. Putrescine could not be reliably measured because of the high endogenous concentration normally present in the hemolymph; spermine showed no consistent changes. After ecdysone injection, the increase in RNA synthesis parallels that of spermidine. Later, however, it diverges; the rate of RNA synthesis begins to decline after 24 hr, whereas the spermidine titer increases until 36 hr. Although this suggests a close relationship between polyamine and RNA synthesis, it does not support the concept of a causal connection with ecdysone action. The fact that the enzyme can be induced by stimuli other than ecdysone suggests that the stimulation may be a mediated response rather than a direct effect of the hormone.

Juvenile Hormone

Vitellogenesis Information on the specific effects of JH during larval intermolt periods is unfortunately not available. Because JH modulates the type of molt that occurs in the presence of ecdysone, it is generally believed that JH modulates ecdysone action. Although juvenile hormone action is, therefore, discussed in the section below, together with ecdysone action, there is one situation in which JH might act by itself and bring about a positive biological response. This was recognized about 40 years ago when Wigglesworth demonstrated the requirement for an active corpus allatum to achieve normal ovarian development in the insect *Rhodnius* (185). Since that time, considerable evidence has accumulated corroborating Wigglesworth's conclusion that the juvenile hormone is essential for vitellogenesis, and there has been rapid progress at the biochemical level in the last few years.

Ovarian maturation occurs during pharate adult development, immediately after adult ecdysis or at some time after adult ecdysis, depending on the species. When maturation is completed after adult ecdysis, the juvenile hormone is generally responsible for eliciting a number of responses critical for normal reproduction (186). These include oogenesis, development of accessory glands, synthesis and release of yolk protein (vitellogenin) by the fat body, release of lipid from the fat body, and uptake of both vitellogenin and lipid by the oocytes, etc. Because most of the available biochemical data pertain to vitellogenin synthesis, release, and uptake, these phenomena are discussed here.

It is now clear that the yolk proteins are synthesized by an extraovarian tissue (fat body), transported via the hemolymph, and incorporated into the oocytes. In this discussion, the yolk proteins are referred to as vitellins and the precursors synthesized by the fat body as vitellogenins. The ease with which the vitellin can be extracted and purified from eggs, where it comprises up to 60–90% of the soluble protein, has aided investigations of the control of vitellogenin synthesis and uptake. In all cases in which they have been purified, vitellins and vitellogenins have been found to be large molecules with molecular weights of the order of 550,000, soluble in high salt, containing phosphate and about 5–10% each of lipid and carbohydrate (Table 1). Invariably, the antibody produced against the vitellin specifically cross-reacts with the corresponding vitellogenin and vice versa. The pattern of vitellogenin appearance in the hemolymph and subsequent accumulation of vitellin in the oocyte during the reproductive cycle indicates a precursor-product relationship. Because of the well known capacity of the insect fat body to synthesize and secrete proteins, it was recognized more than a decade ago that the fat body is the site of synthesis of vitellogenins. Vitellogenin synthesis has been studied in detail in the South American cockroach *Leucophaea maderae*, in which it seems that JH is the sole controlling element of vitellogenin synthesis (190).

Brookes and his colleagues showed that the *Leucophaea* vitellin was composed of 14 S and 28 S components, the 28 S unit being a trimer of the 14 S component. Both had identical amino acid compositions and antigenic characteristics (187, 191–193). Furthermore, it was demonstrated that the fat body synthesized and secreted the 14 S component and that

Table 1. Characteristics of some insect vitellogenins

	Leucophaea maderae (South American cockroach)[a] (187)	*Locusta migratoria* (locust) (188)	*Hyalophora cecropia* (silkmoth) (189)	*Blatella germanica* (German cockroach) (189)
S value	14.5	17.1	15.9	16.8
Molecular weight ($\times 10^5$)	5.59	5.5	5.16	6.59
Isoelectric point	ND	6.9	5.7	5.0
Lipid (%)	6.9	9.6	9.4	15.7
Carbohydrate (%)	4.7	14.2	1.0	4.5

[a] Vitellin extracted and purified from the ovaries.

JH induced the synthesis of this protein and its subsequent deposition in the ovary. In these studies, isolated abdomens were used because they were known to be free of identified endocrine glands and almost totally composed of fat body and ovaries. More recent studies have confirmed and extended these results in the in vitro situation in which JH-stimulated (active) fat body was shown to synthesize and release vitellogenin into the medium. The total protein synthetic activity of fat body stimulated by JH in vivo was about 4–5 times greater than that of inactive fat body, and at maximum stimulation about 50% of the synthesized protein is vitellogenin (190, 194). Dose-response studies with the JH homologues revealed JH I to be the most effective hormone at about 3×10^{-6} M, whereas JH III, the principal hormone of this insect during adult life, was the least effective (194). No firm conclusions on this aspect of the work can be drawn at present because factors such as catabolism, binding proteins, solubility, etc., were not considered.

Studies on RNA and protein synthesis of active fat body have resulted in the following conclusions.

1. Actinomycin D blocked both RNA synthesis and vitellogenin production, thus demonstrating the requirements for transcription and translation.
2. α-Amanitin reduced vitellogenin synthesis as well as the incorporation of precursors into the small (4–18 S) heterogeneous RNA of microsomes, suggesting that these represent the mRNAs for vitellogenin polypeptides.
3. Simultaneously with the increase in protein synthetic activity, heavy polysomes are formed, indicating the involvement of both mRNA and ribosomes.
4. RNA extracted from the microsomes of active fat body stimulates the incorporation of labeled amino acids into vitellogenin (as measured by precipitation with antibody against vitellogenin) in a cell-free protein-synthesizing system containing polysomes from male fat body which presumably does not possess the vitellogenin mRNA.
5. Microsomal membranes isolated from female fat body after ribosome detachment stimulated mRNA-dependent vitellogenin synthesis in a cell-free system (195–197).

The above observations suggest a JH role in transcription leading to increased production of total RNA, mRNA, and membrane components, all of which are required for, or characteristic of, vitellogenin synthesis. However, the specific role of JH is not yet known. Experiments carried out to demonstrate the increased production of vitellogenin mRNA after JH stimulation have yielded equivocal results. Because it has been demonstrated that the 14 S vitellogenin is probably synthesized as two polypeptides (260,000 and 180,000 daltons) (198), the minimum size of the mRNAs for each of these products should be greater than 30 S. However, both short-term and long-term labeling of RNA by orotic acid demonstrated incorporation only into smaller (4–18 S) heterogeneous RNA, and this was inhibited by α-amanitin (197). Furthermore, large quantities of extracted RNA were employed in a homologous in vitro translational system in which the label recovered in vitellogenin precipitated by antibody was very small. Finally, antibody to vitellogenin may react with the separate subunits or even smaller polypeptides which comprise the subunits, and, therefore, it has not been conclusively proven that the product of the in vitro translational system is in fact vitellogenin. This last criticism is supported by the fact that although poly(A)-containing RNA from *Leucophaea* fat body is a good template in a heterologous translational system, the protein product could not be identified as vitellogenin by either electrophoretic or immunochemical techniques (198).

In most of these studies, the induction of vitellogenin synthesis by the fat body has been studied by applying JH to either allatectomized adult females or to isolated adult female abdomens. At various time intervals thereafter, the fat body is placed in culture, and the synthesis and release of vitellogenin are followed with the aid of immunological techniques. This regimen does not rule out the possibility that the ovary secretes a factor that stimulates the fat body to synthesize vitellogenin, particularly because JH has not been demonstrated to act in vitro on the *Leucophaea* fat body. This objection seemed to be nullified by experiments demonstrating that JH applied to ovariectomized abdomens resulted in vitellogenin synthesis (198), but could an ovarian factor have been released prior to ovariectomy?

A major problem with the studies on *Leucophaea* is caused by the presence of symbiotic bacteria in the fat body which make the interpretation of biochemical data tenuous at best. This does not seem to be the situation in the desert locust, *Locusta migratoria*. Egg maturation in *Locusta* is also controlled by JH, but the level of JH needed for the maximum response in allatectomized females is about 2-3 $\times$ 10^{-4} M (Figure 7) as opposed to 3 $\times$ 10^{-6} M in *Leucophaea* (199). As with *Leucophaea*, when adult female locust fat body is exposed to JH, one notes an increased rate of protein synthesis, increased quantities of polysomes, proliferation of the fat body endoplasmic reticulum, and substantial increases in total RNA, all being characteristic of cells actively synthesizing proteins. These metabolic and morphological changes in the production of vitellogenin are blocked by allatectomy and restored by application of JH or its analogue in vivo (Figure 8). In the case of *Locusta,* a certain fraction of RNA [particularly that containing poly(A)], extracted from active fat body, possessed excellent messenger activity in a heterologous system and directed the synthesis of polypeptide products that precipitated with vitellogenin antiserum (188). Although the data (controls, levels of incorporation, etc.) were more convincing than those obtained in the *Leucophaea* system, the products of the cell-free system were not conclusively shown to be vitellogenin poly-

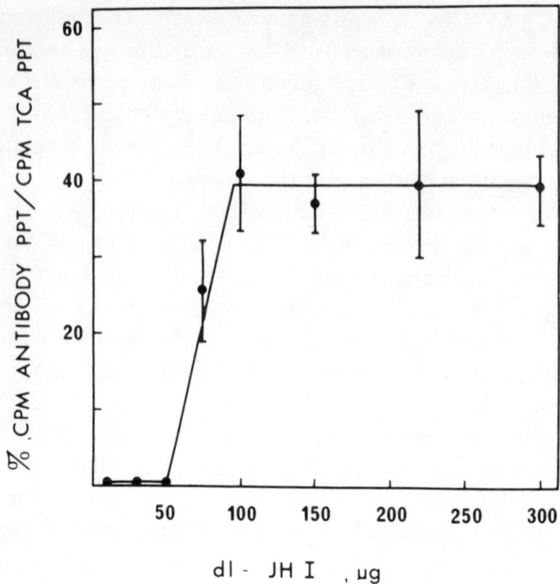

Figure 7. Dose-response curve for induction of vitellogenin synthesis by juvenile hormone in vivo. JH I dissolved in acetone was applied topically to allatectomized adult female *Locusta migratoria* every 8 hr for a period of 32 hr to give the total dose shown. Fat body was dissected out at 36 hr. Vitellogenin synthesis was measured by following [^{3}H]leucine incorporation into proteins precipitated by antibody to vitellogenin. Reproduced from Wyatt et al. (199) with permission of Academic Press Inc.

peptides. However, in the case of *Locusta*, JH induction of vitellogenin synthesis has apparently been demonstrated in vitro (199). The locust fat body may, therefore, be the best system currently available for the study of JH action and is comparable to the chick oviduct as a general, overall model system for the investigation of the hormonal control of transcription and translation.

Analogous studies conducted on the mosquito *A. aegypti* showed that the fat body is the site of vitellogenin synthesis but that the endocrine control of vitellogenesis differs significantly from that of cockroaches and locusts. In the mosquito, a blood meal is required for oogenesis and within a few hours stimulates both RNA and vitellogenin synthesis in the fat body (200). This stimulation is blocked by actinomycin D administered with or just after the blood meal, but the inhibitor is ineffective when administered to stimulated fat body in vitro. Stimulation of vitellogenin synthesis is also prevented by elimination of brain neurosecretory cells or by ovariectomy (201). Further studies have elegantly demonstrated that the stimulus for vitellogenin synthesis is ecdysone derived from the ovary (202) and that physiological concentrations of ecdysone can elicit vitellogenin synthesis in unfed females (203). Indeed, ecdysone induces vitellogenin synthesis in fat body in vitro, suggesting that the mosquito fat body may be the tissue of choice for studies on ecdysone action. Extrapolation would also suggest

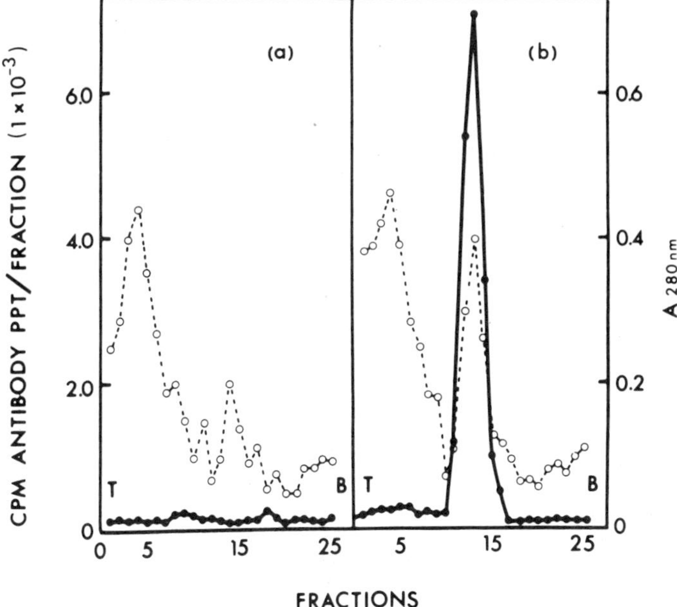

Figure 8. Induced synthesis of vitellogenin in allatectomized female *Locusta migratoria* by juvenile hormone. Hemolymph labeled with [^{3}H]leucine was run on linear sucrose gradients, and fractions were assayed for [^{3}H]vitellogenin by precipitation with antiserum. —•—, radioactivity; ---o---, A280 nm. *Left*, control; *right*, JH treated. Reproduced from Chen et al. (188) with permission of Plenum Press.

that JH and ecdysone act at the transcriptional level in a similar manner, as do gonadotropic agents in the locust and mosquito, respectively.

If the mechanisms of hormone action in insects are to be unraveled, the use of mutants and other genetic techniques might be of significant help. JH seems to control vitellogenesis in *Drosophila* as well, and some female sterile mutants that do not complete oogenesis are induced to do so by application of JH. The appropriate genetic and biochemical analyses should resolve the deficiency at the level of JH biosynthesis, release, or action. Of interest in this regard is a recent report showing that pupal or immature adult ovaries, but not larval ovaries, when transplanted into adult males induce vitellogenin synthesis in the male fat body (204). Whether these ovaries synthesize ecdysone, as in the case of the mosquito, is not yet known; it is of interest that the genes for vitellogenin synthesis are normally repressed in male adult *Drosophila* but are activated as a result of an ovarian transplant.

Thus, two major situations exist as far as vitellogenin synthesis is concerned. Ecdysone, which acts as a molting hormone at preadult stages, acts as a gonadotropic hormone in the female adult mosquito after the insect has passed through an ecdysone-free stage. Juvenile hormone, which acts to maintain the larval state, plays a gonadotropic role in the adults of most species after the insect has passed through a JH-free period. It should be emphasized that in the case of the mosquito, JH is required in addition to ecdysone, albeit at an earlier stage. After adult ecdysis, JH is released and imparts competency to the ovary, so that it can synthesize ecdysone when challenged by a tropic neurosecretory hormone, and to the fat body, so that it can subsequently synthesize vitellogenin in response to ecdysone (203, 205). The observation that a number of insects exhibit high ecdysone titers during adult life is suggestive of a role for ecdysone in oogenesis (or embryogenesis) in those insects in which the primary stimulus now seems to be JH.

Another important aspect of ovarian development in which JH may be involved is the uptake of vitellogenin by the oocytes. Although almost no biochemical data are available on this subject, physiological evidence points to a role for JH in this process (206–209). It seems that uptake involves binding of the vitellogenin to specific regions of the oocyte membrane, i.e., coated pits, and that the protein is then taken into the oocyte by an energy-dependent invagination process (210). This is a process specific for endogenous vitellogenin; vitellogenins of other insect species are rejected (189). How JH may be involved in this process is conjectural, although it has been proposed that JH acts on the follicle cells to allow large macromolecules to move between them and thus be in contact with the oocyte (211).

Juvenile Hormone—Possible Modulator of Translation? Few studies have been conducted at the biochemical level on the effect of JH as a morphogenetic regulator, and most of these have been reviewed recently (6, 9). Other investigations are covered under "Mechanisms of Action of Insect Hormones," but a few comments should be made here on experiments suggesting that JH is a modulator of translation. When actinomycin D was injected into pupae (day 1) of the beetle *Tenebrio molitor*, it did not seem to inhibit the synthesis of adult cuticular proteins as measured during late pharate adult development (day 7) (212), leading to the conclusion that stable mRNA was present and that control could be at the translational level. A series of studies was conducted in which an in vitro protein synthesizing system was used. The major results and conclusions are as follows: 1) the mRNAs for adult cuticular proteins are present on day 1 but are not normally translated until day 7; 2) the synthesis of adult cuticular proteins follows the appearance of new tRNAs and tRNA synthetases and, in particular, a specific leucyl tRNA and its synthetase; 3) a compound which mimics juvenile hormone (dodecyl methyl ether), applied to day 1 pupae, prevented both adult development and the appearance of the specific tRNA and its synthetase; 4) ribosomes from either day 1 or day 7 animals had the same potential for protein synthesis. These data as well as other studies were compatible with the concept of translational control of development by JH in *Tenebrio* (213).

This was consistent with observations revealing the appearance of specific isoacceptor tRNAs in a variety of developmental systems. In some systems, it has been possible to demonstrate a functional adaptation of the tRNAs; i.e., certain isoacceptor tRNAs appear so as to permit the optimum synthesis of specific proteins (214). However, it has not been possible to demonstrate that the appearance of these tRNAs is a causative factor in development rather than a result of the developmental process. Another objection is that actinomycin D is not solely an inhibitor of RNA synthesis; its in vivo effects are very difficult to interpret. An additional problem was that adult cuticular proteins were identified on the basis of a postulated greater concentration of tyrosine than was present in the pupal cuticular proteins. Therefore, in the cell-free system used, the ratio of tyrosine to leucine incorporated into acid-precipitable material was measured. A high ratio was taken as a measure of the synthesis of adult cuticular protein. However, subsequent analysis of cuticular proteins from larval, pupal, and adult *Tenebrio* not only demonstrated a low tyrosine content (3–5% in contrast to the assumed 20%) but also showed no significant difference in the tyrosine content of pupal and adult proteins (215). Despite the claim that the products of the cell-free system differ from pupa to adult, it is still difficult to interpret the data. Finally, the leucyl tRNAs and synthetases of *Tenebrio* have recently been analyzed in detail, and no changes occurred during pharate adult development in either the acylated leucyl tRNAs or isoacceptor leucyl tRNA (216–218). It was concluded that the earlier observations demonstrating changes in the leucyl tRNAs resulted from the employment of suboptimum charging conditions (216). In summary, there is as yet no conclusive evidence that translational control is a critical factor during pupal-adult development, although the possibility remains.

Ecdysone and Juvenile Hormone Interaction

It is the interaction of ecdysone and JH during development that provokes the greatest amount of discussion, speculation, and controversy among insect endocrinologists. This is a consequence of the lack of an experimental system with well-defined products of transcription and translation. The major difficulty is that in most morphogenetic systems JH does not elicit a positive effect. That is, according to the classic picture of the endocrine control of insect development, JH could be considered an agent that prevents ecdysone-stimulated adult development. Such negative effects are also obtained with a variety of biochemical inhibitors affecting metabolic reactions different from those affected by JH. The difficulty in deducing the mechanisms of action of JH is compounded by the use of unphysiological concentrations of JH in most of the past studies (219–221). Although high doses may be defended in in vivo experiments because of the presence of hemolymph binding proteins and other homeostatic mechanisms, their use under in vitro conditions in which these mechanisms may be lacking is unwarranted, and the responses observed are probably more pharmacological than physiological. On the basis of the structure of JH and several analogues, these substances are candidates for membrane destabilization when used in high concentrations. The inhibition of electron transport in mitochondria (222) and the increased activity of Na^+/K^+-dependent ATPase (220), a possible compensatory mechanism for increased membrane permeability after treatment with high doses of JH, are observations consistent with this view. The swelling and lesions in the plasmalemma (223), as well as cell death in insect cell lines (224) treated with JH, provide further evidence. It is, therefore, mandatory that past studies on the in vitro effects of JH on morphogenesis and macromolecular synthesis in which high concentrations of JH were used must be repeated under physiological conditions. One very promising system for the study of ecdysone-JH interactions seems to be epidermal cuticle secretion.

Epidermis The interaction of JH and ecdysone has been studied extensively at the level of the epidermis which undergoes repeated molts when challenged by ecdysone, whereas the nature of the secreted cuticle has been determined by the titer of JH. The evidence for the modulating effect of JH was derived from two types of simple perturbation experiments. First, application of exogenous JH to early last instar larvae resulted in a supernumerary larval-larval molt rather than a metamorphic molt and application of JH to pupae resulted in a pupal-pupal rather than pupal-adult molt (1, 5). Often, developmental progress is only partly prevented and intermediate forms are noted. Secondly, allatectomy (removal of the corpora allata) prior to the last instar resulted in precocious metamorphosis to pupae and adults of minute size. Two chromene derivatives (precocenes) that act as anti-JH produced the same effect as allatectomy in certain insects (225). Thus, the presence of JH results in the repetition of the previous type of molt and, therefore, prevents differentiation to an advanced state. The terminology used in interpreting these experiments has somewhat obscured the issue in that the effect of JH was termed antagonistic, synergistic, permissive, etc. (220, 226, 227). However, it should be emphasized that these are descriptive terms and valid only in the framework of a specific experimental premise. Their usage to describe studies on effects of JH at different levels has led to conceptual confusion.

A system that may clarify the interrelationship of the two hormones at the biochemical level has recently been described. Larval-pupal transformation in *M. sexta* seems to be controlled by two bursts of ecdysone release (Figure 9); after the first burst of ecdysone in the absence of JH, the epidermis is committed (in terms of classic developmental biology) to secrete pupal cuticle (228, 229). This first peak is not sufficient to stimulate the actual expression of this commitment. Synthesis and secretion of a new cuticle require a second and much greater surge of ecdysone (230).

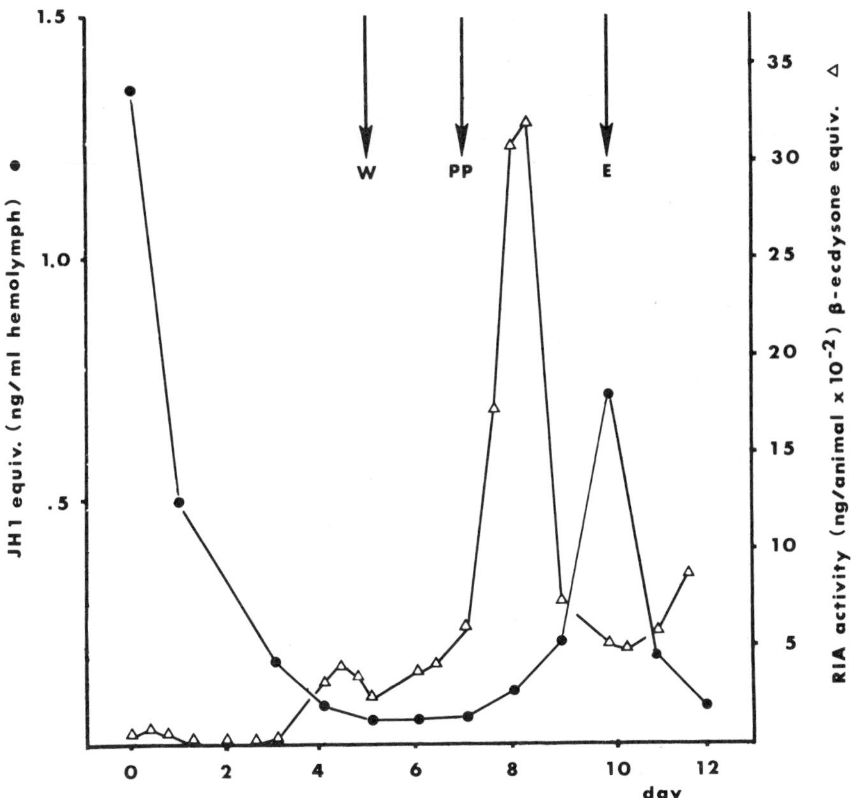

Figure 9. Ecdysone and juvenile hormone titers during fifth (last) larval instar of *Manduca sexta*. Ecdysone values from ref. 19. JH values determined by *Galleria* bioassay (Dr. K. Judy, personal communication). W, PP, and E are defined in legend to Figure 3. RIA, radioimmunoassay.

To elicit a larval molt, experimentally, a critical titer of JH must be present at the time of the first ecdysone peak. The control of commitment and phenotypic expression by two temporally separated phases of hormone release is an interesting phenomenon, but thus far there is no explanation in biochemical terms.

The development of an in vitro culture method for *Manduca* epidermis which yields the same responses as obtained in vivo is a promising first step in exploiting the potential of this system. Dorsal abdominal epidermis extirpated from last instar larvae before the first ecdysone peak produced a pupal cuticle when incubated in a culture medium containing physiological concentrations of β-ecdysone. If there was concomitant exposure to JH I ($\sim 10^{-8}$ M), a new larval cuticle was secreted. However, JH was effective only if applied within 6 hr of exposure of the epidermis to ecdysone and was less effective if the delay between the addition of the two hormones was greater. Actinomycin D or cordycepin similarly prevented the change in commitment (i.e., the synthesis of larval rather than of pupal cuticle), whereas inhibitors of protein synthesis were ineffective (229). Thus, RNA synthesis seems to be a biochemical requisite for the change in commitment.

Further studies of the effects of hormones on macromolecular synthesis in this in vitro epidermal system showed that the incorporation of [^{3}H]uridine into RNA was stimulated to the same degree by β-ecdysone or JH during the first 24 hr of incubation. Simultaneous hormone treatment yielded an enhancement greater than the sum of the individual activities. However, the extent to which changes in the specific activity of precursor pools contributed to the differences in incorporation of label remains to be determined. A critical question in developmental biology is whether this change in commitment requires DNA replication. That is, can the genetic program of a cell be altered without concomitant replication of the genetic material? Most studies in developmental biology suggest a requirement for DNA replication for reprogramming, although the insect has provided some potential exceptions (1). In the case of *M. sexta* epidermis, it has been reported that the change in commitment occurs in the absence of DNA replication (228), but recent studies in the authors' laboratory have demonstrated that a significant percentage of the larval epidermal cells do undergo DNA replication at the time of the first ecdysone surge (J. Wielgus, unpublished observations). The latter observation would agree with the concept that DNA replication is characteristic of the reprogramming process.

The appealing features of the epidermal system are the apparent two-phase mechanism for the induction of cuticle synthesis and deposition and the susceptibility to modification of cuticular character by hormone interaction. The data so far available do not provide insights into the underlying molecular mechanisms, but some discussion of general regulatory processes is appropriate. The above observations fit two basic models. The first assumes a sequence of gene induction in which the products of the early genes (those activated by the first ecdysone peak) have a mediating function in the expression of late genes (those activated by the second ecdysone peak). If JH affects only the primary response to ecdysone, which in itself is not sufficient for cuticle production, the products of the early genes may have regulatory functions. Thus, the RNA molecules synthesized after the small ecdysone peak may support the synthesis of proteins necessary for the amplification of RNA and protein synthesis that occurs during the later stages of steroid hormone action. The second model would invoke post-transcriptional control and require that all genes necessary for molting become derepressed during the first hormonal stimulation. Again, the interaction of ecdysone and JH in some way would determine selectivity. However, translation is delayed by an additional regulatory process which is controlled by the second release of ecdysone. This in turn either eliminates a translational block or stabilizes newly synthesized RNA that was rapidly degraded during the first phase.

It would be important to analyze by hybridization techniques the populations of RNA synthesized during the first and second phases and under different hormonal regimens. Furthermore, the investigation and characterization of induced proteins and their mRNAs are essential for a real understanding of the above observations. Because there seem to be no dramatic differences between larval and pupal cuticle proteins (see under "Juvenile Hormone—Possible Modulator of Translation?"), an analysis of possible differences in the enzymes involved in tanning may be more rewarding.

Effects on Isolated Nuclei The induction of DOPA decarboxylase by ecdysone in blowfly epidermis has been discussed above. Because incubation of nuclei isolated from blowfly larval epidermis with α-ecdysone stimulated the incorporation of labeled precursors into RNA, this was taken as evidence that the nucleus was the primary site of hormone action. These studies were extended by including JH, either alone or in combination with ecdysone; fat body was used, probably because it provided more tissue than epidermis. Both JH and ecdysone seemed to enhance the RNA polymerase activity of isolated nuclei; the extent of stimulation depends on the developmental state of the fat body (231). Hybridization experiments revealed a qualitative difference between the RNA from hormone-stimulated and control nuclei, whereas simultaneous incubation with both hormones eliminated the stimulation of RNA synthesis seen after treatment with either hormone (237).

The data seem to indicate a direct action of the hormones on the nucleus, but the experiments suffer from technical drawbacks (see under "Receptors") typical of experimental regimens used a decade ago. The procedure employed at the time to isolate nuclei clearly causes losses of nuclear RNA polymerases by leakage. The chemical composition of these nuclei is different from that of nuclei isolated recently from other insects and other organisms. The measurements of RNA polymerase activity were carried out before the optimum divalent cation and ionic requirements for the different classes of RNA polymerases were known. Hence, their assay conditions may have been satisfactory for polymerase I, but they were clearly suboptimum for polymerase II and probably for polymerase III. Finally, the procedures employed for hybridization studies were capable of detecting only repetitive sequences. A reassessment of these data and confirmation of the experiments with refined techniques are, therefore, mandatory.

MECHANISMS OF ACTION OF INSECT HORMONES

The most striking feature of the action of growth hormones on their target tissues is that the resultant biochemical changes are remarkably similar, irrespective of the hormone or target tissue. These changes are generally expressed as alterations in either the rate or the extent of DNA, RNA, and protein synthesis. Consequently, in order to understand the ultimate physiological effects of these hormones, the mechanisms by which growth hormones regulate macromolecular syntheses must first be explored. Studies during the past 20 years indicate that the primary cellular response to steroid hormones is an increased rate of RNA synthesis; this may be characteristic of nonsteroid hormones as well.

Action at Nuclear Level

Even prior to the demonstration by Karlson and his colleagues that ecdysone was in fact a steroid, active extracts became available, and Wigglesworth (233) concluded on the basis of histological analysis that the early responses of the epidermis to the active principle included enlargement of the nucleolus and an increased RNA content of the cytoplasm. At about the same time, in the late 1950s, Beermann (234) described profound alterations in the activity of *Chironomus* Balbiani rings that could be correlated with the onset of larval-pupal metamorphosis. On the basis of these observations, it was suggested by Schneiderman and Gilbert (235) about 20 years ago that "perhaps these chromosome changes are, in fact, one of the more immediate actions of ecdysone, changes that reflect the elaboration of specific substances (ribonucleic acid, nucleotide conenzymes?) by the nucleus, substances that are destined to participate personally in the cytoplasmic syntheses that characterize molting. The suggestion that a hormone may act at the nuclear level is admittedly unorthodox. But recent experiments have provided such convincing proof that the nucleus participates actively, not only in reproduction but in the everyday affairs of cells, that the nucleus looks like a strategic place for a hormone to act." This speculative and unorthodox comment of only two decades ago has of course been corroborated by a plethora of studies on both insects and vertebrates. However, it was the observation by Clever and Karlson that injection of ecdysone into *Chironomus* larvae resulted in a specific and rapidly appearing puffing pattern of the salivary gland polytene chromosomes that led Karlson to postulate in the early 1960s that ecdysone acts directly upon the genetic material (236). The steroidal nature of ecdysone was subsequently elucidated, and Karlson then championed the concept that steroid hormones in general may act upon genes, resulting in the stimulation of specific RNA syntheses (237); this was well before the notion of specific cellular receptors for steroid hormones was accepted and the elegant studies on the rat uterus and chick oviduct were conducted. Thus, although the philosophical impetus for studies on gene activation by hormones was provided by studies on dipteran polytene chromosomes, the actual experiments demonstrating the validity of the concept were conducted primarily on vertebrate systems using estrogen and progesterone.

Receptors

In the presently accepted model for steroid hormone action, the steroid binds to a cytosol receptor in the target tissue and is then translocated to the nucleus where the receptor-hormone complex binds to the chromatin (238). This interaction then leads to the stimulation of RNA synthesis. Much of the recent work concerns the isolation and characterization of cytoplasmic receptor molecules and descriptions in molecular terms of the interaction of the steroid hormone-receptor complex with chromatin. In fact, Karlson's original hypothesis envisaged a direct action of ecdysone on the nucleus (Figure 10), and, although a cytoplasmic receptor may not be an obligatory component of hormone action, he has modified his proposal to include a cytoplasmic receptor for ecdysone (239) (Figure 10). The existence of such a receptor is warranted on the basis of two observations. First, nuclei derived from salivary gland cells do not exhibit specific puffing when incubated in vitro with ecdysone (240). Secondly, the microinjection of ecdysone into the cytoplasm of salivary gland cells elicits a normal puffing response but fails to do so if injected directly into the nuclei (241). Consequently, cytoplasmic binding proteins have been investigated in various tissues. When salivary glands from *D. hydei* were pulsed in vitro with [^{3}H]α-ecdysone, both autoradiographic and cell fractionation studies suggested some specificity of localization in the nuclei (242, 243). However, the detection of proteins that specifically bind either α- or β-ecdysone remained elusive, in part because of the unavailability of ecdysone of high specific activity (244, 245). The ambiguity of the biological significance of the nuclear retention is emphasized by the recent studies with α-ecdysone of high specific activity ($\sim$ 60 Ci/mmol), in which the salivary glands were shown to metabolize ecdysone and the metabolites were shown to accumulate within the nuclei (246).

Imaginal discs were isolated en masse from *D. melanogaster* and examined for the presence of ecdysone receptors. The in vitro uptake of β-ecdysone into whole discs was first determined (45), and it revealed that maximum retention is linearly related to the applied hormone concentration over a range of 2×10^{-8}–2×10^{-4} M, indicating a nonsaturable component. There was no detectable metabolism. Competition experiments with cold β-ecdysone revealed the presence of two types of binding although the relatively low specific activity of the β-ecdysone used (6 Ci/mmol) imposed limitations on the detailed characterization of high affinity-low capacity binding sites. The first type was the high affinity-low capacity component that became saturated at a concentration of 7–8×10^{-8} M

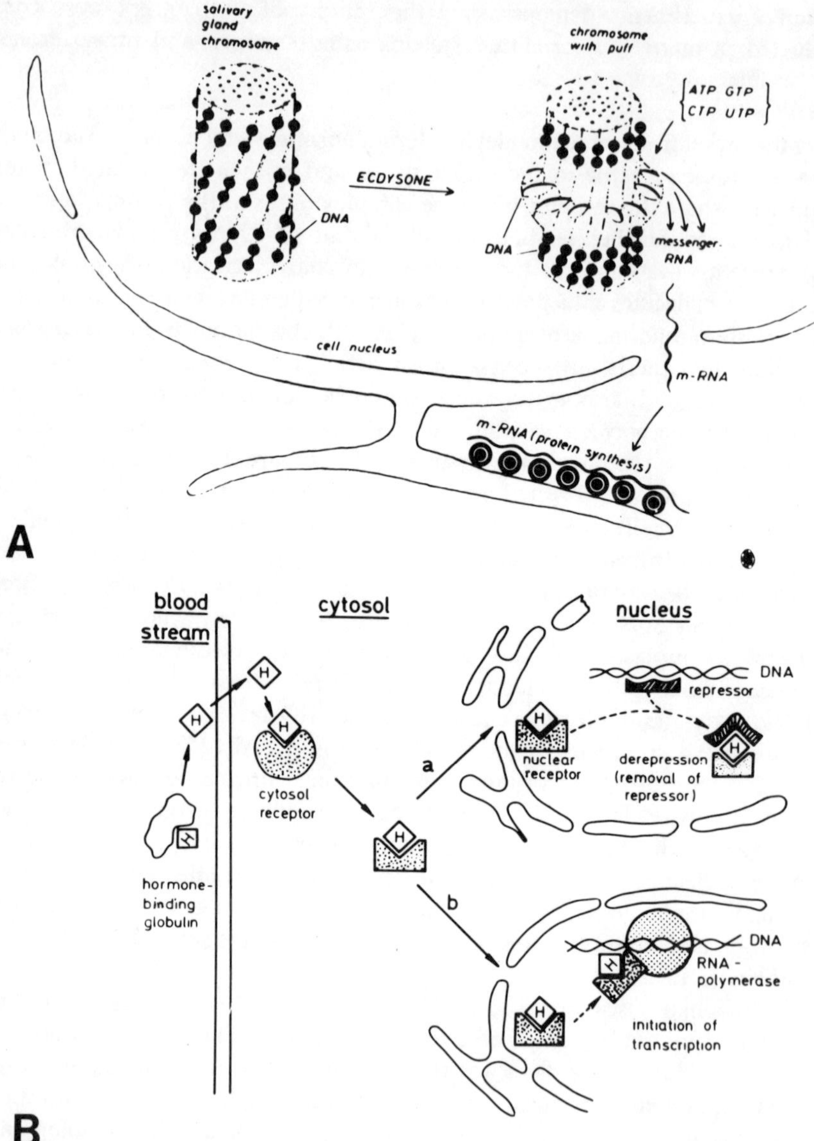

Figure 10. *A*, original hypothesis to explain the mechanism of action of ecdysone. *B*, an alternative model for the regulation of transcription by steroid hormones, including ecdysone. When applied to ecdysone and the insect system, "blood stream" would be replaced by "hemolymph." Reproduced from Karlson (237) with permission of The University of Chicago Press.

ecdysone. This value may be slightly low because the imaginal discs contained endogenous hormone. The second component was of the low affinity-high capacity type and is normally considered nonspecific. That the first component is specific for ecdysone was established by competition with hormone analogues (ponasterone A and α-ecdysone); the data showed a general correlation between concentrations necessary to induce a biological response (i.e., evagination) and the ability to compete. Furthermore, exchange with cold β-ecdysone was strongly decreased at low temperatures, and treatment with *N*-ethylmaleimide eliminated the specific binding. (It should be noted here that unlike the situation in vertebrates, almost all insect tissues are affected by ecdysone during development, and the entire insect could be considered the target for β-ecdysone. Therefore, tissue specificity cannot normally be considered a valid criterion for receptor characterization. On the other hand, specific tissues may become competent to respond to the insect steroid hormone at a specific developmental stage, and, in that sense, tissue specificity may occur at a particular point in time. Indeed, competency may be a function of the ontogeny of the receptor protein.)

Do the high affinity binding sites described above for the imaginal disc represent the ecdysone receptor? The fact that the equilibrium dissociation constant ($K_d = 2 \times 10^{-7}$ M; 25 °C), calculated from the association and dissociation rate constants, is in good agreement with both the in vivo hormone titer at the onset of metamorphosis (2.5×10^{-7} M) and the concentration required for complete evagination in vitro (2×10^{-7} M), is certainly suggestive of a true receptor. The procedure for determining the equilibrium dissociation constant is based on the assumption that the equilibrium is sufficiently described by the two rate constants; this should be confirmed by direct measurement. It will be necessary to determine the subcellular location of the high affinity sites as well as to obtain kinetic data with this subcellular fraction that, ideally, would correlate with the kinetic data derived from whole cells. Even this, however, would not be direct evidence for the existence of a β-ecdysone receptor; perhaps cell lines with varying sensitivity to ecydsone would represent a better model system. For the *Drosophila* Kc cell line, several clones have been isolated representing a continuum of sensitivity (or resistance) to β-ecdysone (50). Preliminary experiments indicated the presence of a class of saturable binding sites in hormone-sensitive clones which was absent in a resistant clone (247). Although these observations require further supporting data, *Drosophila* cell lines, especially when analyzed with the techniques of somatic cell genetics, have great potential for elucidation of the mechanism of ecdysone action.

Hormone-Receptor Action

There is then no definitive evidence for an ecdysone receptor; an even more basic question relates to the controversy involving the direct versus indirect effects of ecdysone (and possibly of JH) on the puffing activity of polytene chromosomes. The direct hypothesis is of course based on analogous observations on vertebrate systems in which the steroid hormone or hormone-receptor complex binds directly to chromosome components. The indirect hypothesis, in contrast, states that the observed puffing is a consequence of changes in the ionic environment of the chromosomes which in turn are brought about by the interaction of ecdysone with the nuclear envelope (248). In summary, the action of ecdysone would be to increase the K^+ to Na^+ ratio, whereas that of JH would be to reduce it. The ionic environment of chromatin no doubt is of critical importance as far as transcription is concerned, particularly when one considers the optimum ion and salt requirements for the activities of the different RNA polymerases, as well as for DNA-protein interactions; there is cogent evidence that some ecdysone puffs could be elicited by artificially manipulating the K^+ to Na^+ ratio. The entire question has recently been analyzed in depth, and the conclusion has been reached that the indirect hypothesis is inadequate to explain puff induction by ecdysone (249). Therefore, at least for the present, the hypothesis that β-ecdysone acts directly on the genome will be accepted, with the reservation, however, that this important growth hormone may have multivalent effects at several levels of cell organization, including effects on permeability. If the role of the hormone-receptor complex is accepted, then what are the nature and consequence of the interaction between the receptor complex and chromatin? Studies with vertebrate steroid hormones indicate that interaction occurs with certain nonhistone (acceptor) proteins of chromatin. Such an interaction has been variously claimed to lead to the activation of RNA polymerases and/or to structural changes in chromatin with a concomitant increase in template capacity, facilitation of elongation of RNA chains, and formation of initiation sites for RNA polymerases (238). These diverse observations in part result from the difficulty encountered in distinguishing between primary and secondary responses. As noted under "Polytene Chromosomes and Puffing Phenomenon," ecdysone or heat-shock elicits specific puffing in polytene chromosomes within 5–10 min of the treatment, independent of protein synthesis. In fact, within 3 min of stimulation, an increase in dye binding at specific regions is noted, the dye being specific for nonhistone proteins (250). This cytologically determined increase in dye binding results from an actual increase in protein mass rather than from just a change in chromosome conformation (251). Further experiments showed that concurrent with the activation of certain puffs (gene activation) certain nonhistone proteins pre-existing in the cytoplasm accumulated within the nucleus (252). This would be in keeping with the concept of the involvement of nonhistone protein in gene regulation (238). However, no studies have been reported on the relationship between ecdysone receptor in the cytoplasm and the nonhistone protein that moves to the nucleus nor on the origin and fate of this protein. Assuming that this specific accumulation of protein in the chromatin causes increased transcription, how might it occur? First, the nonhistone protein may interact at certain regions of the chromatin and elicit new initiation sites for RNA polymerase. Alternatively, it could eliminate or destabilize protein that blocks the movement of RNA polymerase molecules, thereby opening up new genes for transcription. The first possibility is in keeping with the concept developed for the mechanism of action of female sex hormones in the chick oviduct (251). The second model is consistent with the data resulting from in situ RNA polymerase activity measurements (115), localization of endogenous RNA polymerase molecules throughout the chromosomes (116), and the observed changes in template capacity of chromatin after hormone stimulation in other systems (238). It is difficult at present to distinguish between the two possibilities.

At first glance, the polytene chromosome seems to be an ideal system for the evaluation of ecdysone action, but experiments are hampered by the lack of an identifiable ecdysone-induced protein and also by the difficulties of obtaining sufficient quantities of salivary glands for biochemical analysis. The induction of epidermal DOPA decarboxylase (see under "Cuticular Tanning") may be an even better system for analysis than the polytene chromosome, but induction of DOPA decarboxylase by ecdysone has not been demonstrated in vitro, whereas in vivo induction of the enzyme in wing discs requires 2 days.

The preliminary evidence (see under "Receptors") and the precedent established by studies with vertebrate steroid hormones indicate the presence of ecdysone receptors and suggest that the ecdysone-receptor complex may directly stimulate transcription. The problem of receptor involvement in JH action is more difficult to analyze. In part, this is because of the lipoidal nature of JH which results in its adherence to a variety of surfaces (including glass) with relatively high affinity and its tendency to partition out of the aqueous environment into hydrophobic compartments. Furthermore, tissues such as imaginal discs have the capacity to degrade JH at a very rapid rate. Despite these problems, there have been two reports of putative JH receptor proteins in the epidermis (254) and ovary (255) of *Tenebrio*. These receptors were detected by incubating tissues with labeled JH analogues, homogenization in a buffer containing Triton X-100, and subsequent centrifugation, gel filtration, or thin layer isoelectric focusing. The epidermal receptor was regarded as a ribonucleoprotein. The alleged receptor in the ovary showed the rather unusual properties of being heat-stable and resistant to proteolytic and nucleolytic enzymes. Re-examination of these claims revealed that the receptors were actually hormone-Triton X micelles with surprising stability during fractionation procedures (256). The recent development of accurate analytical chemical methods and binding protein assays for JH identification and titer determination can be expected to aid future studies on JH receptors, but the lack of JH of high specific activity may be too difficult a handicap to overcome.

Model Building

The scarcity of specific information on the ultimate mode of action of both hormones makes the extrapolation of results from other systems to the interaction of JH and ecdysone difficult. Some of the problems involved have already been discussed. In the case of insects in which such clear-cut differences between developmental stages are seen in response to ecdysone and JH, one is compelled to assume that the genome is divided into functionally different gene sets specifying the various stages of development. Such an assumption is supported by the study of *Drosophila* mutants, in which defects in disc development result in adult lethality, but have no effect on larval development (257).

As mentioned previously, Karlson proposed a model analogous to the Jacob-Monod model for prokaryotes to explain the action of ecdysone. Since that time, mechanisms proposed to explain the control of gene expression by hormones in vertebrates have been extensions of these models and could probably be adapted to explain the action of ecdysone and JH. Indeed, it was proposed in the case of insect development that a master regulatory gene existed for each gene set and that each of these regulatory gene sets was comprised of its own operator and promotor regions and the corresponding repressors. Which of the gene sets was active at a particular time would be determined by a shift in the affinity of the RNA polymerase for specific promotors, the affinity in turn being determined by σ-like factors. It would follow then that a specific σ factor would exist for each stage of development. JH would act as a negative controlling agent by binding to certain repressors and acting as a co-repressor (258). Unfortunately, it is not known whether such clear-cut gene sets which can be turned on or off in succession actually exist at different developmental stages. More importantly, the original belief that the control of gene expression in higher organisms uses the same mechanisms as occur in bacteria (in which changes in the specificity of a core RNA polymerase have been invoked) has not been sustained by recent work. Suffice it to say that the vast amount of information collected over the past 5 years on eukaryotic RNA polymerases and putative controlling factors indicates that specificity of transcriptional regulation does not occur primarily at the level of RNA polymerase (259).

If gene sets exist in the context discussed above and if ecdysone is the inducing agent or factor essential for the expression of gene sets at a particular developmental stage, then JH can be thought of as differentially suppressing the gene sets that will be activated later in development and perhaps maintaining the ecdysone-induced gene sets. Additionally, ecdysone would also be responsible for turning off the gene sets that were expressed at the previous developmental stage. Although regulatory proteins may have both positive and negative effects, there is no basis at present for suggesting a model by which ecdysone and JH would interact with regulatory proteins to modulate gene expression during insect metamorphosis.

Thus, different models must be formulated which will include different gene sets, circuitry of integrators, activators, inhibitors, receptors, etc. These models will remain conjectural until it is demonstrated in biochemical terms that epidermal cells synthesize stage-specific proteins or mRNA populations that require the activation of different gene sets, ecdysone and JH receptors are characterized, and chromatin acceptor sites are identified. Only then can the morphogenetic action of JH be explained at the biochemical level.

REFERENCES

1. Gilbert, L. I., and King, D. S. (1973). *In* M. Rockstein (ed.), The Physiology of Insecta, Ed. 2, Vol. I, pp. 249–370. Academic Press, New York.
2. Gersch, M. (1976). Actualites Sur Les Hormones D'Invertébrés, pp. 281–305. Colloques Internationaux C.N.R.S. No. 251.
3. Gilbert, L. I., and Chino, H. (1974). J. Lipid Res. 15:439.
4. Tager, H. S., Markese, J., Kramer, K. J., Speirs, R. D., and Childs, C. N. (1976). Biochem. J. 156:515.
5. Doane, W. (1972). *In* S. J. Counce and C. H. Waddington (eds.), Developmental Systems, Vol. II, pp. 291–466. Academic Press, New York.
6. Wyatt, G. R. (1972). *In* G. Litwack (ed.), Biochemical Actions of Hormones, Vol. II, pp. 385–490. Academic Press, New York.
7. Gilbert, L. I., Goodman, W., and Bollenbacher, W. E. (1977). *In* T. W. Goodwin (ed.), MTP International Review of Science, Biochemistry Series Two, (England: Medical and Technical Publishing Co., Ltd.). Vol. 14, pp. 1–50.
8. Gilbert, L. I., Goodman, W., and Nowock, J. (1976). Actualites Sur Les Hormones D'Invertebres, pp. 413–434. Colloques Internationaux C.N.R.S. No. 251.
9. Gilbert, L. I. (1974). Recent Progr. Horm. Res. 30:347.
10. Goodman, W., Bollenbacher, W., Zvenko, H., and Gilbert, L. I. (1976). *In* L. I. Gilbert (ed), The Juvenile Hormones, pp. 75–95. Plenum Press, New York.
11. Tobe, S. S., and Pratt, G. E. (1976). *In* L. I. Gilbert (ed.), The Juvenile Hormones, pp. 147–163. Plenum Press, New York.
12. Kramer, K. J., Dunn, P. E., Peterson, R. C., and Law, J. H. (1976). *In* L. I. Gilbert (ed.), The Juvenile Hormones, pp. 327–341. Plenum Press, New York.
13. Stone, J. V., Mordue, W., Batley, K. E., and Morris, H. R. (1976). Nature (Lond.) 263:207.
14. Kubota, I., Isobe, M., Goto, T., and Hasegawa, K. (1976). Z. Naturforsch. 31c:132.
15. Isobe, M., Hasegawa, K., Kubota, I., and Goto, T. (1976). Agr. Biol. Chem. 40(6):1189.

16. Chino, H., Sakurai, S., Ohtaki, T., Ikekawa, N., Miyazaki, H., Ishibashi, M., and Abuki, H. (1974). Science 183:529.
17. King, D. S., Bollenbacher, W. E., Borst, D. W., Vedeckis, W. V., O'Connor, J. D., Ittycheriah, P. I., and Gilbert, L. I. (1974). Proc. Natl. Acad. Sci. USA 71:793.
18. Bollenbacher, W. E., Goodman, W., Vedeckis, W. V., and Gilbert, L. I. (1976). Steroids 27:309.
19. Bollenbacher, W. E., Vedeckis, W. V., Gilbert, L. I., and O'Connor, J. D. (1975). Dev. Biol. 44:46.
20. Vedeckis, W. V., and Gilbert, L. I. (1973). J. Insect Physiol. 19:2445.
21. Vedeckis, W. V., Bollenbacher, W. E., and Gilbert, L. I. (1974). Zool. Jb. Physiol. 78:440.
22. Vedeckis, W. V., Bollenbacher, W. E., and Gilbert, L. I. (1976). Mol. Cell. Endocrinol. 5:81.
23. Bodnaryk, R. P. (1975). Life Sci. 16:1411.
24. Morishima, I. (1975). Biochim. Biophys. Acta 403:106.
25. Bielinska, M., and Piechowska, M. J. (1975). Bull. Acad. Pol. Sci. 23:1.
26. Castillon, M. P., Catalan, R. E., and Municio, A. M. (1973). FEBS Lett. 32:113.
27. Sharma, R. K. (1973). J. Biol. Chem. 248:5473.
28. Mahaffee, D., Reitz, R. C., and Ney, R. L. (1974). J. Biol. Chem. 249:227.
29. Krishnakumaran, A., Berry, S. J., Oberlander, H., and Schneiderman, H. A. (1967). J. Insect Physiol. 13:1.
30. Timm, U. (1970). Ph.D. thesis, University of Cologne.
31. Pohley, H.-J. (1961). Wilhelm Roux' Arch. 153:443.
32. Löbbecke, E.-A. (1969). Wilhelm Roux' Arch. 162:1.
33. Pohley, H.-J. (1956). Biol. Zbl. 75:86.
34. Muth, F. W. (1961). Wilhelm Roux' Arch. 153:370.
35. Becker, H. J. (1957). Z. Vererb. Lehre 88:333.
36. Garcia-Bellido, A., and Merriam, J. R. (1971). Dev. Biol. 24:61.
37. Garcia-Bellido, A. (1975). *In* D. McMahon and C. F. Fox (eds.), Developmental Biology. pp. 40–59. W. A. Benjamin, Inc., Menlo Park, California.
38. Nöthiger, R. (1972). *In* H. Ursprung and R, Nöthiger (eds.), The Biology of Imaginal Disks, pp. 1–34. Springer, Berlin.
39. Vijverberg, A. J. (1973). Neth. J. Zool. 23:189.
40. Oberlander, H. (1969). J. Insect Physiol. 15:1803.
41. Oberlander, H. (1972). J. Insect Physiol. 18:223.
42. Chihara, C. J., Petri, W. H., Fristrom, J. W., and King, D. S. (1972). J. Insect Physiol. 18:1115.
43. Ohmori, K., and Ohtaki, T. (1973). J. Insect Physiol. 19:1199.
44. Logan, W. R., Fristrom, D., and Fristrom, J. W. (1975). J. Insect Physiol. 21:1343.
45. Yund, M. A., and Fristrom, J. W. (1975). Dev. Biol. 43:287.
46. Galbraith, M. N., Horn, D. H. S., Thompson, J. A., Neufield, G. J., and Hackney, R. J. (1969). J. Insect. Physiol. 15:1225.
47. Borst, D. W., Bollenbacher, W. E., O'Connor, J. D., King, D. S., and Fristrom, J. W. (1974). Biol. 39:308.
48. Courgeon, A.-M. (1972). Nature (New Biol.) 238:250.
49. Courgeon, A.-M. (1972). Exp. Cell. Res. 74:327.
50. Courgeon, A.-M. (1975). Exp. Cell. Res. 94:283.
51. Vijverberg, A. J., and Ginsel, L. A. (1976). J. Insect, Physiol. 22:181.
52. Schooley, D., Judy, K., Bergot, J., Hall, M. S., and Jennings, R. C. (1976). *In* L. I. Gilbert (ed.), The Juvenile Hormones, pp. 101–117. Plenum Press, New York.

53. Karkas, J. D., Margulies, L., and Chargaff, E. (1975). J. Biol. Chem. 250: 8657.
54. Margulies, L., and Chargaff, E. (1973). Proc. Natl. Acad. Sci. USA 70: 2946.
55. Muhammed, A., Goncalves, J. M., and Trosco, J. E. (1967). Dev. Biol. 15: 23.
56. Ashburner, M. (1970). Adv. Insect Physiol. 7:1.
57. Berendes, H. D. (1973). Int. Rev. Cytol. 35:61.
58. Edström, J. E. (1974). *In* H. Busch (ed.), The Cell Nucleus, Vol. II, pp. 293–332. Academic Press, New York.
59. Henning, W. (1974). *In* H. Busch (ed.), The Cell Nucleus, Vol. II, pp. 333–369. Academic Press, New York.
60. Edström, J. E., and Lambert, B. (1975). Prog. Biophys. Mol. Biol. 30:57.
61. Daneholt, B. (1974). Int. Rev. Cytol. 36:417.
62. Daneholt, B. (1975). Cell. 4:1.
63. Daneholt, B., Case, S., Hyde, J., Nelson, L., and Wieslander, L. (1976). Prog. Nucleic Acids Res. 19:319.
64. Beermann, W. (1952). Chromosoma 5:139.
65. Becker, H. J. (1962). Chromosoma 13:341.
66. Clever, U. (1962). Chromosoma 13:385.
67. Clever, U., and Karlson, P. (1960). Exp. Cell. Res. 20:623.
68. Berendes, H. D. (1967). Chromosoma 22:274.
69. Poels, C. L. M. (1970). Dev. Biol. 23:210.
70. Beermann, W. (1967). *In* A. Brink (ed.), Heritage from Mendel, pp. 179–201. University of Wisconsin Press, Madison.
71. Pelling, C. (1964). Chromosoma 15:71.
72. Clever, U., and Romball, C. G. (1966). Proc. Natl. Acad. Sci. USA 56: 1470.
73. Berendes, H. D. (1968). Chromosoma 24:418.
74. Leibovitch, B. A., Belyaev, E. S., Zhimulev, I. F., and Khesin, R. B. (1976). Chromosoma 54:349.
75. Holt, T. K. H., and Kuijpers, A. M. C. (1972). Chromosoma 37:423.
76. Pelling, C. (1970). Cold Spring Harbor Symp. Quant. Biol. 35:521.
77. Ashburner, M. (1971). Nature (New Biol.) 230:222.
78. Ashburner, M. (1973). Dev. Biol. 35:47.
79. Poels, C. L. M. (1972). Cell Differ. 1:63.
80. Richards, G. P. (1976). Wilhelm Roux' Arch. 179:339.
81. Clever, U. (1964). Science 146:794.
82. Ashburner, M. (1972). Exp. Cell. Res. 71:433.
83. Ashburner, M., Chihara, C., Meltzer, P., and Richards, G. (1973). Cold Spring Harbor Symp. Quant. Biol. 38:655.
84. Clever, U. (1966). Am. Zool. 6:33.
85. Richards, G. P. (1976). Dev. Biol. 48:191.
86. Leenders, H. J., and Berendes, H. D. (1972). Chromosoma 37:433.
87. Baudisch, W., and Panitz, R. (1968). Exp. Cell Res. 49:470.
88. Beermann, W. (1961). Chromosoma 12:1.
89. Grossbach, U. (1969). Chromosoma 28:136.
90. Grossbach, U. (1973). Cold Spring Harbor Symp. Quant. Biol. 38:619.
91. Korge, G. (1975). Proc. Natl. Acad. Sci. USA 72:4550.
92. Leenders, H. J., and Beckers, P. J. (1972). J. Cell Biol. 55:257.
93. Leenders, H. J., Derksen, J., Maas, P. M. J. M., and Berendes, H. D. (1973). Chromosoma 41:447.
94. Tissieres, A., Mitchell, H. K., and Tracy, U. M. (1974). J. Mol. Biol. 84: 389.

95. Lewis, M., Helmsing, P. J., and Ashburner, M. (1975). Proc. Natl. Acad. Sci. USA 72:3604.
96. Koninkx, J. F. J. G. (1976). Biochem. J. 158:623.
97. McKenzie, S. L., Henikoff, S., and Meselson, M. (1975). Proc. Natl. Acad. Sci. USA 72:1117.
98. Weinberg, R. A. (1973). Am. Rev. Biochem. 42:329.
99. Gall, J. G. (1973). *In* B. A. Hamkalo and J. Papaconstantinou (eds.), Molecular Cytogenetics, pp. 59–74. Plenum Press, New York.
100. Grigliatti, T. A., White, B. N., Tener, G. M., Kaufman, T. C., Holden, J. J., and Suzuki, D. T. (1973). Cold Spring Harbor Symp. Quant. Biol. 38:461.
101. Lambert, B., Wieslander, L., Daneholt, B., Egyhazi, E., and Ringborg, W. (1972). J. Cell Biol. 53:407.
102. Beermann, W. (1971). Chromosoma 34:152.
103. Egyhazi, E., D'Monte, B., and Edström, J. E. (1972). J. Cell Biol. 53:523.
104. Edström, J. E., and Tanguay, R. (1974). J. Mol. Biol. 84:569.
105. Spradling, A., Penman, S., and Pardue, M. L. (1975). Cell 4:395.
106. Lambert, B. (1973). Cold Spring Harbor Symp. Quant. Biol. 38:637.
107. Lambert, B. (1973). Nature 242:51.
108. Lambert, B., and Daneholt, B. (1975). *In* D. M. Prescott (ed.), Methods in Cell Biology Vol. X, pp. 17–47. Academic Press, New York.
109. Stevens, B. J., and Swift, H. (1966). J. Cell Biol. 31:55.
110. Daneholt, B. (1972). Nature (New Biol.) 240:229.
111. Daneholt, B., and Hosick, H. (1973). Proc. Natl. Acad. Sci. USA 70:442.
112. Zhimulev, I. F., and Kolesnikov, N. N. (1975). Wilhelm Roux' Arch. 178:15.
113. Clever, U. (1972). *In* M. Rockstein and G. T. Baker, II (eds.), Molecular Genetic Mechanisms in Development and Aging, pp. 33–69. Academic Press, New York.
114. Levy, B. W., and McCarthy, B. J. (1975). Biochemistry 14:2440.
115. Pages, M., and Alonso, C. (1976). Exp. Cell Res. 98:120.
116. Plagens, U., Greenleaf, A., and Bautz, E. K. F. (1976). Chromosoma 59: 157.
117. Karlson, P., and Shaaya, E. (1964). J. Insect Physiol. 10:797.
118. Karlson, P., and Sekeris, C. E. (1966). Rec. Progr. Horm. Res. 22:473.
119. Sekeris, C. E. (1974). *In* W. J. Burdette (ed.), Invertebrate Endocrinology and Hormonal Heterophylly, pp. 55–78. Springer-Verlag, Berlin.
120. Sekeris, C. E., and Karlson, P. (1962). Biochim. Biophys. Acta 62:103.
121. Karlson, P., and Sekeris, C. E. (1962). Biochim. Biophys. Acta 63:489.
122. Sekeris, C. E., and Karlson, P. (1964). Arch. Biochem. Biophys. 105:483.
123. Sekeris, C. (1972). Gen. Comp. Endocrinol. 3,(suppl.):149.
124. Sekeris, C. E., and Lang, N. (1964). Life Sci. 3:625.
125. Shaaya E., and Sekeris, C. E. (1971). FEBS Lett. 16:333.
126. Fragoulis, E. G., and Sekeris, C. E. (1975). Arch. Biochem. Biophys. 168: 15.
127. Fragoulis, E. G., and Sekeris, C. E. (1975). Biochem. J. 146:121.
128. Fragoulis, E. G., and Sekeris, C. E. (1975). Eur. J. Biochem. 51:305.
129. Chen, T. T., and Hodgetts, R. B. (1974). Dev. Biol. 38:271.
130. Schlaeger, D. A., and Fuchs, M. S. (1974). J. Exp. Zool. 187:217.
131. Schlaeger, D. A., Fuchs, M. S., and Kang, S. H. (1974). J. Cell Biol. 61: 454.
132. Schlaeger, D. A., and Fuchs, M. S. (1974). J. Insect Physiol. 20:349.
133. Shaaya, E., and Sekeris, C. (1965). Gen. Comp. Endocrinol. 5:35.
134. McCaman, M. W., McCaman, R. E., and Lees, G. J. (1972). Anal. Biochem. 45:242.

135. Karlson, P., and Schweiger, A. (1961). Hoppe Seylers Z. Physiol. Chem. 323:199.
136. Schweiger, A., and Karlson, P. (1962). Hoppe Seylers Z. Physiol. Chem. 329:210.
137. Andersen, S. O., and Barrett, F. M. (1971). J. Insect Physiol. 17:69.
138. Anderson, S. O. (1974). Nature (Lond.) 251:507.
139. Thomson, J. A. (1975). Adv. Insect Physiol. 11:321.
140. Collins, J. V. (1974). Can. J. Zool. 52:639.
141. Koeppe, J. K., and Gilbert, L. I. (1973). J. Insect Physiol. 19:615.
142. Mitchell, H. K., and Lunan, K. D. (1964). Arch. Biochem. Biophys. 106: 219.
143. Seligman, M., Friedman, S., and Fraenkel, G. (1969). J. Insect Physiol. 15:1085.
144. Levenbook, L., Bodnaryk, R. P., and Spande, T. F. (1969). Biochem. J. 113:837.
145. Bodnaryk, R. P. (1970). J. Insect Physiol. 16:919.
146. Munn, E. A., and Greville, G. D. (1969). J. Insect Physiol. 15:1935.
147. Bodnaryk, R. P., and Levenbook, L. (1969). Comp. Biochem. Physiol. 30: 909.
148. Fader, R. G. (1975). Ph.D. thesis, Purdue University, Lafayette, Indiana.
149. Levenbook, L., personal communication.
150. Mandaron, P. (1974). Wilhelm Roux' Arch. 178:123.
151. Fristrom, D., and Fristrom, J. W. (1975). Dev. Biol. 43:1.
152. Tanreuther, G. W. (1910). Wilhelm Roux' Arch. 29:275.
153. Fristrom, J. W. (1972). *In* H. Ursprung and R. Nöthiger (eds.), The Biology of Imaginal Disks, pp. 109–154. Springer-Verlag, Berlin.
154. Raikow, R., and Fristrom, J. W. (1971). J. Insect. Physiol. 17:1596.
155. Chihara, C. J., and Fristrom, J. W. (1973). Dev. Biol. 35:36.
156. Nishiura, J. T., and Fristrom, J. W. (1975). Proc. Natl. Acad. Sci. USA 72:2984.
157. Fristrom, J. W., Logan, W. R., and Murphy, C. (1973). Dev. Biol. 33: 441.
158. Bonner, J. J., and Pardue, M. L. (1976). Chromosoma 58:87.
159. Benecke, B. J., Ferencz, A., and Seifart, K. H. (1973). FEBS Lett. 31:53.
160. Courvalin, J.-C., Bouton, M. M., Baulieu, E.-E., Nuret, P., and Chambon, P. (1976). J. Biol. Chem. 251:4843.
161. Natori, S., Worton, R., Boshes, R. A., and Ristow, H. (1973). Insect Biochem. 3:91.
162. Phillips, J. P., and Forrest, H. S. (1973). J. Biol. Chem. 248:265.
163. Adoutte, A., Clement, J.-M., and Hirshbein, L. (1974). Biochimie 56:335.
164. Gross, R. H., and Beer, M. (1975). Biochemistry 14:4024.
165. Greenleaf, A. L., and Bautz, E. K. F. (1975). Eur. J. Biochem. 60:179.
166. Fristrom, J. W., Brothers, L., Mancebo, V., and Steward, D. (1968). Mol. Gen. Genet. 102:1.
167. Fristrom, J. W., Gregg, T. L., and Siegel, J. G. (1974). Dev. Biol. 41:301.
168. Siegel, J. G., and Fristrom, J. W. (1974). Dev. Biol. 41:314.
169. Siegel, J. (1973). Ph.D. thesis, University of California, Berkeley.
170. Katzenellenbogen, B. S., and Gorski, J. (1972). J. Biol. Chem. 247:1299.
171. Steinberg, R. A., Levinson, B. B., and Tomkins, G. M. (1975). Cell 5:29.
172. Wyatt, G. R. (1968). *In* W. Etkin and L. I. Gilbert (eds.), Metamorphosis, a Problem in Developmental Biology, pp. 143–184. Appleton-Century-Crofts, New York.
173. Wyatt, S. S., and Wyatt, G. R. (1971). Gen. Comp. Endocrinol. 16:369.
174. Greenstein, M. E. (1972). J. Morphol. 136:1.

175. Linzen, B., and Wyatt, G. R. (1964). Biochim. Biophys. Acta 87:188.
176. Bachrach, U. (1973). Function of Naturally Occurring Polyamines. Academic Press, New York.
177. Russell, D. H. (1973). Polyamines in Normal and Neoplastic Growth. Raven Press, New York.
178. Pegg, A. E., Lockwood, D., and Williams-Ashman, H. G. (1970). Biochem. J. 117:17.
179. Russell, D. H., and Taylor, R. L. (1971). Endocrinology 88:1397.
180. Jänne, O., Bardin, W., and Jacob, S. T. (1975). Biochemistry 14:3589.
181. Abraham, K. A. (1968). Eur. J. Biochem. 5:143.
182. Wyatt, G. R., Rothaus, K., Lawler, D., and Herbst, E. J. (1973). Biochim. Biophys. Acta 304:482.
183. Höltta, E., and Raina, A. (1973). Acta Endocrinol. 73:794.
184. Manen, C.-A., and Russell, D. H. (1977). Science 195:505.
185. Wigglesworth, V. B. (1936). Q. J. Microsc. Sci. 79:91.
186. Engelmann, F. (1970). The Physiology of Insect Reproduction. Pergamon Press, New York.
187. Dejmal, R. K., and Brookes, V. J. (1972). J. Biol. Chem. 247:869.
188. Chen, T. T., Couble, P., De Lucca, F. L., and Wyatt, G. R. (1976). *In* L. I. Gilbert (ed.), The Juvenile Hormones, pp. 505–529. Plenum Press, New York.
189. Kunkel, J. G., and Pan, M. L. (1976). J. Insect Physiol. 22:809.
190. Englemann, F. (1974). Am. Zool. 14:1195.
191. Chambers, D. L., and Brookes, V. J. (1967). J. Insect Physiol. 13:99.
192. Brookes, V. J., and Dejmal, R. K. (1968). Science 160:999.
193. Brookes, V. J. (1969). Dev. Biol. 20:459.
194. Engelmann, F. (1971). Arch. Biochem. Biophys. 145:439.
195. Englemann, F. (1972). Gen. Comp. Endocrinol. 3(suppl.):168.
196. Engelmann, F. (1976). Actualite Sur Les Hormones d'Invertébrés, pp. 459–464. Colloques Internationaux C.N.R.S. No. 251.
197. Engelmann, F. (1976). *In* L. I. Gilbert (ed.), The Juvenile Hormones, pp. 470–485. Plenum Press, New York.
198. Koeppe, J. K., and Ofengand, J. (1976). *In* L. I. Gilbert (ed.), The Juvenile Hormones, pp. 486–504. Plenum Press, New York.
199. Wyatt, G. R., Chen, T. T., and Couble, P. (1976). *In* E. Kurstak and K. Marmorosch (eds.), Invertebrate Tissue Culture—Applications in Biology, Medicine and Agriculture, pp. 195–202. Academic Press, New York.
200. Hagedorn, H. H., Fallon, A. M., and Laufer, H. (1973). Dev. Biol. 31:285.
201. Hagedorn, H. H., and Fallon, A. M. (1973). Nature (Lond.) 244:103.
202. Hagedorn, H. H., O'Connor, J. D., Fuchs, M. S., Sage, B., Schlaeger, D. A., and Bohm, M. K. (1975). Proc. Natl. Acad.Sci. USA 72:3255.
203. Hagedorn, H. H. (1974). Am. Zool. 14:1207.
204. Kambysellis, M. P., Craddock, E. M., and Manthos, C. J. (1976). Paper presented at the XV International Congress of Entomology, Washington, D.C.
205. Hagedorn, H. H. (1976). Paper presented at the XV International Congress of Entomology, Washington, D.C.
206. Bell, W. J. (1969). J. Insect Physiol. 15:1279.
207. Bell, W. J., and Barth, R. H. Jr. (1971). Nature (New Biol.) 230:220.
208. Wilhelm, R., and Lüscher, M. (1974). J. Insect Physiol. 20:1887.
209. Bell, W. J., and Sams, G. R. (1975). J. Insect Physiol. 21:173.
210. Roth, T. F., Cutting, J. A., and Atlas, S. B. (1976). J. Supramolec. Struc. 4:527.
211. Davey, K. G., and Huebner, E. (1974). Can. J. Zool. 52:1407.

212. Ilan, J., Ilan, J., and Quastel, J. H. (1966). Biochem. J. 100:441.
213. Ilan, J., and Ilan, J. (1975). Curr. Top. Dev. Biol. 9:90.
214. Garel, J. P. (1976). Nature 260:805.
215. Andersen, S. O., Chase, A. M., and Willis, J. H. (1973). Insect Biochem. 3:171.
216. Lassam, N., Lerer, H., and White, B. N. (1975). Nature (Lond.) 256:734.
217. Lassam, N. J., Lerer, H., and White, B. N. (1976). Dev. Biol. 49:268.
218. White, B. N., Lassam, N. J., and Lerer, H. (1976). *In* L. I. Gilbert (ed.), The Juvenile Hormones, pp. 416–431. Plenum Press, New York.
219. Oberlander, H., and Tomblin, C. (1972). Science 177:441.
220. Fristrom, J. W., Chihara, C. J., Kelly, L., and Nishiura, J. T. (1976). The Juvenile Hormones, pp. 432–448. Plenum Press, New York.
221. Lezzi, M., and Wyss, C. (1976). *In* L. I. Gilbert (ed.), The Juvenile Hormones, pp. 252–269. Plenum Press, New York.
222. Firstenberg, D. E., and Silhacek, D. L. (1973). Experientia 29:1420.
223. Cohen, E., and Gilbert, L. I. (1972). J. Insect Physiol. 18:1061.
224. Courgeon, A.-M. (1975). C. R. Acad. Sci. (D), (Paris) 280:2563.
225. Bowers, W. S. (1976). *In* L. I. Gilbert (ed.), The Juvenile Hormones, pp. 394–408. Plenum Press, New York.
226. Willis, J. H. (1974). Annu. Rev. Entomol. 19:97.
227. Willis, J. H., and Hollowell, M. P. (1976). *In* L. I. Gilbert (ed.), The Juvenile Hormones, pp. 270–287. Plenum Press, New York.
228. Riddiford, L. M. (1976). Nature (Lond.) 259:115.
229. Riddiford, L. M. (1976). *In* L. I. Gilbert (ed.), The Juvenile Hormones, pp. 198–219. Plenum Press, New York.
230. Truman, J. W., and Riddiford, L. M. (1974). J. Exp. Biol. 60:371.
231. Congote, L. F., Sekeris, C. E., and Karlson, P. (1969). Exp. Cell Res. 56: 338.
232. Congote, L. F., Sekeris, C. E., and Karlson, P. (1970). Z. Naturforsch. 25:279.
233. Wigglesworth, V. W. (1957). Symp.Soc. Exp. Biol. 11:204.
234. Beermann, W. (1958). *In* D. Rudnick (ed.)., Developmental Cytology (16th Growth Symposium), pp. 83–103. Ronald Press, New York.
235. Schneiderman, H., and Gilbert, L. I. (1959). *In* D. Rudnick (ed.), Cell Organism and Milieu (17th Growth Symposium), pp. 175–187. Ronald Press, New York.
236. Karlson, P. (1961). Symp. Dtsch. Gesellschaft Endokrinol. 8:90.
237. Karlson, P. (1963). Perspect. Biol. Med. 6:203.
238. Yamamoto, K. R., and Alberts, B. M. (1976). Annu. Rev. Biochem. 45:721.
239. Karlson, P., Doenecke, D., and Sekeris, C. E. (1976). Compr. Biochem. 25:1.
240. Berendes, H. D., and Boyd, J. B. (1969). J. Cell Biol. 41:591.
241. Brady, T., Berendes, H. D., and Kuijpers, A. M. C. (1974). Mol. Cell. Endocrinol. 1:249.
242. Emmerich, H. (1969). Exp. Cell. Res. 58:261.
243. Emmerich, H. (1970). Z. Vergl. Physiol. 68:385.
244. Emmerich, H. (1972). Gen. Comp. Endocrinol. 19:543.
245. Butterworth, F. M., and Berendes, H. D. (1974). J. Insect Physiol. 20:2195.
246. Beckers, C., and Emmerich, H. (1976). Actualite sur les Hormones de'Invertébrés, pp. 351-357. Colloques Internationaux C.N.R.S. No. 251.
247. Best-Belpomme, M., and Courgeon, A.-M. (1975). C. R. Acad. Sci. (D) (Paris) 280:1397.
248. Kroeger, H. (1968). *In* W. Etkin and L. I. Gilbert (eds.), Metamorphosis, a Problem in Developmental Biology, pp. 185–219. Appleton-Century-Crofts, New York.
249. Ashburner, M., and Cherbas, P. (1976). Mol. Cell. Endocrinol. 5:89.

250. Holt, T. K. H. (1970). Chromosoma 32:64.
251. Holt, T. K. H. (1971). Chromosoma 32:428.
252. Helmsing, P. J., and Berendes, H. D. (1971). J. Cell. Biol. 50:893.
253. Schwartz, R. J., Schrader, W. T., and O'Malley, B. W. (1976). *In* L. I. Gilbert (ed.), The Juvenile Hormones, pp. 530–556. Plenum Press, New York.
254. Schmialek, P., Borowski, M., Geyer, A., Miosga, V., Nündel, M., Rosenberg, E., and Zapf, B. (1973). Z. Naturforsch. 28c:453.
255. Schmialek, P., Geyer, A., Miosga, V., Nündel, M., and Zapf, B. (1975). Z. Naturforsch. 30c:730.
256. Akamatsu, Y., Dunn, P. E., Kezdy, F. J., Kramer, K. J., Law, J. H., Reibstein, D., and Sanburg, L. L. (1976). *In* R. Meints and E. Davies (eds.), Control Mechanisms and Development, Vol. 62, pp. 123–149. Plenum Press, New York.
257. Shearn, A., and Garen, A. (1974). Proc. Nat. Acad. Sci. USA 71:1393.
258. Williams, C. M., and Kafatos, F. C. (1971). *In* J. J. Menn and M. Beroza (eds.), Insect Juvenile Hormones—Chemistry and Action, pp. 29–41. Academic Press, New York.
259. Chambon, P. (1975). Annu. Rev. Biochem. 44:613.
260. Daneholt, B., Edström, J. E., Egyhazi, E., Lambert, B., and Ringborg, W. (1970). Cold Spring Harbor Symp. Quant. Biol. 35:513.

International Review of Biochemistry
Biochemistry and Mode of Action of Hormones II, Volume 20
Edited by H. V. Rickenberg

6
Mode of Action of Plant Hormones

J. E. VARNER
Washington University, St. Louis, Missouri

Plant hormones participate in the integration of the various modes of metabolism in the different tissues and organs of the plant. In no case is it known how the hormone interacts at its presumed receptor site to start, stop, or modulate some event. The observations selected for discussion in this chapter represent experimental approaches that seem most likely to lead to the discovery of the chemistry of hormone action.

To list the various responses of cells, tissues, organs, and whole plants to each of the plant hormones is to invite bewilderment. Different hormones cause similar effects in different tissues, and the same hormone causes different effects in different tissues. For example, cytokinin delays senescence in barley leaves, gibberellic acid delays senescence in dandelion leaves, and indoleacetic acid delays senescence in bean endocarp; indoleacetic acid increases the rate of elongation of oat coleoptiles and gibberellic acid increases the rate of elongation of oat internodes; abscisic acid closes stomata, accelerates bean endocarp senescence, prevents the indoleacetic acid-induced extension of oat coleoptiles, and prevents the gibberellic acid-induced synthesis of α-amylase in aleurone cells.

It must be remembered that each response observed is the result of a complex process in which any one of several components might be limiting; clearly, control by different hormones may be exerted at different points in the overall process. Different tissues of the same plant might respond differently to the same hormone simply because the tissues are differentiated and would respond differently to an identical primary action of the hormone.

AUXINS

Occurrence

The demonstration by F. W. Went in 1928 that oat coleoptile tips contain a diffusible substance that promotes elongation of decapitated coleoptiles derived from Darwin's observations of the effect of unilateral light on the bending (growth toward the light) of grass coleoptiles. Kögl's identification in 1934 of the substance in urine that promotes coleoptile growth as indoleacetic acid (IAA) suggested that IAA is the naturally occurring substance of Went's studies. The subsequent demonstration in 1935 of IAA in plant extracts by Kögl and by Thimann seemed to confirm this. In the next three decades, much auxin research dealt with the response of plants to added IAA or to other synthetic auxins with little effort to confirm the assumption that IAA was the only naturally occurring auxin.

However, there are now more and more reports of the unequivocal identification of auxins in plant tissues. These include the mass spectrometric identification of free indoleacetic acid in diffusates of maize coleoptile tips (1), of free IAA in shoots of Douglas fir (2), of free IAA in *Avena* and *Zea* (3), as well as the bound forms of IAA, IAA-glucose, IAA-myoinositols, IAA-myoinositol arabinosides, IAA-myoinositol galactosides (4), and IAA-glucans (5) in maize seeds.

Exogenously introduced, labeled IAA has been recovered as IAA-glucose (6), IAA-myoinositol, IAA-glucans (7), and IAA-aspartate (6, 8). Exogenous labeled IAA appears in bound form within minutes of uptake. Cycloheximide, 6-methylpurine, 2-deoxyglucose, Ca^{2+}, 0.25 M mannitol, and 0.33 M sucrose have little effect on this incorporation (9). The bound forms of auxin might serve for the storage of IAA, permitting the removal of excess hormone from its sites of action or synthesis and yet retaining it for later use.

Sheldrake (10) reviews and revives earlier suggestions that IAA in coleoptile tips comes from the seed tissues via the xylem stream; he shows that the guttation fluid of *Avena sativa, Zea mays, Triticum aestiva,* and *Hordeum vulgare* contains both free and alkali-labile bound IAA. This is a specific example of a more general suggestion that senescent, dying cells are a principal source of auxin (11, 12). Thus the synthesis and/or release of auxins from senescent leaves, endosperm, xylem, etc. may represent an important signal to the rest of the plant.

Biosynthesis

IAA is almost certainly derived from tryptophan (6) (Figure 1). The indole-glucosinolates of the crucifers also arise from tryptophan (13) (Figure 2).

It has been reported that immature pea seeds contain 4-chloroindoleacetic acid (14); 4-chloro-D-tryptophan, also found in the seeds, is considered a likely precursor. Confirmation of this finding would present the possibility that there are separate controls of the biosynthesis of IAA and 4-chloro-IAA (because the biosynthetic pathways are separate and distinct) and further complicate the possible roles of auxins in the integration of plant metabolism.

Phenylacetic acid, an active auxin, also occurs in plants (15–17) and is responsible for about half of the total auxin activity found in extracts of several plant tissues. Phenylalanine is the likely precursor of phenylacetic acid (6) (Figure 3). The occurrence of the two auxins, IAA and phenylacetic acid, raises a number of questions; thus, "Do both compounds affect the same site of action? Are their transport mechanisms different? Are they localized in different tissues or even parts of a cell? Is their biosynthesis so regulated that they can respond to different environmental stimuli?" (17)

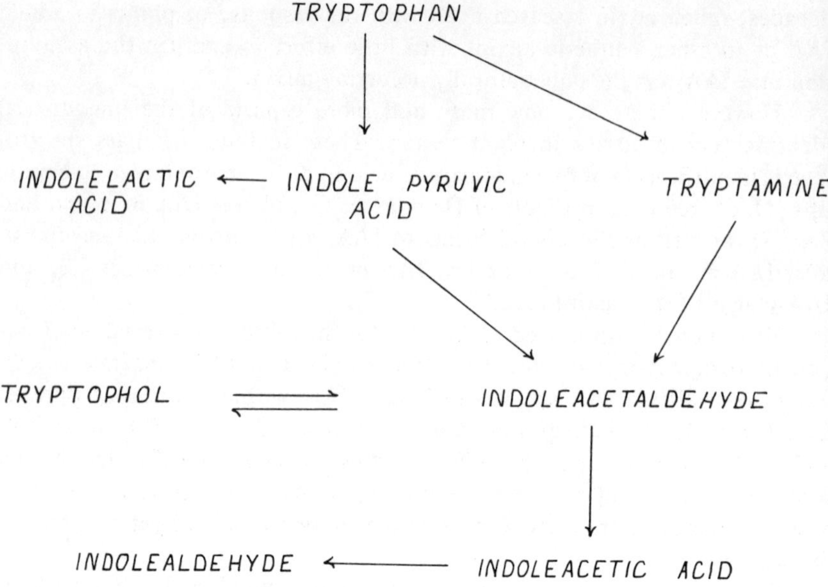

Figure 1. Probable pathway for IAA synthesis in tomato shoots.

Response of Plant to Auxin

Range of Responses The first observed and perhaps most easily observed effect of auxin is to promote cell elongation in stems and coleoptiles. Auxins also induce root formation, vascular differentiation, the formation of β-1,4-glucanases in roots, and the biosynthesis of ethylene; furthermore, they induce (through the mediation of ethylene) fading in flowers, control abscission, prevent senescence in bean endocarp, induce tropic responses, promote growth of ovary into fruit, promote cell division in callus tissue, exert apical dominance, promote cytoplasmic streaming, increase cell wall elasticity and plasticity, and promote hydrogen ion extrusion.

Cell Elongation Added auxins cause an increase in the elongation rate in auxin-sensitive tissues within 10 to 15 min of administration. Auxin-dependent elongation occurs in the presence of cycloheximide. Thus, the primary action must not require protein synthesis. An increased rate of incorporation of labeled glucose into cell wall polymers of pea stem segments can be seen within 15 min, and within an hour of treatment of pea stem segments a 2–4-fold increase in the particulate UDP-glucose:(1,4)-β-glucan glucosyltransferase is observed. This increase is the result of enzyme activation (18), not of enzyme synthesis.

Cytokinins partially inhibit the auxin-promoted elongation of stems and hypocotyls of dicotyledonous plants. A careful examination of this inhibitory effect has shown that there are two overlapping responses of soybean hypocotyl segments to added auxins and that only the first response is inhibited by cytokinins (19) (Figures 4–6). The first response, as observed in the presence of cytokinin, is similar to the elongation response of the hypocotyl segments when incubated in buffers at low pH. Additional evidence for two separable responses of the hypocotyl to auxin comes from the observation that in the presence of the photoaffinity-labeled auxin, 4-azido-2-chlorophenoxyacetic acid (20), the second response is delayed, whereas the first is not.

The first and second responses of the hypocotyl to auxin are affected differently by cycloheximide. The half-lives of the growth-limiting proteins required for the first and second responses have been estimated in experiments in which cycloheximide was used to be 28 min and 11 min, respec-

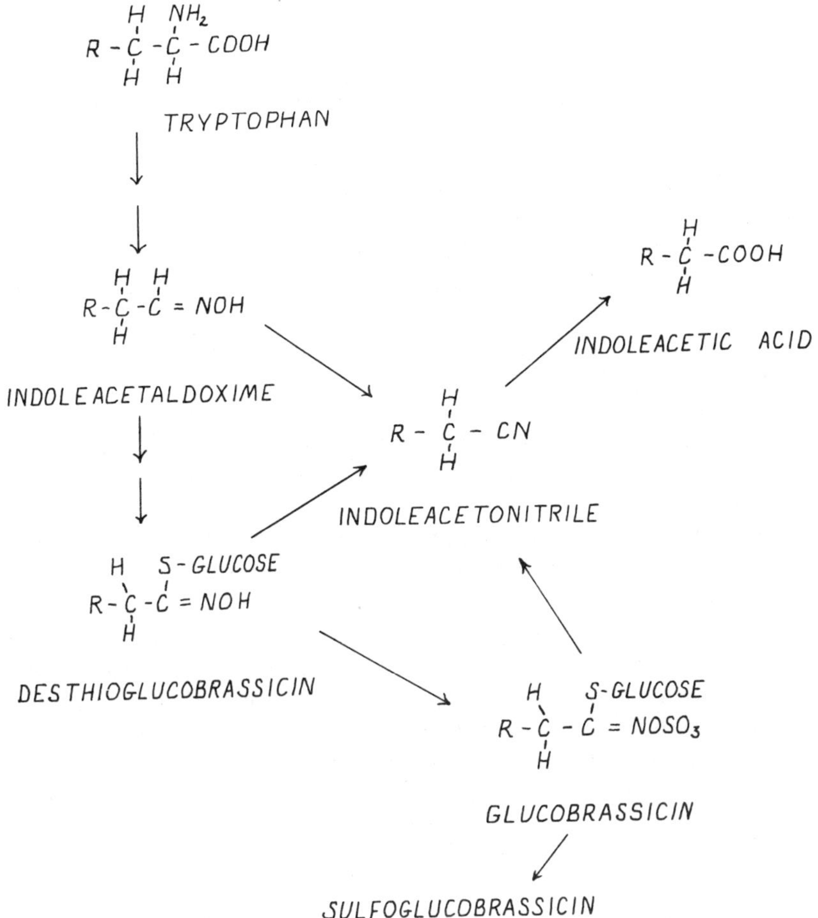

Figure 2. Pathways from tryptophan to the indoleglucosinolates and to IAA in crucifers.

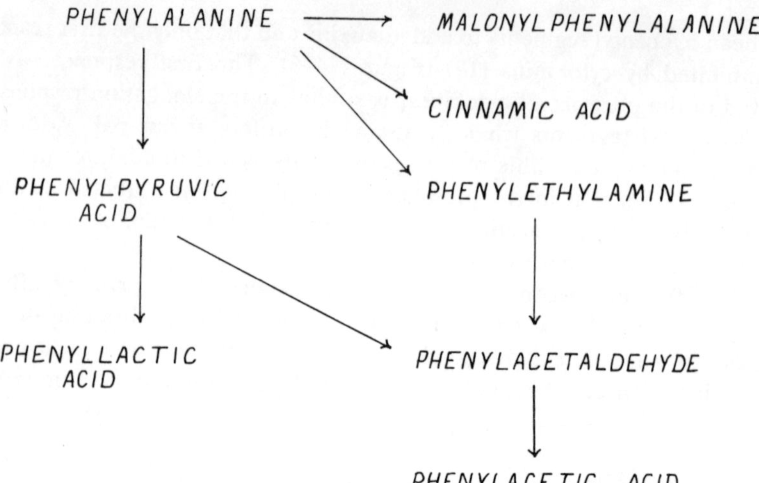

Figure 3. Probable pathways for the metabolism of phenylalanine in tomato shoots.

tively (21); these estimates are based on the assumption that the only short-term effect of cycloheximide is the inhibition of protein synthesis.

The interpretation of published experiments involving the use of cycloheximide is difficult because, at least in wheat and oat coleoptiles, cycloheximide apparently directly inhibits respiration. This inhibition is more effective at low oxygen concentrations and is reversed by continuous diffuse light (22).

Auxin-promoted Acidification Under certain conditions, *Avena* coleoptiles respond to auxins by acidifying the walls and the surrounding medium. This decrease in pH is presumed to permit or cause the loosening of cell walls.

Many cell wall hydrolases have pH optima near pH 5; it seems reasonable, therefore, to invoke their action in hormone-promoted wall extension. There is some evidence for such action (23).

Although it has been known since 1934 (24) that growth rate and plasticity of coleoptiles are increased by acidic buffers, the idea that auxin acts directly by causing a decrease in the pH of the wall solution was not proposed until 1971 (25, 26). Over a range of experimental conditions, acidification, or proton secretion, is proportional to the rate of cell elongation (27). This proton secretion may be accompanied by K^+ uptake (28). In principle, it could be accompanied by anion release.

Exogenous Ca^{2+} causes acidification in *Avena* coleoptiles (29). Most of this effect seems to be a replacement of H^+ by Ca^{2+} at cation exchange sites in the wall (30) because added Ca^{2+} causes the release of protons from frozen-thawed coleoptile sections. In addition to increasing K^+ uptake and H^+ release, exogenous auxin enhances Cl^- uptake (31). It is,

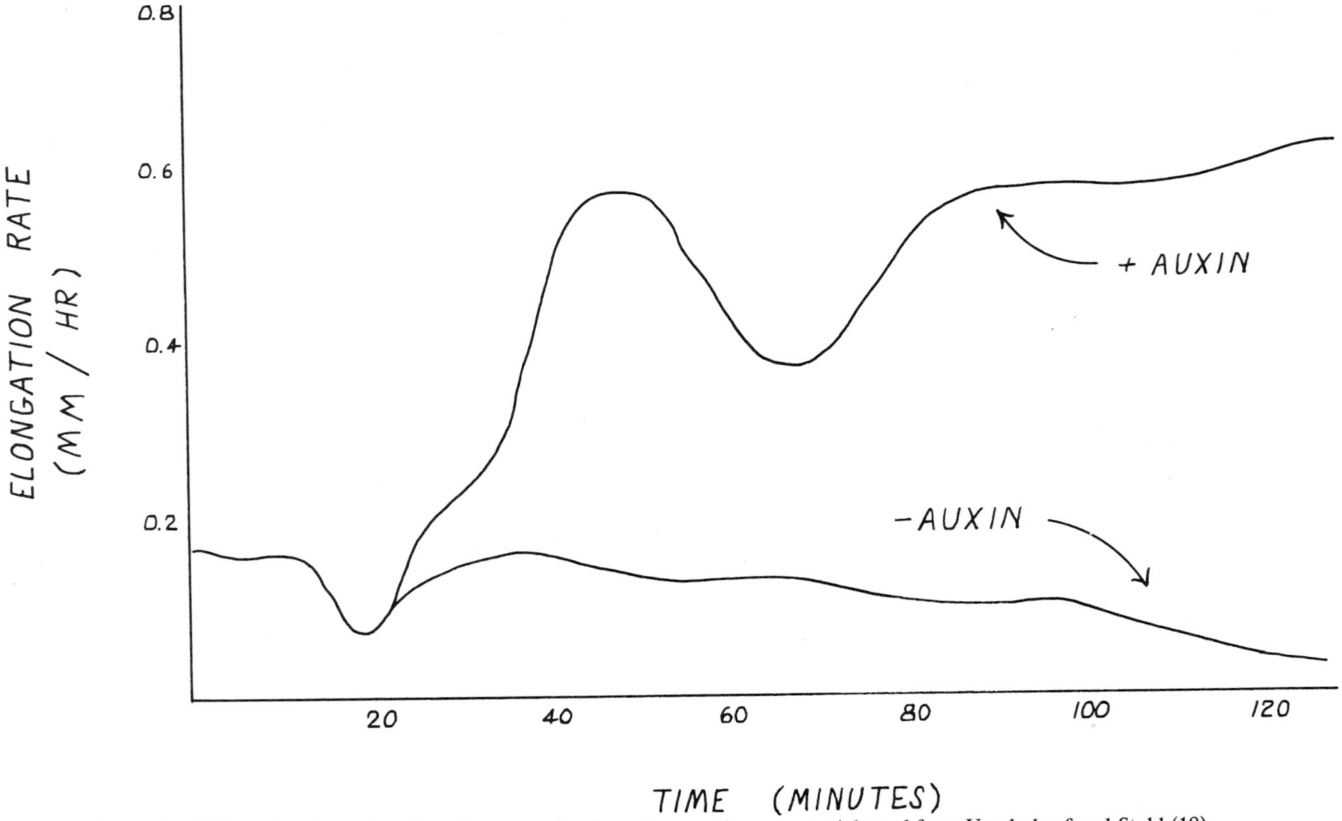

Figure 4. Effect of auxin on the elongation rate of soybean hypocotyl segments. Adapted from Vanderhoef and Stahl (19).

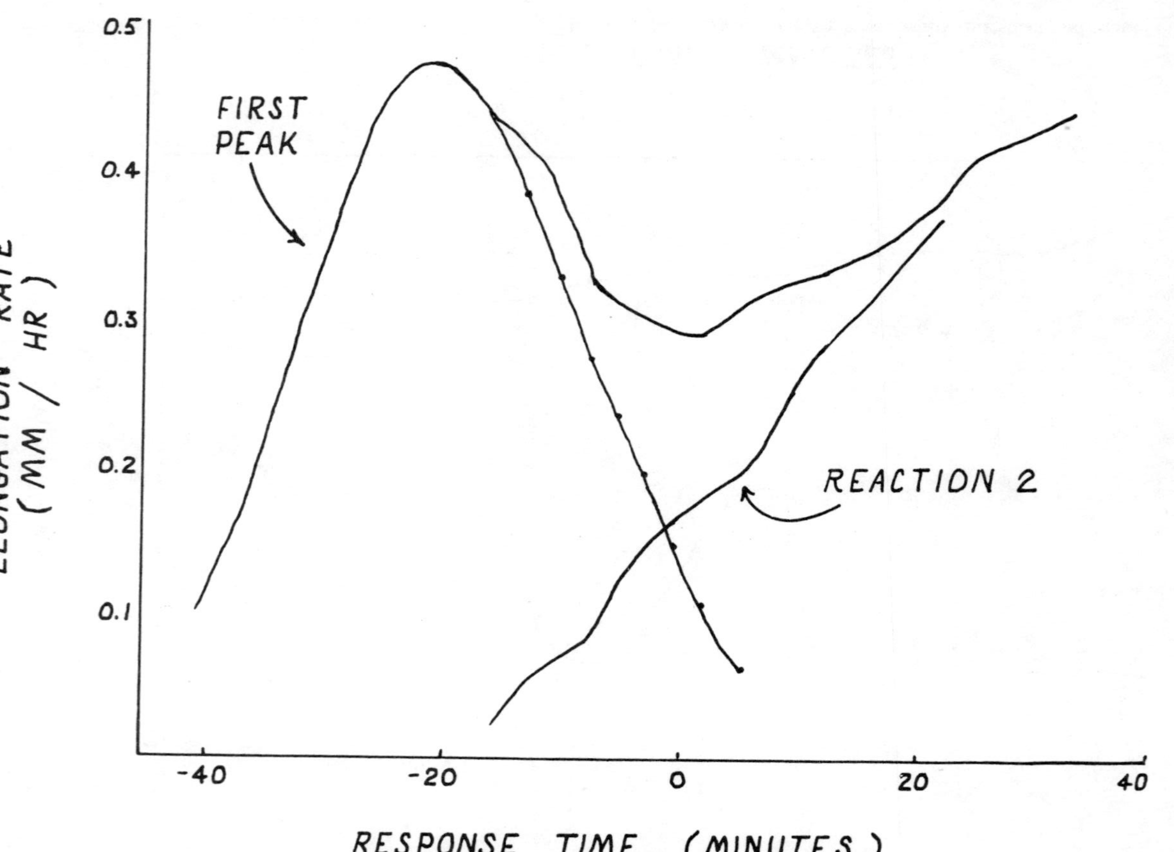

Figure 5 Graphical separation of the

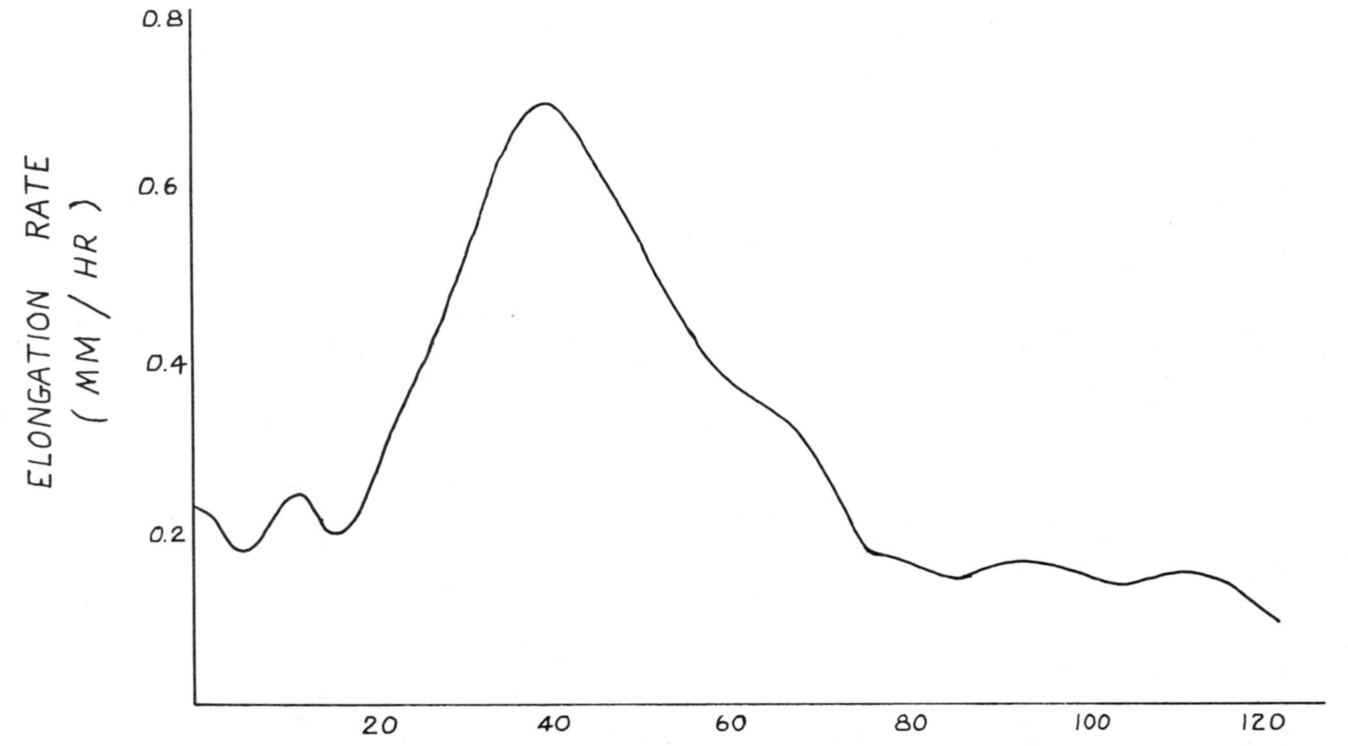

Figure 6. Effect of cytokinin on auxin-induced elongation of soybean hypocotyl segments. The second response is inhibited. Adapted from Vanderhoef and Stahl (19).

therefore, reasonable to conclude that added auxin has a direct effect on the plasmalemma. Coleoptile sections treated sequentially with 0.5 M mannitol and 1 mM sodium phosphate (pH 6.4) at 4 °C, so as to produce shock, show no auxin-dependent growth or acidification although polar auxin transport and the Ca^{2+}-dependent acidification are unaffected (32). It can be assumed that osmotic shock modifies the plasmalemma by selective removal of specific components.

The IAA-dependent acidification and growth that occur in the dark or in far red light are less pronounced in red light (33). Thus, phytochrome and auxin could (but need not) have a common primary site of action in the plasmalemma. Protoplasts treated with IAA immediately (within seconds) acidify the medium (34). These results, as well as the observation that labeled auxins bind to membrane fractions, certainly justify further examination of the effect of auxin on the plasma membrane.

Just as added auxins enhance growth rates and acidification in oat and wheat coleoptiles and in pea stem segments, gibberellic acid increases the rates of growth and acidification of oat stem segments (35). It should be pointed out that the claim that auxin-induced growth is invariably accompanied by a decrease in wall pH (36) or that acid-induced growth is identical with auxin-induced growth (37) is not universally accepted.

Auxin Transport Inhibitors of auxin transport cause, inter alia, growth retardation, loss of apical dominance, loss of tropic responses, parthenocarpy, and shortened internodes. The auxin transport inhibitors, 2,3,5-triiodobenzoic acid, *N*-1-naphthylphthalamic acid, morphactin, and the 2-(3-aryl-5-pyrazolyl)benzoic acids all have a free carboxyl group. Because the free carboxyl group is also one of the structural requirements for auxin activity, it can be presumed that this similarity in some way reflects the mechanism of action of the inhibitors (38).

Auxin-promoted Synthesis of RNA and Protein Auxins applied to soybean hypocotyls cause, in addition to the rapid responses already described, a large increase in the rate of synthesis of RNA, especially rRNA. The activity of the α-amanitin-resistant nucleolar RNA polymerase, RNA polymerase I, is increased 5–8-fold after 2,4-dichlorophenoxy acetic acid (2,4-D) treatment (39, 40). There is no effect of 2,4-D on the α-amanitin-sensitive RNA polymerase II. It is not known whether RNA polymerase I activity increases as a result of activation of an inactive form or as a result of an increased rate of synthesis.

Application of an auxin to the cut surface of a pea seedling decapitated below the apical hook causes the knob-like swellings that are accompanied by a many-fold increase in cellulase activity. This increased cellulase activity reflects parallel increases in two cellulases (41). The poly(A) RNA isolated from the polysomes of auxin-treated pea epicotyls, when translated in a cell-free system, yields an antigen identical with one of the cellulases. The mRNA for this cellulase is located on membrane-bound polysomes and is not detectable before treatment of the seedlings with auxin (42). The increase in translatable mRNA for cellulase begins immediately after the auxin treatment, whereas the increase in cellulase activity is not detected until after 24 hr. Thus, in this system, there may be controls at the levels of both transcription and translation.

Auxin and Senescence The senescence of *Rhoeo* leaf sections in darkness is associated with marked increases in ribonuclease and acid phosphatase. These increases in hydrolases are characteristic of many detached leaves and fruits. In *Rhoeo* leaf sections, the increases are prevented by auxins and promoted by abscisic acid. Because the increase in hydrolases results from de novo synthesis (43), the *Rhoeo* leaf system may be useful for the study of hormonal control of the synthesis of specific enzymes.

GIBBERELLINS

Discovery

The gibberellins are a large (> 50 members) family of diterpenoid acids, members of which can probably be found in all plants (44). The discovery of the gibberellins resulted from studies over the period 1912–1926 of the "bakanae" disease of rice, the symptom of which is the appearance of tall thin plants. Crystalline, biologically active gibberellins were first obtained from culture filtrates of *Fusarium moniliforme* (*Gibberella fujikoroi*) in Japan in 1938. Western researchers did not become aware of the gibberellins until the early 1950s. In 1956 substances that were chemically similar to, and caused the same physiological effects in plants as, the fungal gibberellins were found in higher plants. Since then, over 50 naturally occurring gibberellins have been identified.

Biosynthesis and Metabolism

Although the pathway for the biosynthesis of gibberellins is fairly well understood (acetate→mevalonate→geranylgeranyl pyrophosphate→copalyl pyrophosphate→kaurene→kaurenoic acid→gibberellins) (45), there is little to report about the control of gibberelin biosynthesis.

The development of methods for the preparation of tritiated gibberellins of high specific activity (46) has made it possible to observe the in vivo binding of gibberellins (47). In dwarf peas, labeled gibberellins accumulated only in the gibberellin-responsive regions of the stem, and gibberellins of low biological activity in dwarf peas failed to accumulate. It may be that this accumulation represents formation of the physiologically active gibberellin-receptor complex. Considerable interconversion of gibberellins occurs in dwarf peas (48), but the physiological reason for this is not clear.

Barley aleurone layers also metabolize gibberellins (49), principally by glucosylation (50, 51). Although the uptake and metabolism of gibberellins are enhanced by abscisic acid, these effects are probably too slow to explain the inhibitory action of abscisic acid on the response of aleurone tissue to gibberellins.

Responses to Gibberellins

The dramatic response of dwarf pea stems to added gibberellins is accompanied by an increase in chromatin-associated RNA polymerase activity (52). It is not known how these responses are related.

English ivy (*Hedera helix*) is changed from the juvenile form to the adult form by added gibberellins. Abscisic acid causes the reverse transition. The transition does not involve the transcription of new kinds of RNA, as determined by DNA-RNA hybridization (53).

Avena stem segments show a change in plasticity within about 90 min of the application of gibberellins. This is about the time required for the stem elongation response to become visible (54). As already mentioned, *Avena* stem segments show a change in the rate of acidification of the medium in response to gibberellins (35). A decreased rate of acidification is observed in 15 min, and after 90 min there is a marked increase in the rate of acidification of the medium. These responses seem deserving of much more attention because they occur quickly and in nondividing cells.

Synthesis and Secretion of Hydrolases in Cerebral Grain Aleurone Layers

Because it is an easily isolated tissue in which all cells respond to added gibberellins, abscisic acid, and ethylene, the aleurone layer is frequently used to study hormone action. The response of the isolated tissue to added gibberellins is either identical with, or at least similar to, its response in situ, i.e., in the seedling to gibberellins synthesized by the embryo.

The barley aleurone layer synthesizes and secretes, in response to added gibberellins, α-amylase, protease (55), ribonuclease, β-glucanase (56), α-glucosidase, acid phosphatase, and limits dextrinase (57). There is a basal level of ribonuclease, β-glucanase, and α-glucosidase in the aleurone layers prior to the addition of gibberellins.

This response is specifically evoked by GA_1, GA_3, GA_4 and GA_7, but not by GA_8 (58). Gibberellic acid, GA_3, has been used in most investigations. The response to gibberellins is prevented by the presence of abscisic acid and permitted if ethylene is present at the same time as abscisic acid (59). In contrast to many reports that α-amylase is produced in response to a variety of compounds, Clutterbuck and Briggs (58) found no α-amylase produced by aleurone tissue in response to kinetin, benzylaminopurine, hydroxylamine, glutamine, ornithine, ent-kaurene, ent-kaurenol, ent-kaurenoic acid, phorone, isophorone, phenobarbitone, ATP, APD, or cyclic-3′,5′-AMP. High concentrations (80 mg/l) of helminthosporal and of helminthosporic acid evoke the production of only a fraction of the α-

amylase that is produced in response to gibberellins. Studies with helminthosporin analogues may be useful in determining the minimum structure required to evoke a response to gibberellin-sensitive cells. In particular, it has been shown that the carbonyl group must be present in a certain spatial relationship, but the hydroxymethyl group is not required (60).

Acid phosphatase activity in the isolated aleurone layers increases in the absence of added gibberellins and increases somewhat more with added GA_3 (61). There is some acid phosphatase localized in the walls of the aleurone cells of the dry grain; further secretion of acid phosphates into the walls occurs during imbibition and incubation in the absence of gibberellins. Release of this acid phosphatase depends, however, upon the production by the cell of cell wall-degrading enzymes in response to added gibberellins (61). The release from the cell wall of other enzymes also depends upon cell wall degradation (61). The aleurone cell walls consist almost entirely of arabinoxylan. Therefore, it is likely that gibberellin-dependent increases in pentosanases (62) are largely responsible for cytolysis. In addition, at least for α-amylase, a certain ionic concentration is required for rapid release from the walls (63). Because the only source of these ions is the aleurone cell itself, the cell has an opportunity to control the rate of release of already secreted α-amylase by controlling the secretion of its ions. This secretion of ions is also determined by the concentration of added gibberellins (64).

The increases in the synthesis of several enzymes could result from the translation of already existing mRNAs or the translation of newly transcribed mRNAs, with such transcription being gibberellin-dependent. Addition of GA_3 does not cause any marked increase in protein synthesis; so there is neither reason to expect, nor evidence for, a dependence on rRNA, 5 S RNA, and tRNA synthesis (55, 65). There is a gibberellin-dependent increase in the rate of labeling of a poly (A)-rich, mRNA-like fraction (66, 67), and there is a parallel increase in translatable mRNA for α-amylase (68). This increase in translatable mRNA could result from either a decreased rate of degradation of mRNA, an increased rate of synthesis of mRNA, or some kind of processing or activation of an inactive form of mRNA.

In considering possible controls of the translation of amylase mRNA, it is interesting to note that further amylase synthesis is prevented when abscisic acid is added 12 hr after the addition of GA_3 (at this time, amylase synthesis is not susceptible to inhibition by cordycepin). Thus, abscisic acid added at this time, directly or indirectly, inhibits translation (69). Cordycepin added with abscisic acid, or within a few hours after abscisic acid, prevents or relieves the inhibition by abscisic acid of amylase synthesis. If, indeed, cordycepin specifically inhibits RNA synthesis, then the inhibition of amylase translation by abscisic acid must involve the synthesis of new RNA (69).

It should not take long to resolve at what point abscisic acid inhibits translation of amylase mRNA and whether amylase mRNA is synthesized de novo or activated after the addition of GA_3. The outcome may or may not reveal much about the primary action of GA_3 and of abscisic acid. The earliest observed responses of aleurone tissue to GA_3 are related to phospholipid metabolism and do not depend on protein synthesis. These responses include changes within 30 min in the rate of labeling of soluble nucleotides, particularly CTP (70), and increases (beginning in a few minutes and prevented if abscisic acid is added at the same time as GA_3) in the activity of phosphorylcholine cytidyltransferase and phosphorylcholine glyceridetransferase as measured in cell-free membrane preparations (71). These increases in enzyme activity apparently do not require RNA synthesis or protein synthesis because increases in activities are not prevented by cordycepin or the presence of amino acid analogues (72). The mechanism of these responses to GA_3 and to GA_3 and abscisic acid may be quite different from that responsible for the changes in the rate of translation of amylase mRNA.

These changes in the activities of enzymes of phospholipid metabolism as measured in vitro do not have a strict parallel in observable in vivo phospholipid metabolism. Although there is a marked GA_3-dependent (and abscisic acid-inhibited) increase in the incorporation of $^{32}P_i$ into phospholipid, this does not begin until 4 hr after the addition of GA_3 (73). This lack of parallelism between the in vitro and in vivo observations, as well as the failure to observe any net increase in the quantity of phospholipid in responses to added gibberellins (74), raises questions about, but does not yet eliminate, the attractive idea that the proliferation of rough endoplasmic reticulum observed by electron microscopy is the result of a gibberellin-enhanced phospholipid synthesis.

Gibberellic acid changes the permeability of model membranes (liposomes) composed of plant components, presumably as a result of direct interaction of gibberellic acid with the phospholipid of the membrane (75). Although these observations are the basis of a proposal that some of the in vivo effects of gibberellic acid might be explained by such interactions (76), it is difficult to see how such a model could account for the great specificity of cells for the different gibberellins. In this context, a report claiming the phospholipids act as ionophores (77) might be mentioned.

Even though the synthesis of various hydrolases by aleurone layer tissue depends upon certain gibberellins, it is not at all certain whether gibberellins play any direct role in the secretion of these hydrolases. In barley, appearance of rough endoplasmic reticulum depends upon added gibberellins (55), but in wheat, formation of the endoplasmic reticulum occurs during inhibition of the half-seeds before any addition of gibberellins (78). Amylase synthesis in wheat, as in barley, does not occur in the absence of added gibberellins. Immunohistochemical localization of α-

amylase in barley shows that the enzyme first appears in the rough endoplasmic reticulum associated with the nuclear envelope and the perinuclear region (79). How amylase proceeds from the rough endoplasmic reticulum to a position beyond the plasmalemma is uncertain. Some evidence suggests that it is released from the endomembrane system into the cytoplasm and moves across the plasmalemma molecule in "molecular form" rather than in packaged form (79). Other evidence indicates that amylase moves from the endoplasmic reticulum to the outside of the plasmalemma by way of secretory organelles (80, 81). Whatever the mechanism for the movement of amylase molecules across the plasmalemma, such movement requires respiratory energy and has not been shown to be directly under the control of any hormones (63). Release of amylase from the wall into the medium requires a certain ionic strength and some degree of degradation of the wall. The cell wall degradation apparently occurs only in response to gibberellins (61), beginning and being most extensive around the plasmalemma.

In spite of the many published claims that cyclic AMP mediates or potentiates the action of gibberellins in the barley endosperm system, this author sees no evidence that would modify an earlier conclusion that cyclic AMP does not mediate the action of gibberellins (82).

CYTOKININS

Structure-activity Correlations

Kinetin, 6-(furfuryl amino)purine, the first chemically defined cytokinin, does not occur naturally (see Figure 7 for the naturally occurring cytokinins). Its production in the autoclave as a degradative product of DNA has no known physiological parallel. There are now hundreds of biologically active synthetic cytokinins. The most effective are N^6-substituted adenines. Any substitution of 1 atom for another in the adenine ring or for the nitrogen in the N^6 position results in a loss of activity (in the tobacco callus assay). In the N^6-alkylaminopurines, the optimum length of the side chain is 5 carbon atoms, and N^6-pentylaminopurine has the same activity as kinetin. The activity of N^6-(3-methylbutyl)aminopurine is the same as that of N^6-pentylaminopurine. Introduction of a double bond to form N^6-(3-methyl-2-butenyl)aminopurine increases the activity 10-fold. Introduction of the hydroxyl group to form the 4′-hydroxy derivative, N^6-(4-hydroxy-3-methyl-*trans*-2-butenyl)aminopurine, further increases the activity. Many N^6-(methylbutyl)aminopurine compounds have been tested, and it seems that substituents tending to make the side chain more planar increase biological activity, whereas those tending to make it less planar decrease it. Modification of the purine ring with the 2-methylthio and/or the 9-β-D-ribofuranosyl group leads to decrements in biological activity.

Ring substituents in the N^6 position also confer cytokinin activity

FREE BASES OF NATURAL CYTOKININS

NH—R_1 (purine ring: N, N, N, N—H; R_2)

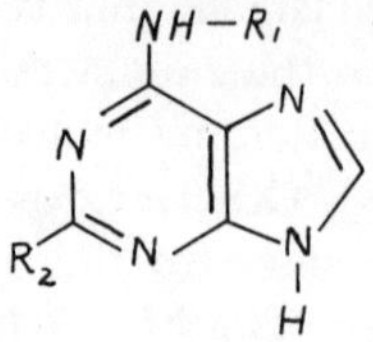

	STRUCTURE R_1	R_2	COMMON NAME
I	$-CH_2-CH=C(CH_3)-CH_2OH$	H	TRANS-ZEATIN
II	$-CH_2-CH=C(CH_2OH)-CH_3$	H	CIS-ZEATIN
III	$-CH_2-CH_2-CH(CH_3)-CH_2OH$	H	DIHYDROZEATIN
IV	$-CH_2-CH=C(CH_3)-CH_3$	H	ISOPENTENYL ADENINE (IPA)
V	$-CH_2-CH=C(CH_2OH)-CH_3$	CH_3S-	CH_3S-ZEATIN (CIS OR TRANS)
	$-CH_2-CH=C(CH_3)-CH_2OH$	CH_3S-	
VI	$-CH_2-CH=C(CH_3)-CH_3$	CH_3S-	CH_3S-IPA

Figure 7. Free bases of natural cytokinins.

on adenine. The benzyl ring is effective; the furfuryl, phenyl, and thienyl rings are less effective, and the cyclohexyl ring is much less effective. Other ring substituents produce cytokinins, which have still less activity. (For a review of the extensive literature on the relationships between the chemical structure and biological activity of cytokinins, see ref. 83.)

Cytokinins in tRNA

The finding of cytokinins in tRNAs raised the possibility that cytokinins might act by virtue of their incorporation into the structure of certain tRNAs. However, it now seems that exogeneous cytokinins are not incorporated in significant amounts into tRNAs but rather that the substitution of the various side chains on the N^6 of adenine is accomplished after the tRNA molecule is formed. In addition, 9-methylbenzyladenine, which cannot be converted to the riboside triphosphate for incorporation into tRNA, is nonetheless active as a cytokinin (84).

Tobacco callus cells synthesize N^6-(3-methyl-2-butenyl)adenosine of the tRNA by attachment to a side chain to an adenosine residue of preformed tRNA (85). A similar synthetic pathway occurs in other organisms. Thus, the turnover of the tRNAs is a possible source of cytokinins. However, the level of free cytokinins in pea root tips is 27 times higher than that which could be obtained by hydrolysis of the tRNA of the cells. One must, therefore, assume either a high turnover rate of tRNA or a second path for the biosynthesis of cytokinins, presumably the addition of the appropriate side chain to adenine (86).

Cytokinins apparently occur in certain tRNAs of all organisms. The first cytokinin identified as a constituent of tRNA was N^6-isopentenyl adenine (IPA). It was shown to be present only once in each of two $tRNA^{Ser}$ species and was adjacent to the 3′ end of the anticodon in both species. A short time later, $tRNA^{Tyr}$ was shown to contain CH_3S-IPA adjacent to the 3′ end of the anticodon.

The presence of the cytokinin base in this position helps to determine the conformation of the anticodon loop. Modification of the IPA of yeast $tRNA^{Ser}$ by treatment with aqueous iodine impaired its binding by ribosomes but did not alter its ability to accept serine (87). Mutants of *Escherichia coli* suppressor $tRNA^{Tyr}$ that contain an adenosine in place of the CH_3S-IPA accept amino acids as well as the normal $tRNA^{Tyr}$ but are ineffective in in vitro tests of suppression and are poorly bound by ribosomes (88). Thus, it is clear that cytokinin bases have an important role in certain tRNAs. However, this role does not seem to be associated with the role of cytokinins as hormones.

Isopentyl adenine and its derivatives occur only in those species of tRNA that respond to codons beginning with uridine (89). Another modified adenosine, 6-(threonylcarbamoyl)purine, is found exclusively at the 3′ end of the anticodon of those species of tRNA that respond to codons beginning with adenosine (89). Neither this base nor its riboside has growth-promoting activity in the soybean callus assay or in the tobacco callus assay. This lack of activity may result from lack of uptake into the cells because of the polarity of the side chain. Less polar ureidopurine analogues do have activity in the soybean callus assay; those analogues most resembling 6-(threonylcarbamoyl)purine in three-dimensional and electronic configuration have the most activity (90).

Metabolism

The main features of cytokinin metabolism are shown in Table 1.

Table 1. Metabolic reactions of cytokinins in plant tissues

Adenine in tRNA $\xrightarrow[\text{(EC 2.5)}]{\text{isopentenyl transferase}}$ IPA in tRNA
IPA-riboside → zeatin riboside
IPA-riboside → adenosine → adenine
IPA-riboside → IPA
IPA → N^6-(3-methyl-3-hydroxybutyl)aminopurine
Zeatin → zeatin riboside → zeatin ribotide
Dihydrozeatin → dihydrozeatin riboside → dihydrozeatin ribotide
Zeatin riboside → adenosine ↘ Zeatin
Zeatin → 7-glucosyl zeatin
Zeatin → 9-glucosyl zeatin
Zeatin → β-zeatinyl alanine

From Varner and Ho (44) and MacLeod et al. (91).

Antagonists of Cytokinins

The availability of cytokinin antagonists is relevant to the eventual study of cytokinins in tissues that produce their own cytokinins endogenously. A series of such antagonists has been synthesized (92) by modification of the side chain, by interchanging the carbon and nitrogen atoms of positions 8 and 9 of the purine ring, and by substitution with a methyl group of the equivalent of position 9 of the purine nucleus. One of the more effective antagonists is 3-methyl-7-(3-methylbutylamino)pyrazolo-(4,3-D) pyrimidine.

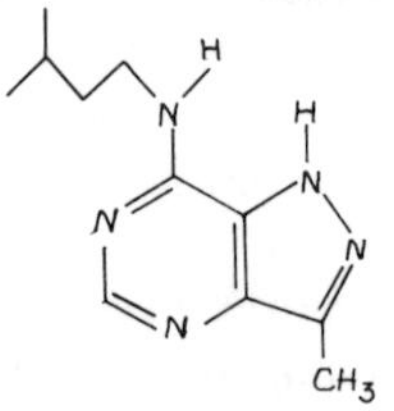

3-METHYL-7-(3-METHYLBUTYLAMINO)-PYRAZOLO-[4,3-d] PYRIMIDINE

This compound is not only devoid of cytokinin activity but also inhibits the growth of tobacco callus supplied with optimum concentrations of IPA or N^6-benzylaminopurine or (BAP). It also inhibits the growth of a strain of tobacco callus that requires no exogenous cytokinin.

Physiological Responses to Cytokinins

The response of excised tobacco pith tissue to exogenous cytokinins is one of many test systems employed to assay the response of plant tissue to added cytokinins; the growth of cultured tissue from soybean cotyledons is about equally sensitive. Suspension cultures of some strains of tobacco cells and of *Acer pseudoplatanus* require added cytokinins for growth and prove to be advantageous for kinetic studies of growth. Soybean tissue in liquid culture synthesizes measurable and proportionate quantities of two deoxyisoflavones after only 24 hr of treatment with cytokinins. This response is not known to be related to the growth response; nonetheless, it deserves further attention.

Cytokinins enhance DNA synthesis and elongation in the hypocotyl of rootless soybean seedlings, induce tuberization in excised potato stolons, overcome the inhibitory effect of abscisic acid on the growth of *Lemna minor*, promote the formation of tyramine methylpherase activity in roots of germinating barley and the formation of isocitrate lyase and protease activity in excised squash cotyledons, remove the thiamine requirement for growth of tobacco callus (Wis. #38) by inducing thiamine synthesis, induce auxin synthesis in tobacco tissue cultures, and enhance the activities of carboxydismutase and NADP-glyceraldehyde phosphate dehydrogenase in etiolated rice seedlings; they also promote bud development, germination of some seeds, and the accumulation of nitrate reductase in some embryos.

Cytokinins, in the absence of added nitrate, increase the activity of nitrate reductase (NR) in excised embryos of *Agrostemma githago* (93). Benzyladenine (BA), 10 μM, increases nitrate reductase maximally 3- to 5-fold over the control. The increase is detectable 30–60 min after BA treatment and is, therefore, one of the fastest biochemical responses to cytokinin so far observed. NADH-cytochrome *c* reductase and flavin mononucleotide (FMN)-nitrate reductase activities are associated with NADH-nitrate reductase; these activities increase in response to either nitrate or benzyladenine (94). The physicochemical properties of nitrate-induced and benzyladenine-induced nitrate reductase seem to be identical as demonstrated by sedimentation velocity through a sucrose gradient, differential precipitation with $(NH_4)_2SO_4$, gel filtration, and gel electrophoresis (94).

When excised embryos of *A. githago* are incubated with a mixture of nitrate and benzyladenine, an additive effect on nitrate reductase is observed. Benzyladenine-mediated nitrate reductase induction is not the consequence of leakage of nitrate from a storage pool to a metabolically active pool. Unlike the induction by nitrate, the induction of nitrate reductase by benzyladenine does not depend on the level of endogenous nitrate, suggesting that benzyladenine and nitrate act by different mechanisms to induce nitrate reductase activity.

Although benzyladenine does not have a significant effect on the degradation of pre-existing protein in general, in *Agrostemma* embryos (95) it is still possible that benzyladenine preferentially slows down the degradation of nitrate reductase. Unlike nitrate, benzyladenine does not increase nitrite reductase activity in *Agrostemma* (96).

The cytokinins kinetin and zeatin promote respiration of soybean callus tissue beginning at about 3 hr after treatment with the hormones (97). Addition of adenine does not affect respiration, and the effects of various cytokinin concentrations on oxygen uptake parallel effects on overall growth. This finding also seems worthy of more attention because it occurs in a relatively short time.

ETHYLENE

General

It seems likely that every plant tissue at some stage in its development both produces, and responds to, ethylene. This appraisal of the importance of ethylene is of recent origin. Yet, since 1901, ethylene has been known to regulate plant growth and development.

Ethylene can regulate ripening, senescence, anthocyanin synthesis, abscission, epinasty, mitosis, swelling and elongation, hypertrophy, dormancy, hook closure, leaf expansion, flower induction, sex expression, and exudation (98). As judged by the close similarities in the dose-response relationships, the action of ethylene homologues, and the competitive action of carbon dioxide, each of these responses involves an identical ethylene receptor site.

Biosynthesis of Ethylene

L-Methionine is the only generally accepted precursor of ethylene in the tissues of higher plants. On the basis of experiments in which methionine, labeled with ^{14}C in various positions, was administered to apple slices, it was shown that the C-1 of methionine is converted to CO_2, C-2 to several metabolites, C-3 and C-4 form ethylene, the methyl C and some of the sulfur appear in *S*-methylcysteine, the methyl C in pectin, and some of the sulfur in the sulfur-containing amino acids. Prior to the acceptance of methionine as an ethylene precursor, several alternative models to account for the formation of ethylene were studied (44). In the first of these, peroxidized linolenic acid in reactions catalyzed by Cu^+ (produced from Cu^{2+} and ascorbate) yielded ethylene as well as several other unsaturated and saturated hydrocarbon gases.

In a second model system (44), cupric ions and ascorbate catalyzed the degradation of methionine to ethylene. In this system, methional and 2-keto-4-methylthiobutyrate were more effective as ethylene precursors than was methionine. In a third model system (44), FMN and light catalyzed the conversion of methionine to ethylene. In a fourth model system (44), purified horseradish peroxidases catalyzed ethylene formation from methional or 2-keto-4-methylthiobutyrate. This system required either H_2O_2 or Mn^{2+}, a phenol, and sulfite as co-factors; orthodiphenols were inhibitory and methionine itself was not converted to ethylene.

Although these model systems are instructive regarding the possible chemistry of the biosynthesis of ethylene, the details of the in vivo conversion by higher plants of methionine to ethylene remain to be established. In addition to the possible ethylene precursors already mentioned, ethanol, acetate, and acrylic acid have been given serious consideration (98). None of these is generally accepted as an ethylene precursor in higher plants. However, fungi, notably *Penicillium digitatum,* also produce ethylene, probably from glutamate rather than from methionine (99).

In many tissues, treatment with auxins increases ethylene production, apparently by a stimulation of the formation of the enzymes involved in the biosynthesis of ethylene. The increased production of ethylene evoked by indoleacetic acid in subapical pea stem sections quickly (within 1-4 hr) ceases when the sections are transferred to indoleacetic acid-free medium. The rate of ethylene production by subapical pea stem tissue closely parallels the level of free auxin in the tissue, which is determined by the balance between the rate of uptake of auxin from the external medium and the rate of conjugation of the auxin in the tissue. Although both the auxin conjugation and the auxin decarboxylase systems increase in response to an increased external concentration of auxin, the conjugation system seems to be more important in determining the tissue concentration of free auxin (101). The rate of ethylene production may be a useful indicator of the tissue levels of free auxin. For example, more ethylene is produced by the dark side than by the light side of unilaterally illuminated plants and more by the lower side than the upper side in horizontal stem tissue (98).

Tissue Response

The concentration of ethylene required to produce threshold effects with respect to a variety of physiological responses is 0.01 ppm; half-maximum responses occur at 0.1 ppm; saturation of the responses occurs at 10 ppm and higher concentrations are generally not toxic. In addition, relative effectiveness of ethylene homologues is approximately the same in the various responses.

The order of physiological activity of olefins (ethylene, propylene, butylene, etc.) active in the ethylene bioassay is similar to their order in forming complexes with silver ions, suggesting that the receptor site includes a metal ion. This possibility is consistent with the fact that carbon monoxide, at concentrations well below those required to inhibit cytochrome oxidase, mimics the physiological effects of ethylene. Carbon monoxide characteristically binds to (and usually inhibits) only those enzymes that include a metal at their active site. The most likely metals at the ethylene receptor site would seem to be Cu^{2+}, Fe^{2+}, or Zn^{2+}.

There is as yet no clue as to the intracellular localization of the ethylene receptor site(s). No one has yet reported a response to ethylene by a cell-free enzyme, organelle, or subcellular fraction. If there is a direct effect of ethylene on a membrane or membranes, it is most likely a highly specific effect that involves only a small fraction of the solutes which normally move through the membrane.

Progress toward an understanding of the mode of action of ethylene may well depend on the choice of system to be used for further study. Even though the response is slow, ripening fruit is an attractive material because a relatively large amount of tissue is available in each fruit and all, or nearly all, cells respond to the presence of ethylene.

The specialized cells involved in fruit and leaf abscission offer the advantage of the expression of the ethylene effect, i.e., abscission, apparently involving the synthesis and secretion of only a few cell wall-dissolving enzymes, the principal one being cellulase.

Just as fruit tissue responds to ethylene by producing the enzymes characteristic of ripening fruit and abscission layer cells respond to ethylene by producing enzymes involved in abscission, a great variety of other tissues produce, in response to ethylene, enzymes characteristic for each tissue. In many cases, the increase in enzyme activity is most probably a result of enzyme synthesis rather than activation. In some tissues, inhibitors of RNA synthesis block the expression of ethylene action. It is reasonable to suppose that at least a part of the RNA synthesis required is that RNA specifically required for the synthesis of the ethylene-evoked, tissue-specific enzymes; however, this has not yet been shown. Alternatively, protein synthesis and RNA synthesis may be required to maintain cells in a state competent to respond to ethylene.

The etiolated pea seedling is an attractive experimental tissue because all parts of the seedling respond to ethylene; "Stem growth slows, the hook tightens, the sub-apex swells and nutates horizontally, root growth slows and the zone of elongation swells, root hairs form, lateral root formation is inhibited, and the root tip bends plageotropically" (101, 102). The subapical swelling occurs 3–4 hr after the application of ethylene, requires RNA and protein synthesis, and is accompanied by a marked decrease in the incorporation of hydroxyproline-containing peptides into the wall and by an alteration in the birefringence pattern of the wall.

In many respects, the tropistic and epinastic responses are meaningful starting points for the search for the primary site of action of ethylene. These effects, i.e., hook tightening in etiolated seedlings, leaf epinasty, horizontal nutation in stems, and plageotropism in roots, apparently result from the inhibition by ethylene of the lateral transport of auxin and are visible within minutes after the application of ethylene. Because the immediate effects of changed auxin concentrations on elongation are visible in 0–10 min and do not require RNA or protein synthesis, the effect of ethylene on lateral transport of auxin may well be close to the primary site of action of the hormone.

The regulation by ethylene of aging in the flowers of *Ipomoea tricolor* provides an exceptionally convenient system for studies of the mechanism of action of ethylene. In this tissue, ethylene promotes ethylene production, RNase synthesis, and senescence of the corolla (103).

Elucidation of the mechanisms by which the plant responds to ethylene should be aided by the observation that Co^{2+} specifically inhibits the synthesis of ethylene from methionine (104) and by the observation that Ag^{2+} specifically inhibits the response of the plant to both endogenous and exogenous ethylene (105).

ABSCISIC ACID

Catabolic Metabolism

When abscisic acid (ABA) is applied to a tomato plant, both plus and minus enantiomers are conjugated with glucose to form polar, water-soluble abscisyl-β-D-glucopyranoside (44, 106).

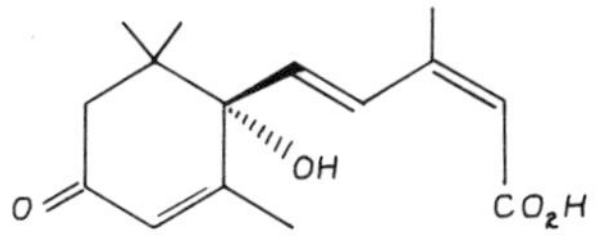

(+) – ABSCISIC ACID

The glucose ester of abscisic acid (Figure 8) is as inhibitory as abscisic acid on a molar basis when it is tested in a rice seedling bioassay, probably as a result of the hydrolysis of the ester to release free abscisic acid. As in the case of the glucose derivatives of gibberellins and auxins, the function of the glucose ester of abscisic acid is still unknown. Because the glucose ester can be rapidly hydrolyzed to form free abscisic acid, a function of the glucose ester as a storage form seems likely. Although both plus and minus ABA are converted to the glucose derivatives, only the plus enantiomer is hydroxylated on one of the terminal methyl groups to form the stable metabolite C, which rearranges to give phaseic acid by a nucleophilic attack of the 6′-hydroxymethyl group on C-2′ to form a saturated furan ring. In bean embryo, phaseic acid can be further converted to 4′-dihydrophasic acid (107).

Physiology and Mode of Action

One of the findings leading to the isolation of abscisic acid was based on the discovery that a growth-inhibiting substance, then called dormin, was able to cause bud dormancy in several woody plant species.

Abscisic acid is an effective inhibitor of seed germination; many dormant seeds contain abscisic acid, suggesting that abscisic acid is able to maintain seed dormancy, although it may not be the only factor controlling this state. As is the case in dormant buds, the abscisic acid content in some dormant seeds decreases after stratification. Apparently, growth-promoting substances such as gibberellins and cytokinins also play a role in seed germination. Therefore, the balance between abscisic acid and growth-

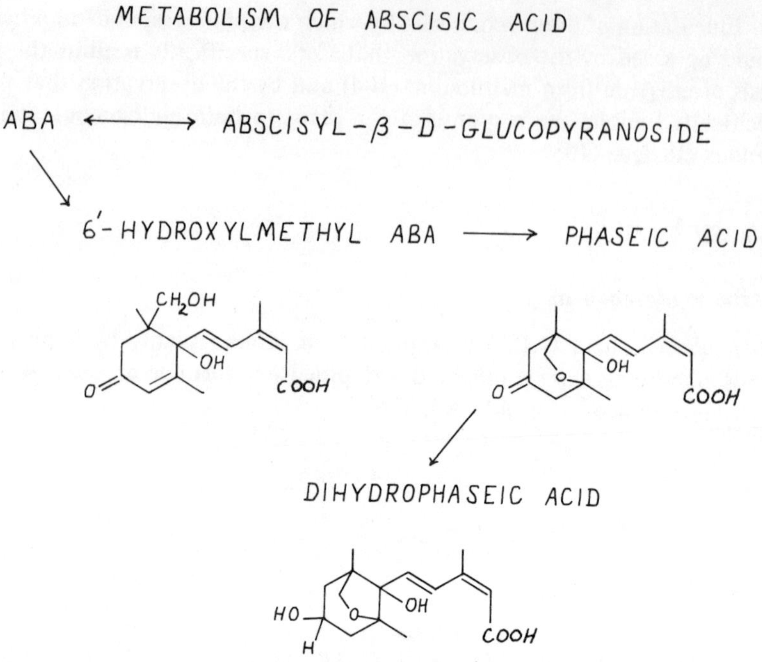

METABOLISM OF ABSCISIC ACID

ABA ⟷ ABSCISYL-β-D-GLUCOPYRANOSIDE

6′-HYDROXYLMETHYL ABA ⟶ PHASEIC ACID

CH₂OH OH COOH O

O OH COOH

DIHYDROPHASEIC ACID

HO OH COOH H

Figure 8. Metabolism of abscisic acid.

promoting substances may control the release of seed dormancy. For those seeds containing abscisic acid, the content of abscisic acid starts to increase during the development of seed and fruit. Many fruits, the cotyledon of avocado seeds, and, to a lesser extent, the endosperm of developing wheat seeds are capable of incorporating labeled mevalonate into abscisic acid, indicating that the increase in abscisic acid in these tissues is not attributable to transport from other parts of the plant.

The effect of abscisic acid on the germination of cotton and wheat seed has been studied in detail. The content of abscisic acid in cotton fruit (mostly in the ovary wall) increases during seed development. Isolated developing embryos can be germinated precociously if the abscisic acid is washed away. This precocious germination can be prevented if abscisic acid or ovary extract is added again to the washed embryos. The inhibitory effect of abscisic acid can be abolished by germinating the seeds in the presence of inhibitors of transcription such as actinomycin D, indicating that the effect of abscisic acid depends on continuous synthesis of RNA (108).

Abscisic acid and structurally related compounds reduce the growth of wheat coleoptiles. The incorporation of amino acids into proteins by polysomes isolated from coleoptiles treated with abscisic acid is strongly inhibited. Because total RNA synthesis is not altered by the hormone treatment of the coleoptile, the effect of abscisic acid in this system may also be post-transcriptional (109).

The level of abscisic acid in leaves increases dramatically when the plant is subject to water stress. When excised wheat leaves are kept in a wilted state, the content of abscisic acid increases by as much as 40-fold in 4 hr (110). Similar observations have been made with intact plants of several other species. The most rapid increase in abscisic acid occurred when 5–10% of tissue fresh weight was lost. The increase has been shown by Milborrow and Noddle (111) to result from synthesis and not the release of a bound form. The principal role of abscisic acid in wilted plants is to close the stomata, thus preventing further loss of water. Stomata open up when the guard cells are under turgor, which is caused by the accumulation of K^+ ion in the guard cells (112). Abscisic acid prevents and reverses the accumulation of K^+ within minutes of application. Abscisic acid also affects K^+ uptake in slices taken from expanding leaves (113) and Na^+, K^+, and Cl^- uptake in discs in beet roots (114).

Abscisic acid can reverse the effect of growth promoters such as auxins, gibberellins, and cytokinins in certain plant tissues. The effect of gibberellic acid on hydrolase synthesis and secretion in the aleurone layer of barley seed has been extensively studied. Abscisic acid reverses the gibberellin-enhanced amylase and protease synthesis, membrane-bound polysome formation, incorporation of ^{32}P into membrane phospholipids, poly A RNA synthesis, and the activation of phosphorylcholine transferases (55, 69). Abscisic acid has no effect on general cellular metabolism as measured by oxygen consumption. After 12 hr of gibberellin treatment, α-amylase synthesis is no longer sensitive to inhibitors of transcription such as cordycepin (3′-deoxyadenosine) (67), whereas abscisic acid at this stage still effectively prevents the further synthesis of α-amylase. It seems that abscisic acid selectively inhibits the synthesis of α-amylase, although the synthesis of other proteins remains normal.

Abscisic acid prevents the auxin-enhanced cell elongation in *Avena* coleoptile with a lag period of a few minutes (115).

XANTHOXIN

Xanthoxin is a possible precursor of abscisic acid and is itself an endogeneous inhibitor (having effects similar to or identical with those of abscisic acid) in many plant tissues (116).

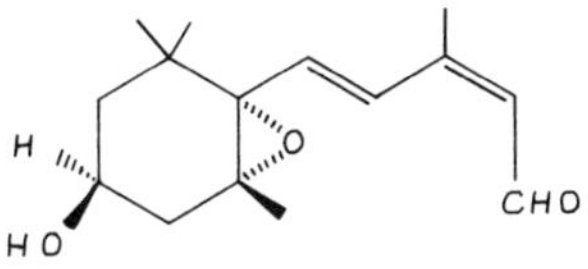

H O HO CHO

XANTHOXIN

The natural occurrence of xanthoxin throughout the plant kingdom seems to follow a pattern similar to that of ABA. However, within the plant the distribution of xanthoxin seems to be limited; young shoots are the best source. When present, the levels of xanthoxin seem to be no lower than those of abscisic acid (115).

The biological activity of xanthoxin has been tested in several systems, including the elongation of wheat coleoptile and of lettuce hypocotyl, cress seed germination, tobacco callus tissue growth, and bean petiole abscission formation (117). Although 2-*trans*-xanthoxin is much less effective than the 2-*cis* isomer, the latter seems to be as effective as abscisic acid in most of these tests. When xanthoxin was applied to cut tomato shoots, the level of abscisic acid was drastically increased (117), suggesting that the inhibitory effect of xanthoxin may result in this instance from its conversion into abscisic acid. Xanthoxin added to the transpiration stream of detached leaves closes stomata rapidly (118). However, when tested with epidermal strips, xanthoxin had no effect on the stomatal aperture. This suggests that xanthoxin, which probably has no direct effect on guard cells, can be rapidly converted into abscisic acid by plant tissues. It was estimated that more than 50% of the xanthoxin could be converted into abscisic acid in less than 5 min (118).

FATTY ACIDS

Short chain free fatty acids seem to be present in dormant seeds and buds in sufficiently high concentrations to account for the dormancy (119) (Table 2). Exogenous short chain fatty acids at 5×10^{-4} inhibit lettuce seed germination. This inhibition is overcome by red light, gibberellic acid, or kinetin (119) (Table 3). Added short chain fatty acids also inhibit the gib-

Table 2. Concentrations of short chain fatty acids in various plant materials (mg/kg)

	C-6	C-7	C-8	C-9	C-10	C-11	Total
Wild oat grain	8.93	23.92	22.48	19.61	13.98		88.92
Dock seed	11.17	18.79	17.76	17.49	20.8		86.01
Ash seeds	9.48	10.87	12.46	11.71	8.99		53.51
Sycamore buds	4.69	48.77	5.10	8.06	7.55	1.51	75.68
Spruce buds	9.54	35.78	463.87	12.52	5.67	3.18	530.56
French bean pods	0.4	0.49	0.9	0.95	0.45		3.19
Broad bean seeds			0.20	0.19	0.08		0.47
Ivy leaves	0.15	0.09	0.11				0.35
Willow catkins	0.24	3.16	1.57	0.32	1.15		6.55
Potato peelings	0.1	0.85	1.27	1.18	1.62		5.02
Holly berry flesh	0.09	0.09	0.42	0.26			0.86
Moss	0.5	1.35	2.75	1.35	0.13		6.08
Liverwort	0.11	0.1	0.17	0.06	0.11		0.55

Table 3. Percentage of germination of lettuce seed

Acid	Light treatment	Concentration			
			6×10^{-4}	5×10^{-3}	10^{-2}
Hexanoic (C-6)	Dark	92	50	3	0
	R[a]	94	95	41	0
	R + FR	87	45	0	0
	R + FR + R	96	94	46	0
Heptanoic (C-7)	Dark	92	39	0	0
	R	94	94	39	0
	R + FR	87	35	0	0
	R + FR + R	96	94	27	0
Octanoic (C-8)	Dark	92	35	0	0
	R	94	92	2	0
	R + FR	87	31	0	0
	R + FR + R	96	93	5	0

[a] R and FR, short exposures to red and far red light, respectively.

berellin-dependent amylolysis of barley endosperm (120); this inhibition is overcome by high (10^{+5} M) concentrations of gibberellic acid (B. Johnson and J. E. Varnar, unpublished observations). These free fatty acids have two interesting properties as possible natural growth regulators: as free fatty acids, they are volatile, and their pK values are about 5. Thus, within a cell or tissue they would move up a pH gradient, and they would tend to be lost to the atmosphere at pH values below pH 7.

HORMONE BINDING IN PLANTS

In spite of the availability of plant hormones of high specific radioactivity and of good methods for the study of hormone receptors, no plant hormone receptor has been isolated. Work in this area is discussed in an excellent review (121), and alternative suggestions to the idea that plant hormones must act by first forming a hormone-receptor complex with a highly specific macromolecule (usually assumed to be a protein) are made.

REFERENCES

1. Greenwood, M. S., Shaw, S., Hillman, J. R., Ritchie, A., and Wilkins, M. B. (1972). Planta 108:179.
2. DeYoe, D. R., and Zaerr, J. B. (1976). Plant Physiol. 58:299.
3. Bandurski, R. S., and Schulze, A. (1974). Plant Physiol. 54:257.
4. Ueda, M., Ehmann, A., and Bandurski, R. S. (1970). Plant Physiol. 46:715.
5. Piskornik, Z., and Bandurski, R. S. (1972). Plant Physiol. 50:176.
6. Schneider, E. A., and Wightman, F. (1974). Annu. Rev. Plant Physiol. 25:487.

7. Kopcewicz, J., Ehmann, A., and Bandurski, R. S. (1974). Plant Physiol. 54:846.
8. Andreae, W. A., and Good, N. E. (1955). Plant Physiol. 30:380.
9. Davies, P. J. (1976). Plant Physiol. 57:197.
10. Sheldrake, A. R. (1973). New Phytologist 72:433.
11. Sheldrake, A. R., and Northcote, D. H. (1968). Nature (Lond.) 217:195.
12. Sheldrake, A. R., and Northcote, D. H. (1968) New Phytologist 67:1.
13. Mahadevan, S., and Stowe, B. B. (1972). *In* D. J. Carr (ed.), Plant Growth Substances, p. 117. Springer-Verlag, Heidelberg.
14. Hattori, A., and Marumo, S. (1972). Planta 102:85.
15. Isogai, Y., Okamoto, T., and Koizumi, T. (1967). Chem. Pharm. Bull. Tokyo 15:151.
16. Abe, H., Uchiyama, M., and Sata, R. (1974). Agr. Biol. Chem. 38:896.
17. Milborrow, B. V., Purse, J. G., and Wightman, F. (1975). Ann. Bot. 38: 1143.
18. Ray, P. M. (1973). Plant Physiol. 51:601.
19. Vanderhoef, L. N., and Stahl, C. A. (1975). Proc. Natl. Acad. Sci. USA 72:1822.
20. Vanderhoef, L. N., Stahl, C. A., Williams, C. A., Brinkman, K. A., and Greenfield, J. C. (1976). Plant Physiol. 57:817.
21. Vanderhoef, L. N., Stahl, C. A., and Lu, T.-Y. S. (1976). Plant Physiol. 58:402.
22. Dheidah, M., and Black, M. (1976). Physiol. Plant. 37:83.
23. Jacobs, M., and Ray, P. M. (1975). Plant Physiol. 56:373.
24. Bonner, J. (1934). Protoplasma 21:406.
25. Hager, A., Menzel, H., and Krauss, A., (1971). Planta 100:47.
26. Rayle, D. L., and Cleland, R. (1972). Planta 104:282.
27. Cleland, R. (1975). Planta 127:233.
28. Haschke, H. P., and Lüttge, U. (1975). Plant Physiol. 56:696.
29. Cohen, J. D., and Nadler, K. D. (1976). Plant Physiol. 57:347.
30. Rubenstein, B., Johnson, K. D., and Rayle, D. L. (1976). Proceedings of the Workshop in Regulation of Cell Membrane Activities in Plants.
31. Rubinstein, B., and Light, E. N. (1973). Planta 110:43.
32. Rubinstein, B. (1977). Plant Physiol. 59:369.
33. Lürssen, K. (1976). Plant. Sci. Lett. 6:389.
34. Bayer, M. H., and Sonba, J. (1976). Z. Pflanzenphysiol. 78:271.
35. Hebard, F. V., Amatangelo, S. J., Dayanandan, P., and Kaufman, P. B. (1976). Plant Physiol. 58:670.
36. Penny, P., Dunlop, J., Perley, J. E., and Penny, D. (1975). Plant Sci. Lett. 4:35.
37. Perley, J. E., Penny, D., and Penny, P. (1975). Plant Sci. Lett. 4:133.
38. Beyer, E. M., Johnson, A. L., and Sweetser, P. B. (1976). Plant Physiol. 57:839.
39. Guilfoyle, T. J., Lin, C. Y., Chen, Y. M., Nagao, M. A., and Key, J. L. (1975). Proc. Natl. Acad. Sci. USA 72:69.
40. Lin, C.-Y., Chen, Y.-M., Guilfoyle, T. J., and Key, J. L. (1976). Plant Physiol. 58:614.
41. Byrne, H., Christou, N. V., Verma, D. P. S., and Maclachlan, G. A. (1975). J. Biol. Chem. 250:1012.
42. Verma, D. P. S., Maclachlan, G. A., Byrne, H., and Ewings, D. (1975). J. Biol. Chem. 250:1019.
43. Sacher, J. A., and Davies, D. D. (1974). Plant Cell Physiol. 15:157.
44. Varner, J. E., and Ho, T. T. H. (1976). *In* J. Bonner and J. E. Varner (eds.), Plant Biochemistry. Academic Press, New York.

45. West, C. A. (1973). *In* B. V. Milborrow (ed.), Annu. Proc. of the Phytochem. Soc., Number 9. Academic Press, New York.
46. Kende, H. (1967). Plant Physiol. 42:1612.
47. Musgrave, A., Kays, S. E., and Kende, H. (1969). Planta 89:165.
48. Railton, I. D., Durley, R. C., and Pharis, R. P. (1974). Plant Physiol. 54:6.
49. Musgrave, A., Kays, S. E., and Kende, H. (1972). Planta 102:1.
50. Nadeau, R., Rappaport, L., and Stolp, C. F. (1972). Planta 107:315.
51. Nadeau, R., and Rappaport, L. (1974). Plant Physiol. 54:809.
52. McComb, A. J., McComb, J. A., and Duda, C. T. (1970). Plant Physiol. 46:221.
53. Rogler, C. E., and Dahmus, M. E. (1974). Plant Physiol. 54:88.
54. Adams, P. A., Montague, M. J., Tepfer, M., Rayle, D. L. Ikuma, H., and Kaufman, P. B. (1975). Plant Physiol. 56:757.
55. Yomo, H., and Varner, J. E. (1972). Curr. Top. Dev. Biol. 6:111.
56. Bennett, P. A., and Chrispeels, M. J. (1972). Plant Physiol. 49:445.
57. Hardie, D. G. (1975). Phytochemistry 14:1719.
58. Clutterbuck, V. J., and Briggs, D. E. (1973). Phytochemistry 13:45.
59. Jacobsen, J. V. (1973). Plant Physiol. 51:198.
60. Coombe, B. G., Mander, L. N., Paleg, L. G., and Turner, J. V. (1974). Aust. J. Plant. Physiol. 1:473.
61. Ashford, A. E., and Jacobsen, J. P. (1974). Planta 120:81.
62. Taiz, L., and Honigman, W. A. (1976). Plant Physiol. 58:380.
63. Varner, J. E., and Mense, R. M. (1972). Plant Physiol. 49:187.
64. Jones, R. L. (1973). Plant Physiol. 52:303.
65. Jacobsen, J. V., and Zwar, J. A. (1974a). Aust. J. Plant Physiol. 1:343.
66. Jacobsen, J. V., and Zwar, J. A. (1974b). Proc. Natl. Acad. Sci. USA 71: 3290.
67. Ho, D. T.-H., and Varner, J. E. (1974). Proc. Natl. Acad. Sci. USA 71: 4783.
68. Higgins, T. J. V., Zwar, J. A., and Jacobsen, J. V. (1976). Nature (Lond.) 260:166.
69. Ho, D. T.-H., and Varner, J. E. (1976). Plant Physiol. 57:175.
70. Collins, C. G., Jenner, C. F., and Paleg, L. G. (1972). Plant Physiol. 49:398. 49:398.
71. Johnson, K. D., and Kende, H. A. (1971). Proc. Natl. Acad. Sci. USA 68:2674.
72. Ben-Tal, Y., and Varner, J. E. (1974). Plant Physiol. 54:813.
73. Koehler, D. E., and Varner, J. E. (1975). Plant Physiol. 52:208.
74. Firn, R. D., and Kende, H. (1974). Plant Physiol. 54:911.
75. Wood, A., and Paleg, L. G. (1972). Plant Physiol. 50:103.
76. Wood, A., Paleg, L. G., and Spotswood, T. M. (1974). Aust. J. Plant Physiol. 1:167.
77. Tyson, C. A., Zande, H. V., and Green, D. E. (1975). J. Biol. Chem. 251: 1326.
78. Varty, K., and Laidman, D. L. (1976). J. Exp. Bot. 27:748.
79. Jones, R. L., and Chen, R. F. (1976). J. Cell. Sci. 20:183.
80. Gibson, R. A., and Paleg, L. G. (1975). Aust. J. Plant Physiol. 2:41.
81. Firn, R. D. (1975). Planta 125:227.
82. Keates, R. A. B. (1973). Nature (Lond.) 244:355.
83. Leonard, N. J. (1974). Recent Adv. Phytochem. 7:21.
84. Kende, H., and Tavares, J. E. (1968). Plant Physiol. 43:1244.
85. Chen, C. M., and Hall, R. H. (1969). Phytochemistry 8:1687.
86. Short, K. C., and Torrey, J. G. (1972). Plant Physiol. 49:155.

87. Fittler, F., Kline, L. K., and Hall, R. H. (1968). Biochemistry 7:940.
88. Gefter, M. L., and Russell, R. L. (1969). J. Mol. Biol. 39:145.
89. Skoog, F., and Armstrong, D. J. (1970). Annu. Rev. Plant Physiol. 21:359.
90. Dyson, W. H., Hall, R. H., Hong, C. I., Dutta, S. P., and Chedda, G. B. (1972). Can. J. Biochem. 50:237.
91. MacLeod, J. K., Summons, R. E., Parker, C. W., and Letham, D. S. (1975). J. C. S. Chem. Comm. 19:809.
92. Skoog, F., Schmitz, R. Y., Bock, R. M., and Hecht, S. M. (1973). Phytochemistry 12:25.
93. Kende, H., Hahn, H., and Kays, S. E. (1971). Plant Physiol. 48:702.
94. Dilworth, M. F., and Kende, H. (1974). Plant Physiol. 54:821.
95. Kende, H., and Shen, T. C. (1972). Biochim. Biophys. Acta 286:118.
96. Dilworth, M. F., and Kende, H. (1975b). Plant Physiol. 54:826.
97. Moore, T. S., and Miller, C. O. (1972). Plant Physiol. 50:594.
98. Abeles, F. B. (1973). Ethylene in Plant Biology, p. 302. Academic Press, New York.
99. Yang, S. F. (1974). Recent Adv. Phytochem. 7:131.
100. Kang, B. G., Newcomb, W., and Burg, S. P. (1971). Plant Physiol. 47:504.
101. Burg, S. P., Apelbaum, A., Eisinger, W., and Kang, B. G. (1971). Hort. Sci. 6:359.
102. Burg, S. P. (1973). Proc. Natl. Acad. Sci. USA 70:591.
103. Kende, H., and Baumgardner, B. (1974). Planta 116:279.
104. Lau, O.-L., and Yang, S. F. (1976). Plant Physiol. 58:114.
105. Beyer, E. M. (1976). Plant Physiol. 58:268.
106. Milborrow, B. V. (1974). Recent Adv. Phytochem. 7:57.
107. Tinelli, E. T., Sondheimer, E., Walton, D. C., Gaskin, P., and MacMillan, J. (1973). Tetrahedron Lett. 1973:139.
108. Ihle, J. N., and Dure, L. (1970). Biochem. Biophys. Res. Commun. 38: 995.
109. Bonnafous, J. C., Mousseron-Canet, M., and Olive, J. L. (1973). Biochim. Biophys. Acta 312:165.
110. Wright, S. T. C., and Hiron, R. W. P. (1969). Nature (Lond.) 224:719.
111. Milborrow, B. V., and Noddle, R. C. (1970). Biochem. J. 119:727.
112. Raschke, K. (1975). Annu. Rev. Plant Physiol. 26:309.
113. Horton, R. F., and Bruce, K. R. (1972). Can. J. Bot. 50:1915.
114. Van Stevenick, R. F. M. (1972). Z. Pflanzenphysiol. 67:282.
115. Rehm, M., and Cline, M. G. (1973). Plant Physiol. 51:93.
116. Firn, R. D., Burden, R. S., and Taylor, H. F. (1972). Planta 102:115.
117. Taylor, H. F., and Burden, R. S. (1972). Proc. R. Soc. Lond. (Biol.) 180: 317.
118. Raschke, K., Firn, R. D., and Pierce, M. (1975). Planta 125:149.
119. Berrie, A. M. M., Don, R., Buller, D., Alam, M., and Parker, W. (1975). Plant Sci. Lett. 6:163.
120. Buller, D. C., Parker, W., and Reid, J. S. G. (1976). Nature (Lond.) 260: 169.
121. Kende, H., and Gardner, G. (1976). Annu. Rev. Plant Physiol. 27:267.

International Review of Biochemistry
Biochemistry and Mode of Action of Hormones II, Volume 20
Edited by H. V. Rickenberg
Copyright 1978 University Park Press Baltimore

7
Replacement of Serum in Cell Culture by Hormones

G. SATO and L. REID[1]
John Muir College, The University of California, San Diego, La Jolla, California

---

[1] Present Address: Department of Molecular Pharmacology, Albert Einstein College of Medicine, Bronx, New York.

It is the authors' conviction that the cell culture technique has matured sufficiently over the past several years so as to be a major tool in endocrine physiology with the promise of providing insights unique to this experimental approach. For this reason, this chapter deviates slightly from the major theme of this volume; it is hoped that the reader will accept this presentation as an early attempt to describe a newly emerging field of inquiry.

The authors have previously proposed that the major role of serum in cell culture media is to provide groups of hormones required for the growth and survival of cells (94). In the present text and in previous publications (52, 53), evidence was offered in support of this hypothesis. Our studies raise the issue of the definition of a hormone. (The hypothesis that serum can be replaced by groups of hormones involves circular reasoning if the factors are defined as hormones because they are active in culture.) Developing rules for defining a hormone becomes further complicated by considerations of conditioning factors secreted by cells in culture and required by the same or other cells for optimum growth. Nevertheless, one can argue for the hormonal status of these factors. They are synthesized by living cells and, when purified, are effective at doses so small as to exclude any but a regulatory role.

The classical definition of a hormone is that it is a molecule synthesized by glandular cells and secreted directly into the bloodstream which transports it to target tissues on which the molecule exerts specific regulatory effects. It seems to us that the definition is unduly restrictive. It is likely that hormonal physiology has evolved from regulatory mechanisms found in single cell organisms and lower metazoans. These regulatory mechanisms have become associated with a spectrum of physiological processes, among which are a subset classically defined as hormonal. A case in point is the common origin of the sympathetic nervous system and the adrenal medulla. Norepinephrine secreted at a synaptic junction is a neurotransmitter, whereas norepinephrine secreted into the bloodstream by the adrenal medulla is considered a hormone although it activates tissues innervated by the sympathetic nervous system. Evolution has taken the same basic regulatory mechanism and the same molecule and varied the mode of delivery—one through an axonic process and the other through the bloodstream. It is likely that an intermediate mode, the delivery of non-neural regulatory secretions from one cell to neighboring cells, will also be found to be widely prevalent. It seems to these authors that, while nature may have evolved many modes of delivery, the basic mechanisms of action of regulatory molecules will fall into a small number of discrete classes.

The authors would prefer to redefine hormones as regulatory molecules secreted by living cells, effective at low concentrations, and which bring about their final results by one of a few discrete classes of mechanisms, such as steroid mechanisms or cyclic AMP mechanisms. The definition of these mechanisms remains to be elucidated in the future. We confidently expect that the factors active on cultured cells will operate via mechanisms identical with those effected by classical hormones in vivo. In this light, our hypothesis could be restated as follows: the function of serum in cell culture is to provide groups of hormones, a few accessory factors such as the iron transport protein, transferrin, and growth factors whose hormonal status is likely but not yet established. Whether or not growth factors prove to be hormones, we believe the hypothesis has important implications for integrative physiology and cell culture technology.

TISSUE CULTURE MEDIUM

Development of Defined Synthetic Media

The formulation of cell culture media has been under intensive investigation since the first medium, frog lymph, was used by Arnold and later by Harrison to sustain explants (7, 51). At the present time, cell culture medium consists of two parts: 1) a defined synthetic medium, and 2) a biological fluid, commonly serum. Attempts to fabricate synthetic medium began before the discovery of some of the vitamins and the amino acids. This required that the early investigators such as the Lewises had to resort to using substances such as bouillion as a source of these compounds (31, 64, 67). With the advent of modern biochemistry, attempts were made to incorporate the new knowledge into the preparation of synthetic media. Notable among the investigators studying synthetic media were P. White and later the Connaught laboratories group which formulated Medium 199 (76, 110–112). Medium 199 was noteworthy in that it contained most of the vitamins, amino acids, salts, etc. which are now known to be necessary for cell growth.

The modern era in cell culture nutrition started with H. Eagle. His laboratory determined the essentiality of each component of the medium for the growth of the cells in the presence of dialyzed serum (28, 29). This type of work has continued in various laboratories around the world. A useful series of media has been devised by R. G. Ham, who has carefully formulated complex media in which all the components are balanced and can support the growth of several cell lines in completely synthetic media, i.e., in the absence of serum (48). One of Ham's recent contributions is his discovery of the need for trace elements in the media; selenium, for example, is needed for a number of cell lines (74). A definitive review of the subject by Charity Waymouth is a useful source of information on the nutrition of cells in culture (109).

Development of Use of Serum in Cell Culture

Defined synthetic media are usually inadequate for supporting the growth of cell lines and must be supplemented with an undefined component, serum. The use of such biological fluids dates back to the origins of cell culture. In the early days, cells were grown in plasma clots or in medium of salt solutions supplemented with embryo extract. There were a number of technical difficulties in the use of plasma clots (e.g., retraction of the clots and interference with cell counting) which were eliminated by substituting with serum. This was first done by W. R. Earle and was a significant advance (27, 93).

Attempts to Define Serum

Extensive efforts have been made to define the components in serum which are essential for the growth of cells. These efforts are detailed in a recent review by R. F. Brooks, who presents a thorough discussion not only of the many efforts, but also of the confusion and lack of coherence in the field (12). This lack of coherence in the field of serum analysis can be traced to two dominant ideas which have prevailed throughout these investigations. The first is the finding by A. Fisher and then by H. Eagle that dialyzed serum is sufficient for support of the growth of cells in defined media, suggesting that the active components of the serum are macromolecular (28, 29). The second idea is that serum contributes one or at most a few components, probably macromolecular. This bias is expressed in many contemporary papers and by statements such as "factor X substitutes completely for serum." The authors believe that the many attempts to purify the serum factors have suffered primarily from these biases, i.e., the concept that one or a few macromolecular factors isolated from the serum can replace whole serum for the growth of the cells in culture. Instead, we believe that most of the serum in cell culture can be replaced by groups of known hormones in addition to a few accessory factors such as transferrin. When these are provided either no, or very little, serum is required. In some cases, this residual serum requirement can be met by novel growth factors. In other cases, the residual serum provides as yet unknown factors.

This hypothesis clarifies not only why those trying to isolate active factors from serum have found the task so complex but also indicates a systematic route by which to circumvent the difficulties, that is, to initially define the needs of a cell culture for conventional hormones and pure growth factors and then to extract from serum or from tissues, the components responsible for the residual requirement for serum.

REPLACEMENT OF SERUM WITH HORMONES

Use of Hormones in Cell Culture

Hormones have been incorporated into cell culture medium since the early works of Baker and Carrel (8) and Baker and Ebeling in 1939 (9). These investigators used pituitary extracts and adrenocortical extracts in their medium because the hormones produced by these tissues had not yet been purified. Insulin was first used by Gey and Thalhimer shortly after its discovery (41). These early investigators had considered the possibility that the role of serum in cell culture was to provide hormones. However, because of the poor technical development of hormone chemistry at the time, this idea could not be pushed to its logical conclusion. In our laboratory, we have resurrected this idea, having arrived at the notion by a circuitous route. Over the last eight years, we have been attempting to establish cell cultures which depend on trophic hormones active on the parent tissue. This is still an important area badly in need of further development. We began the work by following an old technique in which ovarian cells are injected into the spleens of ovariectomized rats (10). The rationale is that if the ovarian cells are injected at any other site they would produce steroid hormones which would enter the circulation and eventually inhibit secretion of pituitary trophic hormones by the endocrine feedback loop mechanisms. Thus, the transplant of ovarian tissue would be self-limiting. The spleen's blood supply is drained by the hepatic portal veins traveling directly to the liver. Therefore, steroids produced by the ovarian cells implanted into the spleen would first traverse the liver, where they would be metabolically inactivated by the liver's enzymes. Thus, gonadal steroids do not appear in the circulation beyond the liver, and the hypothalamus, sensing the deficiency of ovarian steroids, hypersecretes releasing factors for the gonadotropins, causing enlargement of the splenic implant of ovarian tissue. After six months, such splenic implants develop into sizable masses, whose growth is presumably driven by the high levels of gonadotrophins in the circulation.

Such tissue is ideal for selecting cells in culture dependent for their growth on gonadotrophins. J. L. Clark in our laboratory established cell cultures from such ovarian cells implanted into the spleens of rats (14). The first experiments with the cultures were very disappointing. No response to gonadotrophins could be detected in cell culture, although the cell cultures were known to be hormone dependent because they only grew in animals if they were implanted in the spleens of ovariectomized animals. We quickly realized that we were unable to demonstrate hormone dependency because the serum component of the medium was contributing hormones. Therefore, methods were developed to deplete serum of the hormones. H. A. Armelin in our laboratory developed methods for depleting the serum of steroids (5, 6), and K. Nishikawa developed a CMC column chromatographic method which depleted the serum of basic peptides (78). Such sera would not support the growth of the ovarian cells unless the medium was supplemented with a crude luteinizing hormone (LH) preparation or with steroids. We were able to show that purified LH is not active in this system, and, with the collaboration of D. Gospodarowicz at the Salk Institute, we were able to isolate the active component from the crude LH preparation. This was a novel basic peptide, ovarian growth factor (OGF), which was distinct from any of the known pituitary hormones (43). Subsequently, H. A. Armelin used 3T3 cells as control cells for the OGF experiments because it was assumed that 3T3 cells would not require OGF. Much to our surprise, Armelin (3–6) and his co-workers showed that 3T3 cells also require crude LH, insulin, and hydrocortisone for their growth. This requirement led Gospodarowicz to pursue the purification of the factor in crude LH which was necessary for 3T3 cells; in so doing, he was able to purify another new factor, fibroblast growth factor (FGF) (42).

At this point several conclusions could be drawn.

1. It is possible to establish cell cultures which are subject to in vivo growth regulatory mechanisms. For example, growth of transplants of ovarian cells requires hormonal conditioning of the host.
2. In order to demonstrate hormonal dependency in culture, it is necessary to deplete the serum of those hormones.
3. Some of the hormones required by cells in culture might be novel peptides.
4. Multiple hormones are necessary for the growth of cells.

From these observations, the hypothesis which most easily follows is that the role of serum in cell culture is to provide hormones, and that it should be possible to replace the serum with hormones, some of which would be novel factors. This hypothesis was published in 1974 (94). Some objected that, if it were true, it should have long been the practice to combine synthetic media with commercially available hormones. To counter this criticism, the argument was advanced that many of the hormones were yet to be discovered. In fact, as discussed under "Growth Factors," many novel hormones are being discovered by the use of cell culture technique.

Initial Attempts to Replace Serum with Hormones: Studies with GH_3

I. Hayashi, in our laboratory, began work on the replacement of serum by hormones in cultures of GH_3 cells, a rat pituitary tumor line which secretes growth hormone and prolactin (53). Hayashi diluted the serum of the medium to the point at which growth no longer took place. She then added various hormones to the medium in an attempt to restore the growth of the cells. When a hormone was found which was active, the serum was diluted further until growth again ceased and another hormone was sought which would again restore growth. In this manner, Hayashi was eventually able to eliminate all the serum. In these first experiments, GH_3 cells grew as well in serum-free F12 medium supplemented with transferrin, a somatomedin preparation, parathyroid hormone, triiodothyronine, and thyrotropin-releasing hormone as they did in serum-supplemented medium (Figure 1).

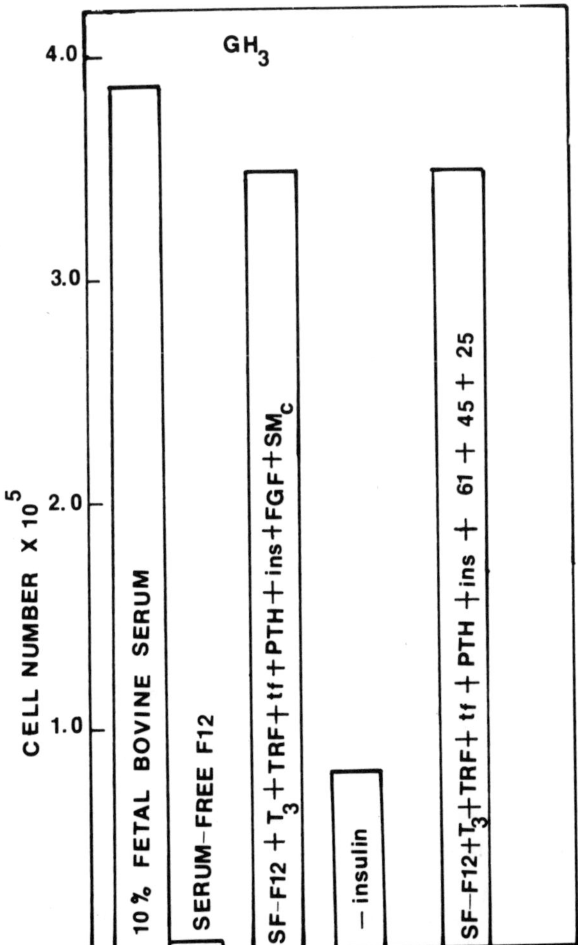

Figure 1. Growth of GH_3 cells in serum-supplemented and hormone-supplemented serum-free F12 medium. Cells were inoculated onto 60-mm Falcon plastic petri dishes at a seeding density of 10^5 cells per plate. After 4 days of incubation at 37 °C in a 5% CO_2, 95% air atmosphere, the cells were trypsinized, and the number of cells assayed with a Coulter counter. In serum-free medium, most of the cells disintegrated after 4 days. If the serum-free medium is supplemented with the highly purified hormones insulin, triiodothyronine, TSH-releasing hormone, parathormone, fibroblast growth factor, somatomedin C, and transferrin, the growth is almost equivalent to that in serum-supplemented media. If insulin is omitted, the cells barely survive. Isoelectric focusing fractions (fraction numbers given; refer to Figure 8) from bloodmeal can substitute for FGF and somatomedin C.

A partially purified preparation (5,000 X) of somatomedin was donated to us by K. Uthne and used in these first experiments. Dr. J. Larner, as a visiting scientist in our laboratory, was puzzled by the apparent lack of an insulin requirement by GH_3 cells and pointed out the general significance of this hormone on cellular metabolism and its influence on many tissues. Subsequently, he and Hayashi determined that insulin was indeed a requirement of GH_3 cells (the crude somatomedin preparation was shown to be replaceable by insulin, FGF, and somatomedin C). Since then, all cell lines studied have been shown to require insulin. These studies provided substantial support for the hypothesis that the serum in cell culture could be replaced by hormones and led to investigations on the hormone requirements of a number of cell lines.

CURRENT STATUS OF STUDIES DEFINING HORMONE REQUIREMENTS OF CELLS GROWN IN VITRO

Cell Lines Whose Requirements Are Defined

HeLa cells exemplify an in vitro cell line for which the complete hormonal requirements for growth have been elucidated. Previously, Hayashi and Sato (53) showed that HeLa and BHK_{21} cells could grow in a serum-free medium supplemented with transferrin, ceruloplasmin, and a number of other factors or hormones. The importance of each component to the growth of the cell lines was not demonstrated at that time. S. Hutchings in our laboratory has now demonstrated that for HeLa cells the essential substances are insulin, transferrin, hydrocortisone, epidermal growth factor (EGF), and FGF (52, 95). When Ham's F12 medium is supplemented with these hormones and transferrin, it supports single cell colony formation as well as does serum-supplemented medium (Figure 2). This medium also supports the long-term continuous culture of these cells (Figure 3). In fact, HeLa cells have been maintained continuously in serum-free media supplemented with these hormones for over six months. A similar medium has been found adequate for BHK_{21} cells (71).

J. Mather (73) in our laboratory has developed a hormone-supplemented, serum-free medium which supports the growth of a B16 mouse melanoma cell line developed by P. Roberts. For B16 melanoma, the media must be supplemented with insulin, transferrin, testosterone or progesterone, partially purified follicle-stimulating hormone (FSH), and luteinizing hormone-releasing hormone (LHRH) (Figure 4).

A newly established normal rat ovarian cell line (72) has been studied by Ohasa and Sato (80). They have found that for growth these require insulin, transferrin, hydrocortisone, and triiodothyronine. The remaining need for serum is after subculturing. The enzymic dissociation of the cells from the plates affects the cells in some way and prevents survival of the cells unless they are suspended in a serum-supplemented medium for at least 6 hr. It has been found that cells grown in serum-free medium attach more tightly to tissue culture plates and require longer than usual enzymic treatment to detach them. The prolonged enzymic treatment may damage the cells, a damage which can be overcome by treatment in serum for a

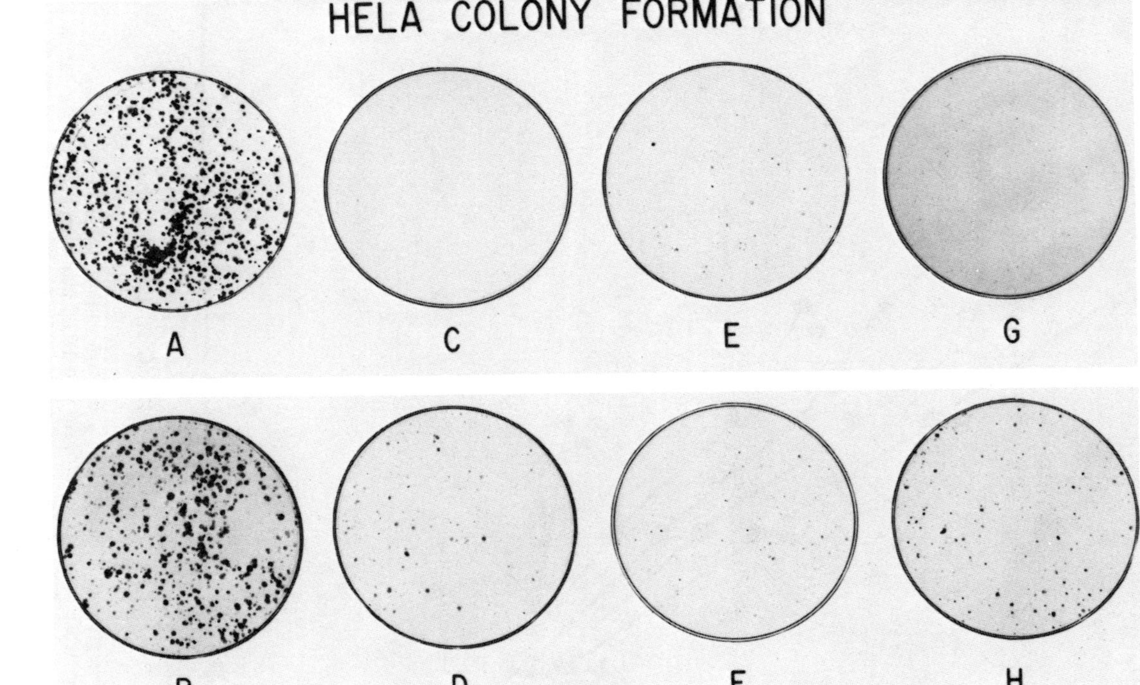

Figure 2. Colony formation by HeLa cells in serum-supplemented and hormone-supplemented F12 media. Cells were plated in 10 ml of media in 100-mm Falcon petri dishes and inoculated at seeding densities of 5×10^3. After 10 days of incubation, the plates were fixed with buffered formaldehyde and stained with 1% gentian violet in 70% ethanol. One medium change occurred after 5 days of growth. *A*, 10% serum-supplemented medium; *B*, serum-free F12 medium supplemented with insulin, transferrin, hydrocortisone, NIH-LH (FGF) and EGF; *C*, serum-free F12 medium; *D*, as in *B* with the omission of NIH-LH (source of FGF); *E*, as in *B* with the omission of transferrin; *F*, as in *B* with the omission of EGF; *G*, as in *B* with the omission of hydrocortisone; *H*, as in *B* with the omission of insulin.

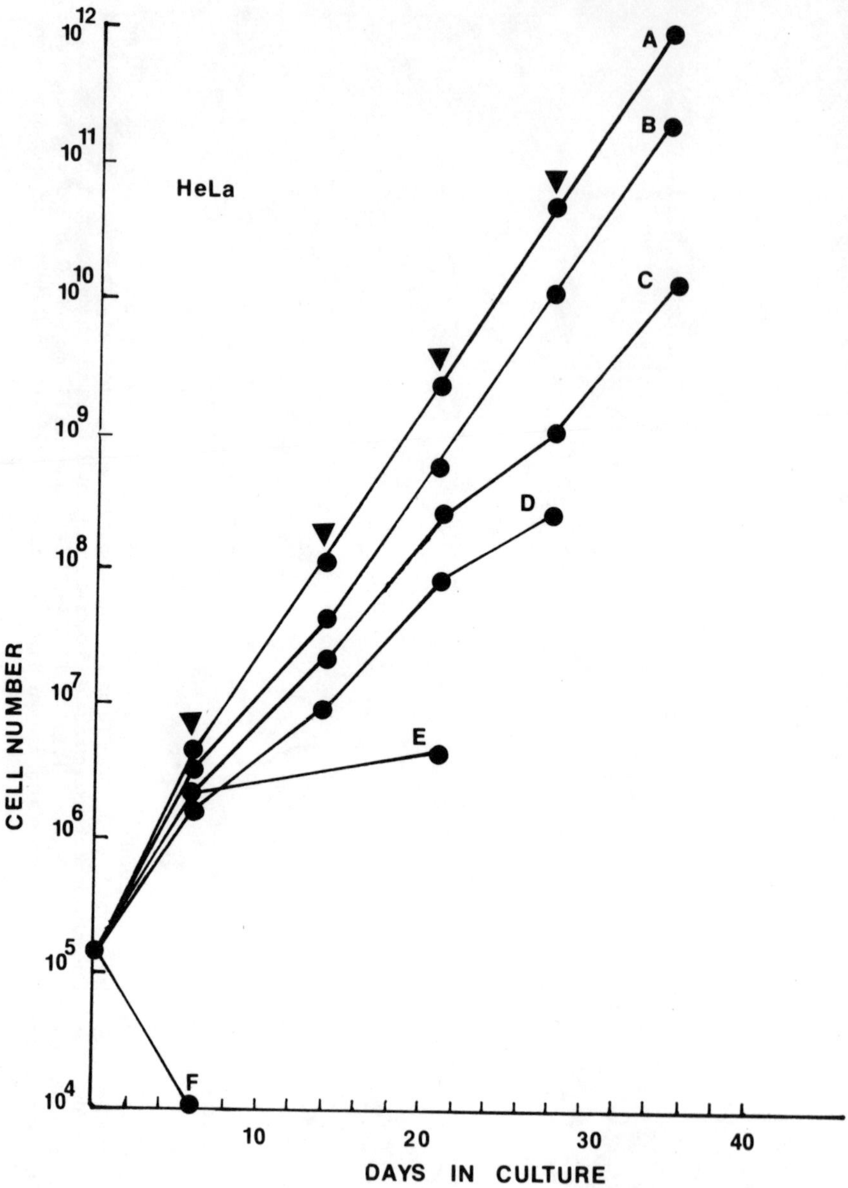

Figure 3. Long-term growth of HeLa cells in hormone-supplemented media. Cells were inoculated at seeding densities of 1.5×10^5 cells into 10 ml of medium in 100-mm Falcon plastic tissue culture plates. Every 7 days the cells were subcultured at a ratio of 1:20. The serum-free media were supplemented with insulin, transferrin, hydrocortisone, EGF, and NIH-LH (FGF). Curves C, D, and E represent growth in media equivalent to that in B except that insulin (C), EGF (D), or hydrocortisone (E) were omitted. Curve F shows the fate of the cells in serum-free F12 medium. Curve A represents the growth in F12 medium supplemented with 10% fetal calf serum. Although NIH-LH served as the source of FGF in this experiment, equivalent results are obtained with highly purified FGF.

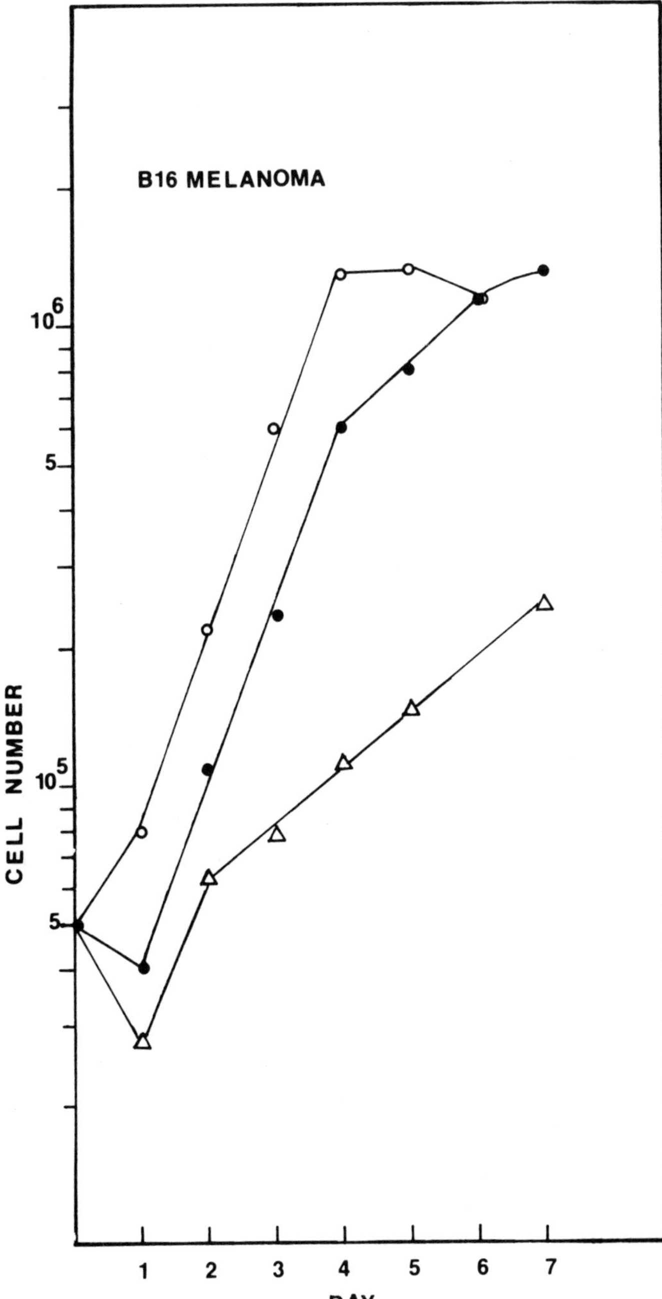

Figure 4. Growth of B16 melanoma cells in serum-supplemented and hormone-supplemented medium. The synthetic medium consisted of a 50:50 mixture of F12 and DME. ○, medium supplemented with 10% serum, •, medium supplemented with insulin, transferrin, progesterone, NIH-FSH, and LH-releasing hormone; △, synthetic medium (DME/F12).

few hours. Other cell lines such as HeLa and B16 melanoma, which we now grow in serum-free medium supplemented with hormones, can be grown on bacteriological plates to which they attach less tightly than to tissue culture plates.

Cell Lines with Partially Defined Hormone Requirements

With some cell lines, the serum cannot be completely eliminated despite strenuous efforts to entirely replace the serum with hormones. It is useful to list some of these cases. In Table 1 are given some of the cell lines currently under investigation. Table 2 lists their known requirements. In those cases in which the medium is only partially defined, a small amount of serum, 1% or less, must be added. If the exact concentration of required hormone has been determined, it is given in parenthesis after the name of the hormone.

Balb-3T3 cells undergo only two rounds of replication in a serum-free medium supplemented with insulin, transferrin, EGF, FGF, and hydrocortisone. If the medium is supplemented with only 0.1% fetal calf serum, the growth rate is equivalent to that in 10% serum. Serum, therefore, contains unidentified factors necessary for the growth of 3T3 cells. It is noteworthy that very low concentrations of serum can satisfy this requirement. Larner and Hayashi (unpublished results) have found that human fibroblasts have similar requirements.

An intimal cell line developed by Levine and Buonassisi (63) from the subendothelial cell layer of rabbit aorta (63) is currently being studied by G. Serrero (unpublished results), who finds that the cells can grow in a serum-free medium supplemented with insulin, transferrin, NIH-LH, growth hormone (GH), progesterone, and bovine serum albumin. The cells grow, but at a lower rate than that attained in serum-supplemented medium. To date, no substitute has been found for serum which will allow a high growth rate.

A. Rizzino in our laboratory can grow two different lines of mouse embryonal carcinoma cells, PCC 4-aza-1 and F9, indefinitely in a serum-free medium supplemented with insulin, transferrin, β-mercaptoethanol, and partially purified fetuin (88). PCC4 cells grown in this medium for long periods (4 weeks and 20 generations) retain their capacity for differentiation when they are injected into animals (Figures 5 and 6). In this case, it has not yet been possible to replace the fetuin by more defined substances.

It should be emphasized that although our ultimate goal is to develop defined media for many cell lines, we have the more immediate goal of creating conditions, as with low serum or with fetuin supplementation, in which many of the defined substances become essential. This permits the definition of further hormonal requirements and provides a degree of control over the cells not possible previously.

Generalizations

From the results of our current research, several generalizations can be set forward in a tentative manner.

1. The hypothesis that serum can be replaced by hormones and a few accessory factors such as transferrin and can maintain cell growth in vitro is essentially correct.
2. Insulin and transferrin are requirements for almost all cell lines studied and should always be among the first substances to be tested in attempts to replace serum with defined components.
3. Most of the cells studied require at least one hormone of the class which localizes in the nucleus, such as steroids and thyroid hormone.

GROWTH FACTORS

Use of Cell Culture to Purify Hormones

Progress in our studies has been made possible by the recent discoveries and purification of novel growth factors, such as EGF, the somatomedins, OGF, FGF, multiplication-stimulating activity (MSA), nonsuppressible insulin-like activity (NSILA), platelet factor, and the basic peptide discovered by Antoniades et al. (2). Future progress in this field will certainly depend on the discovery of factors as well as on the characterization of some of the more novel factors. Many of the novel factors are likely to be found by cell culture assays. This is apparent if one examines the methods by which hormones are discovered. They are discovered by one of two classical routes: 1) extirpation of an organ and observation of the physiological effects of its removal (e.g., removal of the pituitary causes cessation of growth of the animal, so it can be deduced that the gland is producing a growth hormone); 2) injection of extracts of the storage organ of the hormone into animals (e.g., injection of pituitary extracts makes the animal grow larger than normal). This type of experiment also leads to the conclusion that the pituitary produces a growth hormone. If the organ or tissue cannot be extirpated because it would lead to the death of the animal, or if the hormone is produced by, but not stored in, a tissue, then neither method works for the discovery of that hormone. A classical precedent for this can be seen in the discovery of the somatomedins (92). In the 1950s, Salmon and Daughaday studied the effects of growth hormone in growing bone. Because cartilage is a sulfated polysaccharide, growth of the bone can be monitored by the incorporation of radioactive sulfate into the epiphyseal regions of long bone. They incubated young growing bones in sera of different animals. If the bones were incubated in normal serum, radioactive sulfur was incorporated as was expected. Also as expected,

Table 1. Cell lines used in studies

Designation	Origin	Morphology	Tumorigenic in nude mice	Comments and references
GH_3	Rat pituitary carcinoma	Epithelioid	Yes	GH_3 was established by Yasumura et al. (113) from a transplantable pituitary tumor developed by Takemoto et al. (100). Secretes prolactin and GH.
GC	Rat pituitary carcinoma	Epithelioid	Yes	A subclone of GH_3.
HeLa	Human cervical carcinoma	Epithelioid	Yes	Established by Gey et al. (40).
B16	Mouse melanoma	Epithelioid	Yes	Cloned line established by Roberts (unpublished results) from a transplantable tumor developed from a spontaneous tumor in C57B1/6 mice.
RF1	Rat ovarian cells	Epithelioid	No	Cloned line established by Masui et al. (72) from primary cultures of rat ovary. Can be stimulated by LH to secrete testosterone and other steroids.
31A	Rat ovarian cells	Epithelioid	No	Cloned cell line from hyperplastic rat ovary grown in the spleen. Established by Clark et al. (14).
3T3	Balb/c mouse embryos	Fibroblast-like	No	Cloned cell line from disaggregated 14–17-day-old mouse embryos (1).
BHK_{21}	Hamster kidney cells	Fibroblast-like	Yes	Cloned cell line from kidneys of 1-day-old Syrian hamster. Established by MacPherson and Stoker (70).

$DS8_i$	Mouse mammary carcinoma	Epithelioid	Yes	Established by Desmond et al. (24) from the Shionogi transplantable tumor carried in nude mice.
PCC4-aza-1	Mouse teratocarcinoma	Epithelioid	Yes	Established by Jacob et al. (55) from a teratocarcinoma carried in ascitic transplantation in 129 mice. Differentiates when injected subcutaneously into 129 mice or nude mice.
F9	Mouse teratocarcinoma	Epithelioid	Yes	A subline of PCC4. Does not differentiate in vivo but does differentiate to some extent in vitro.
BRL 6278	Human melanoma	Epithelioid	Yes	A cloned line established by Grace Cannon at Bionetics Research Lab, Rockville, Md.
HT 29	Human colon carcinoma	Epithelioid	Yes	A cloned line developed by Fogh and Trempe (33).
	Rabbit intimal cells		No	A cloned line established by Buonassisi (unpublished results) from the subendothelial cell layer of rabbit aorta. Suspected of being smooth muscle cells.
PC3	Human prostatic carcinoma	Epithelioid	Yes	A cloned line established by Kaighn (unpublished results). Derived from a bone metastasis. The cell line is anaplastic and grows in both male and female nude mice.

Table 2. Hormone requirements of various cell lines

Cell line	Insulin	Transferrin	Hormone requirements for growth in vitro	Investigators primarily responsible for work; references
In completely defined medium				
GH_3	5 μg/ml	5 μg/ml	PTH (0.5 ng/ml), T_3(10^{-11} M), TRH (10^{-9} M), somatomedin C[a] (1 ng/ml), FGF[a] (1–10 ng/ml)[3]	Hayashi (52, 53)
GC	5 μg/ml	5 μg/ml	Requirements are the same as for GH_3 except that SM_C is not needed	Wu (unpublished results)
HeLa	5 μg/ml	5 μg/ml	Hydrocortisone (10^{-7} M), EGF (1 ng/ml), FGF (now supplied by NIH-LH at 1 μg/ml)	Hutchings (52, 95)
B16	5 μg/ml	0.5 μg/ml	NIH-FSH (0.4 μg/ml), LH-RH (10 ng/ml), and testosterone (10^{-9} M). Testerosterone can be replaced by progesterone (5×10^{-8} M)	Mather (73)
RFl	2 μg/ml	5 μg/ml	Hydrocortisone (40 ng/ml) and T_3 (3×10^{-8} M). After trypsinization, at least 6 hr of preincubation in serum are needed prior to plating in serum-free medium plus hormones	Ohasa (80)

BHK_{21}	5 μg/ml	5 μg/ml	EGF (5 μg/ml), FGF (1 μg/ml), and hydrocortisone (10^{-7} M)	Maciag (71) and Hutchings
In partially defined medium				
31A	2 μg/ml	5 μg/ml	EGF and LH	Nishikawa (78)
Balb-3T3	2 μg/ml	5 μg/ml	Hydrocortisone (0.1 μg/ml), FGF (25 ng/ml), EGF (30 ng/ml)	Armelin (3) and Nakamura (77)
$DS8_i$	insulin	transferrin	Hydrocortisone (10^{-8} M), dihydrotestosterone (10^{-9} M), prolactin	Desmond (24)
PCC4-aza-G and F9	10 μg/ml	5 μg/ml	Fetuin (500 μg/ml), β-mercaptoethanol (0.5×10^{-5} M)	Rizzino (88)
BRL 6278	5 μg/ml	5 μg/ml	The same as the requirements for B16 plus 1% fetal calf serum	Breitman (unpublished results)
HT 29	5 μg/ml	5 μg/ml		John Elovson (unpublished results)
Chick primary myoblast cultures	10 μg/ml		Bovine serum albumin (7 mg/ml), GH (5 μg/ml), and 1% horse serum	Rizzino (89)
Rabbit intimal cells	10 μg/ml	1 μg/ml	NIH-LH (1 μg/ml), GH (5 μg/ml)	Serrero (unpublished results)
PC3	5 μg/ml	5 μg/ml	T_3(10^{-11} M), TRH (10^{-9} M), NIH-LH (1 μg/ml)	Russell and Reid (unpublished results)

[a]Can be supplied by bloodmeal peptides: pI 8.5, 9.5, and 10.5. If concentration not listed, dose-response studies have not been done.

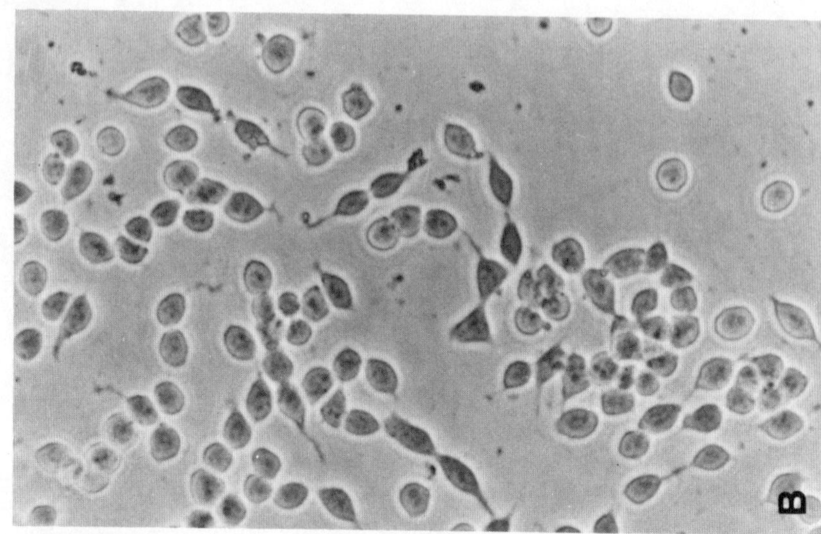

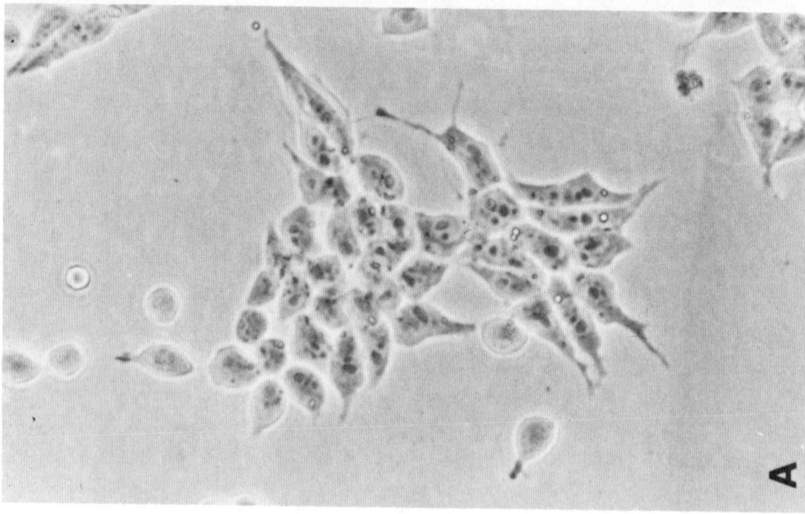

Figure 5. Appearance of PCC4-Aza-1 cells, an embryonal carcinoma cell line, in medium supplemented with 10% fetal calf serum (*A*) and in serum-free F12 supplemented with insulin, transferrin, β-mercaptoethanol, and fetuin (*B*).

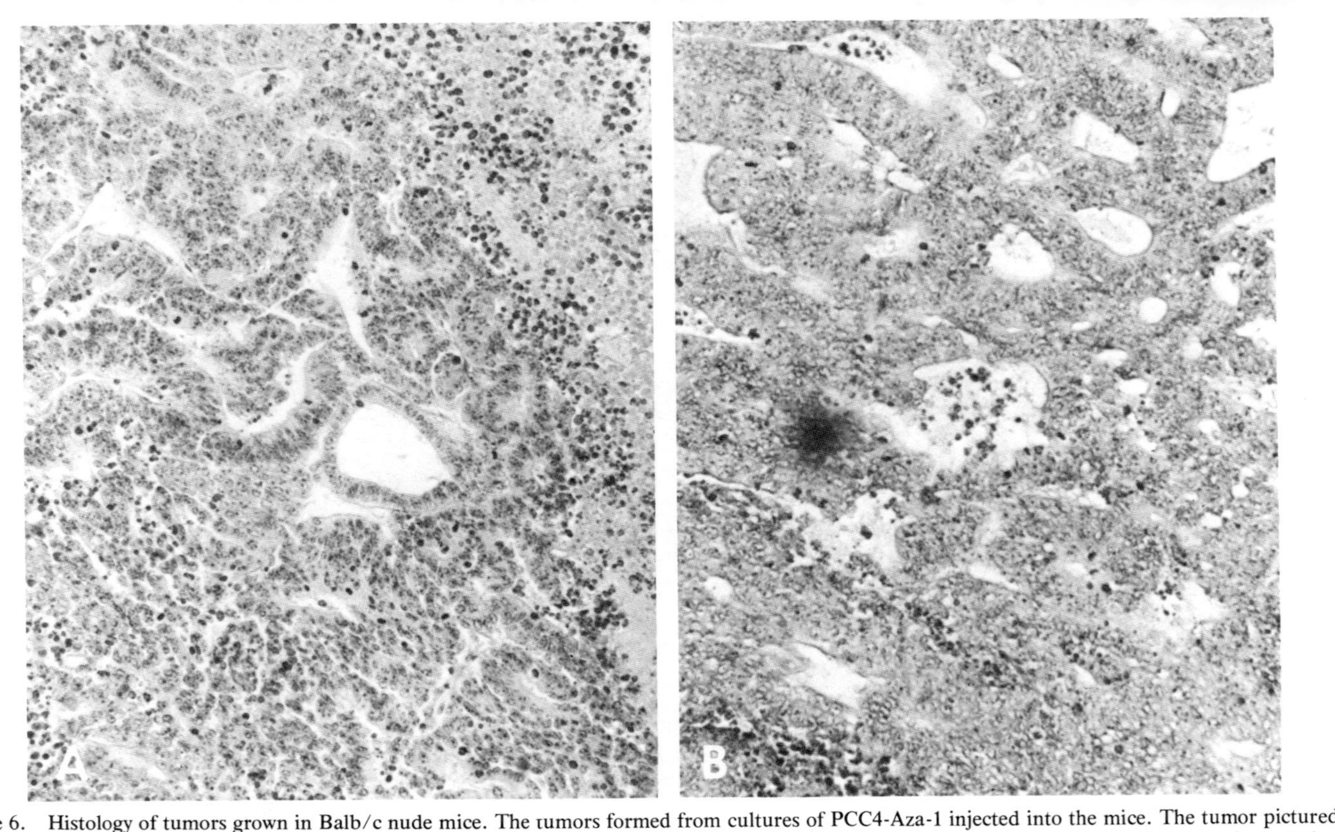

Figure 6. Histology of tumors grown in Balb/c nude mice. The tumors formed from cultures of PCC4-Aza-1 injected into the mice. The tumor pictured in *B* derived from cultures grown in serum-supplemented medium; that pictured in *A* derived from cultures grown in serum-free medium supplemented with the substances listed in the legend to Figure 5. Note that the cultures grown for weeks in the serum-free, hormone-supplemented media have produced a tumor which differentiated much as the one derived from cultures maintained in serum-supplemented media.

when bones were incubated in serum from hypophysectomized animals, no incorporation of radioactive sulfate was observed because the lack of a pituitary meant a deficiency of GH. But, to their surprise, when they added growth hormone to the serum from hypophysectomized animals, no stimulation of sulfate uptake was observed. If hypophysectomized animals were injected with growth hormone 6 hr prior to taking their blood, the serum from such animals supported the incorporation of radioactive sulfur. From these experiments, Salmon and Daughaday concluded that growth hormone does not act directly on bones to stimulate sulfate incorporation; instead, it acts via a sulfation factor. Today sulfation factor is known to be a complex of peptides which are collectively called somatomedins. It is presently believed that somatomedins are secreted by the liver in response to growth hormones, and that the liver does not store somatomedins. Therefore, the two classical routes for discovering somatomedins were not available. First, hepatectomized animals would not survive. Second, extracts of liver could not be injected into animals to watch superstimulation of growth because the liver does not store somatomedins. Therefore, this class of hormones was discovered by the use of culture methods. Tissue culture studies will be an increasingly important means of discovery of hormones in the future.

Purified Growth Factors

Strictly speaking, many substances, including trophic hormones, vitamins, essential amino acids, and other nutrients, can be considered growth factors, but the term "growth factor" is loosely used to refer to those endogenously produced substances which have a marked trophic effect on cells but have not been labeled hormones as yet. The best known group of growth factors are the peptide growth factors including FGF, OGF, EGF, nerve growth factor (NGF), the somatomedins, MSA, NSILA, and others. A review of these factors has been written by Van Wyk and Hintz (105). The most frequently used growth factors in defining hormone requirements have, thus far, come from this class of trophic substances. Table 3 lists some of the factors and relevant references.

Two of the substances, OGF and FGF, were discovered as the active principles of a pituitary extract with LH activity and were referred to as NIH-LH. They are both basic peptides with molecular weights of approximately 10,000 and can trigger DNA synthesis in a variety of cell types. Because there are a number of similarities between OGF and FGF, they may be chemically identical or closely related molecules.

Nerve growth factor was discovered by Levi-Montalcini et al. (63) in snake venom. It has marked trophic activity on nerve cells both in vivo and in vitro. It is produced by mouse salivary glands (62), glioma cells (54), and chick fibroblast cultures (54).

Epidermal growth factor is also found in mouse salivary glands (16–

18). S. Cohen and his colleagues discovered and purified EGF and found that testosterone dramatically increases the amount of immunoreactive EGF in vivo. As can be seen in Table 2, EGF demonstrates mitogenic activity on a number of different cell types, including both fibroblast-like and epithelial cells.

Other growth factors such as thrombin, erythropoietin, and thymosin have not been shown to be active on any of the cells under investigation in our laboratory. There is, of course, the distinct possibility that they will prove to be significant in the growth regulation of cell types not yet studied. A discussion of their origins and their chemical nature is given in a review by Van Wyk and Hintz (105).

Partially Purified Factors

Insulin-like Factors Multiplication-stimulating activity, nonsuppressible insulin-like activity, and somatomedin C are closely related and may be identical. All have insulin-like activity, stimulate the incorporation of sulfate into cartilage, and are under growth hormone control. Somatomedin C competes for insulin-binding sites on adipocytes. All are considered now to be members of the family of growth factors known as the somatomedins. As discussed previously, the somatomedins were discovered in culture experiments testing for the sulfation factor in bone cartilage. No storage organ for this material has been found, and it is laboriously purified from ethanolic fractions (Cohn fraction IV) (19) of human serum. Progress has been slow, and only small amounts of highly purified material have been available. Recent reviews by Hall et al. (47) and by Van Wyk et al. (107). detail the present knowledge about the somatomedins.

Multiplication-stimulating activity is a growth factor purified from calf serum by Pierson and Temin (83). It stimulates DNA synthesis in serum-depleted cultures of chick fibroblasts (25, 26, 101). It has since been found that four similar peptides with MSA activity are secreted by a cell line of rat liver cells developed by H. Coon (20, 97).

Nonsuppressible insulin-like activity was found as an insulin-like activity in serum which is not suppressed by guinea pig antiserum against insulin (35). NSILA consists of two components: NSILA-S, which is soluble in acid ethanol, and NSILA-P, which is insoluble in acid ethanol. NSILA-S has been shown by Froesch and his colleagues to fulfill all the criteria for a somatomedin and to stimulate mitogenic activity in fibroblast cultures (36–38). It has been purified (87) and its primary structure determined, showing extensive homology of its amino acid sequence with that of insulin (86).

Platelet and Antoniades Factor Plasma prepared carefully so as to be free of platelets and substances released during clot formation does not support the proliferation of arterial smooth muscle cells (90), most normal fibroblast-like cells (61, 85), or some normal epithelial cells (85). In fact,

Table 3. Growth factors and accessory factors frequently used in defining hormone requirements

Name	Origin	Sources	Comments and references
Nerve growth factor (NGF)	Snake vemon Mouse salivary gland	Burroughs Welcome, Research Triangle Park, Raleigh, N.C.	Discovered by Levi-Montalcini et al. (11, 34, 54, 62, 63) and found active on sympathetic and sensory neurons (102).
Fibroblast growth factor (FGF)	Bovine pituitary Brain	Collaborative Research, Waltham, Mass.	Purified by Gospodarowicz, et al. (42). May be similar to or identical with platelet factor. Active on many cells (see Table 2).
Platelet factor	Platelets		Being purified by R. Ross (90). Necessary for the growth in vitro of most normal fibroblast-like cell lines (60, 85) and for a number of epithelioid cell lines (85). Chemically similar to FGF.
Antoniades factor	Human serum		Purified by Antoniades et al. (2). Active on fibroblast-like cells. Chemically similar to FGF.
Epithelial growth factor (EGF)	Salivary gland	Collaborative Research, Waltham, Mass.	Purified by Cohen et al. (16–18). Active on a number of cell lines. (See Table 2.) (98)
Fetuin	Fetal calf serum	Calbiochem	Partially purified extract shown to facilitate attachment of cells in vitro.
NIH-luteinizing hormone (NIH-LH)	Bovine pituitary	Hormone Distribution Program, NIH, Bethesda, Md.	Contains FGF and OGF activity (42, 43, 78).

Follicle-stimulating hormone (NIH-FSH)	Bovine pituitary	Hormone Distribution Program, NIH, Bethesda, Md.	
Growth hormone (NIH-GH)	Bovine pituitary	Hormone Distribution Program, NIH, Bethesda, Md.	
Ovarian growth factor (OGF)	NIH-LH (bovine pituitary)		Purified by Gospodarowicz et al. (43) from NIH-LH. Active on ovarian cells. May be the same as FGF.
Somatomedins	Liver, serum		
Somatomedin C(SM_c)	Cohn fraction IV	Collaborative Research, Waltham, Mass.	Purified by several groups (15, 21, 22, 104, 107) and found to mediate the effects of GH. Active on various cell lines (see Table 2). Can be replaced by bloodmeal peptides (39, 46–67, 69, 77, 83, 92, 106).
Multiplication-stimulating activity (MSA)	Cohn fraction IV		Purified by Temin's group (83, 97). Chemically similar to SM_c (25, 26, 101).
Nonsuppressible insulin-like activity (NSILA)	Human serum		Purified by Froesch's laboratory (36). Chemically similar to SM_c (35,37,38,79,86,87).
Transferrin	Serum	Sigma	Iron-carrying protein. Found by Tormey et al. (103) to stimulate the growth of lymphocytes.

Reid et al. (85) have shown that, at least for fibroblast-like cells, the ability to grow in plasma is a property highly correlated with the tumorigenicity of the cells in nude mice. Isolated platelets can be made to discharge factors which will permit the growth of cells cultured in plasma (90). Work is under way in R. Ross' laboratory at the University of Washington in Seattle to purify the platelet factor which is mitogenic in vitro.

Antoniades et al. (2) have isolated a highly basic peptide from human serum which stimulates DNA synthesis in serum-starved fibroblast cultures. This peptide, like platelet factor, is found in serum and not in plasma, and the two may be closely related or identical. Moreover, the factor may also be chemically similar or the equivalent of FGF because both are found in the pituitary and share common physical characteristics such as pI and molecular weight.

As discussed earlier, many trophic substances purified or partially purified from serum have not been identified as distinct entities. Extensive or complete homologies among the insulin-like peptides have been demonstrated; other growth factors may also prove to be identical with or fall within a chemically distinct class of trophic substances.

Bloodmeal as Source of Hormones The difficulty of isolating factors from blood, when it is necessary to do so for lack of a storage organ in the body, is readily apparent in the case of the somatomedins. Progress in its purification has been slow mainly because of the absence of a concentrated source material. It has recently been discovered that bloodmeal is a rich concentrate of several biologically active, heat-stable blood peptides (78). It is produced by the furnace drying of whole blood and is used primarily as a fertilizer. Because most of the blood components have been irrevocably denatured in the furnace drying process, the purification of the heat-stable peptides has been greatly facilitated. The major contaminants, salts, are easily eliminated in the first step by washing the bloodmeal powder with water. The relative ease of isolation of these peptides and the low cost and ready availability of the material make bloodmeal a particularly ideal source for these peptides.

Our procedure for purification of bloodmeal peptides is given in Figure 7, and an isoelectric focusing profile of the growth activity of bloodmeal peptides on GH_3 cells is given in Figure 8. At this time, it is premature to assign identities to the various peaks of activity, but, as shown in Figure 1, several of the bloodmeal peptides can replace the activity of FGF and somatomedin C in the GH_3-defined medium. Furthermore, the bloodmeal peptide BGF2 is extremely potent in a somatomedin receptor assay (M. Grumbach, personal communication).

Epithelial-Fibroblast Interactions: A New Class of Hormonal Interactions in Adult Tissues?

There is much evidence indicating the importance of secretory products of cells which influence neighboring cells. In embryological induction, a great effort has been made to isolate the mesenchymal inducers which influence the differentiation of epithelial cells (13, 23, 32, 45, 96). A renewal of interest in this field is typified by the recent work of Green et al. (44) and of Vaughan and Bernstein (108), who have studied the influence of fibroblasts on the growth and the differentiation of epithelial cells. It is also typified by the recent investigations showing that in mammary tissue hormones act directly on the mesenchyme and through them influence the behavior of the epithelial elements (91). Moreover, other workers have found that the constituents of basement membrane, e.g., certain mucopolysaccharides, increase the survival of explanted epithelial cells (57–59, 81, 99).

Thus, certain cell-cell relationships within the tissue architecture are probably critically responsible for the differentiative state of cells. The epithelial-mesenchymal relationship is likely to be one of the most primary since it is ubiquitously present in tissues. In vivo all normal epithelial cells capable of proliferation or long-term survival are attached to a basement membrane which is, in turn, associated with mesenchymally derived cells, most commonly fibroblasts. The basement membrane, a layer of secretion located between and produced by the epithelium and the mesenchymal cells is likely to contain components which effect a portion of the epithelial-mesenchymal interactions. This cell-cell relationship is typically altered in the malignant state so that there is proliferation of either epithelial cells alone (carcinomas) or mesenchymal cells alone (sarcomas), a fact probably significant in the genesis of malignancy.

There may be multiple types or classes of fibroblasts which are interdependent with specific classes of epithelial cells. The possible classes range from each epithelial cell type having its distinct type of associated fibroblast to two prototype classes: 1) a class of fibroblasts interdependent with a class of epithelial cells involved in a differentiated function, and 2) a class of fibroblasts interdependent with a class of epithelial cells involved in growth. In the first class, differentiated cells which slowly or rarely mitose in vivo are attached to the basement membrane. Mitotic activity occurs rarely, except in the malignant condition, in which the epithelial cells become dissociated from the basement membrane and the cells usually lose their differentiated functions. In the second class, undifferentiated epithelial cells are anchored to the basement membrane, and mitotic activity among these stem cells gives rise to daughter cells which are detached from the basement membrane and gradually differentiate as they become distant from the basal epithelium.

The possibility of an interdependency of epithelial cells and fibroblasts for survival, growth, and/or differentiation should have relevance for in vitro studies. Culture techniques such as organ culture which retain the tissue architecture maintain tissue-specific functions and hormone and pharmacological responses. Cell culture procedures use explanted tissues which are disaggregated into single cells and adapted to grow as a mono-

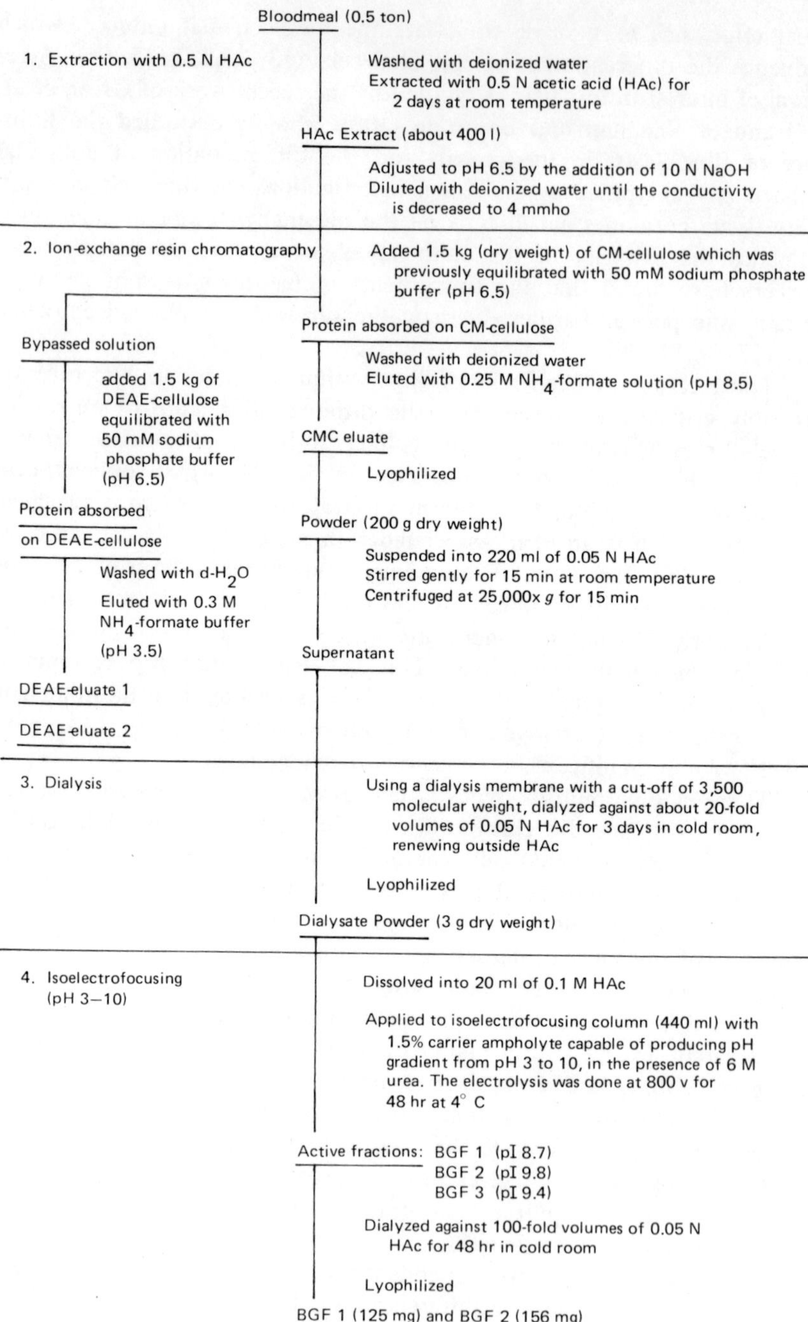

(continued on next page.)

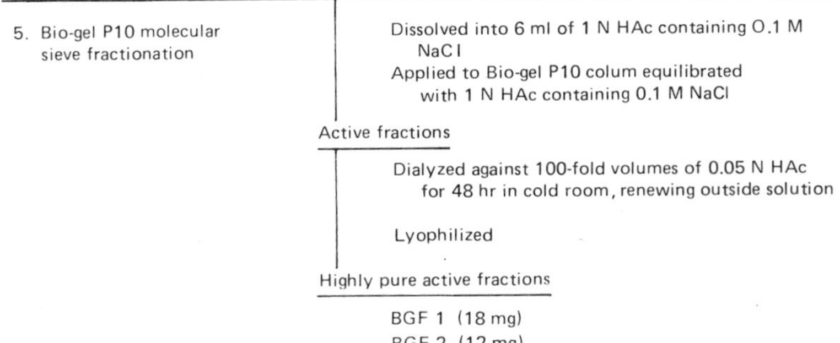

Figure 7. Purification procedures for bloodmeal peptides with mitogenic activity. Adapted from Nakamura and Sato (unpublished results).

layer on treated plastic or as a cell suspension. The mechanical and enzymatic dissociation into single cell suspensions not only may cause damage of cell-cell junctions perhaps important in cellular physiology (30, 68, 75, 82, 114) but also disrupts cell-cell relationships which also may be relevant to growth and/or functioning of cells. In order to establish in culture normal epithelial cells or fibroblasts dissociated from one another, the cells must be provided in the media with those factor(s) which maintain their interdependency in vivo. Because malignant cells have become qualitatively or quantitatively independent of their need for these factor(s), it is easy to understand why they can be established more easily in vitro. By using hormone-supplemented media, there should be enhanced opportunities to discover and purify some of these factors. Studies on the epithelial-fibroblast interaction may facilitate the establishment of some normal epithelial cells in vitro as well as the retention of differentiated functions in either normal or neoplastic cell types.

FUTURE PERSPECTIVES

It is clear that different cell types in culture require different combinations of hormones for growth and probably also for their differentiated functions. Eventually, as the complex for each cell type is elucidated, patterns of requirements will emerge which should be meaningful for physiology and developmental biology. It is possible that the complex of hormones required by a cell line could serve as a "fingerprint" to identify the physiological role and embryonic origin of the cell from which the culture was derived. The patterns of hormone requirements could also shed new light on the subtle modulations of cellular activity made possible by a multiplicity of interacting hormonal signals, each of which can vary in intensity. Clarification of this concept will require that we first identify the hormones required by many kinds of cells.

The hormones required by established cell cultures probably represent a minimum in terms of the hormones effective on the normal cell of origin

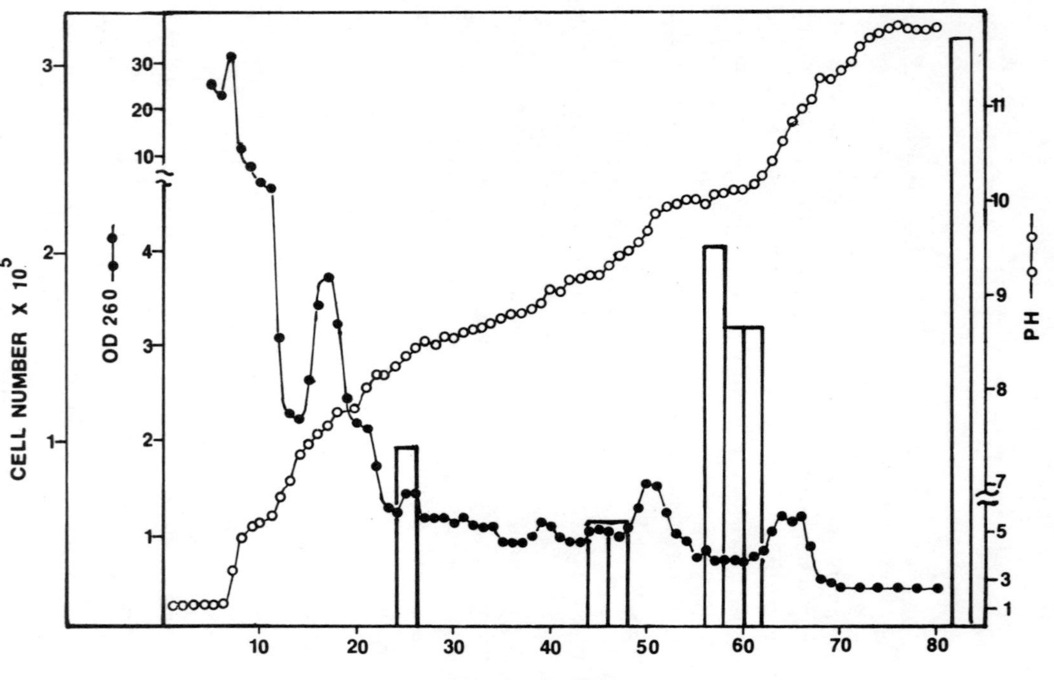

Figure 8. Isoelectric focusing fractionation of bloodmeal extracts. ○, pH; ●, optical density (OD) at 260. The *bars* represent cell numbers of cultures grown in media supplemented with insulin, transferrin, TSH-releasing hormone, triiodothyronine, and parathormone, plus the isoelectric focusing fraction at a concentration of 0.5% v/v. Cells were inoculated at a seeding density of 10^5 cells into serum-free F12 medium plus the hormones. The *bar* at the *extreme right* represents the cell number when the starting material (before fractionation) of the ammonium formate eluate from the CMC column was used in media at a concentration of 10 μg/ml.

in the animal. At the present time, these requirements are detected as requirements for growth and survival in culture, whereas in the animal they might be revealed as modulators of metabolic activity. Despite the possible discrepancy between culture and animal experiments, the determination of hormone requirements in vitro should be used as a first approximation to the in vivo situation, because the most convenient and radical "endocrine ablation" which can be performed is the removal of serum from cells in culture.

The hormones required by an established cell line should form a basic core from which one can expand to the larger set operative on the cell type in vivo. This may have important technical consequences for cell culture. At the present time, many cell types have never been established in vitro, and for those which grow, there is always a lag period before the primary cultures begin vigorous growth. We surmise that this lag often results from selection of variants with decreased hormonal requirements. Primary cultures could be used as a tool for finding those hormones above and beyond those required for established cell lines and could help determine the hormone requirements of the normal cell of origin. In turn, this knowledge could be used to formulate media with hormone instead of serum supplementation, which would be more suitable for establishing those cells in culture.

Our present ability to formulate serum-free, hormone-supplemented media is possible because of the discoveries of novel growth factors. Their hormonal status is not yet established, but it is likely that they will prove to have important physiological roles. Culture technology played an important role in their discovery, and it is likely that cultures will play an increasingly important role in the future discovery of hormones. A case in point is the list of cells which can grow in low serum concentrations when the medium is hormone-supplemented but for which we have not yet found the correct factors or hormones to totally substitute for the serum requirement. If this residual serum provides unknown hormones or if novel hormones can be derived from tissues, then they can be detected and purified by standard purification procedures. The likelihood of finding and purifying them has been greatly enhanced because many of the hormone requirements have been determined and provided in precise amounts.

Of these yet-to-be-discovered hormones, a significant group is likely to derive from epithelial-fibroblast interactions. Discovery of these hormones or factors should permit the establishment of normal and benign cells, as well as of certain malignant cells in vitro. For the first time, studies will be done on established cell cultures which closely resemble their in vivo counterparts, and it will be feasible to have true model systems of normal versus malignant cells. Hormone mechanisms which involve the epithelial-fibroblast interactions can be studied and elucidated. Moreover, there are a number of pathological conditions other than malignancy which may involve an aberration of the epithelial-fibroblast relationship. For example, benign prostatic hyperplasia may be an abnormality of the stromal matrix rather than of the epithelial cells embedded in that matrix. The proliferation of the fibroblasts may stimulate a concomitant growth in the epithelial cells which results from the fibroblast-epithelial cell interactions. Such a hypothesis can be tested in vitro in serum-free, hormone-supplemented medium by comparing the fibroblasts from normal prostate with those from benign hyperplastic or hypertrophic prostates.

The use of serum-free, hormone-supplemented media can contribute to a better understanding of hormonal mechanisms. The elimination of serum reveals that cells are responsive to a surprising variety of hormones whose effects can now be studied in these defined systems. Because the requirements of any given cell are most likely to be multiple, the interaction of their effects can be dissected. Moreover, the search can start for those hormones and factors which affect the differentiated functions of the cells. Because the hormones are required for growth, the isolation of mutants which are independent of the need for certain hormones becomes a possibility, paving the way for genetic studies of hormone mechanisms.

These approaches can be applied to the cancer cell. Knowing the multiple hormone requirements of a given cancer cell opens the possibility of sophisticated approaches to hormone therapy. Hormones whose elimination would not be lethal to the patient can be withdrawn by surgical ablation of the hormone source. Analogues which are nonfunctional can be administered. Toxins can be coupled to hormones. Hormones can be administered and withdrawn along with chemotherapeutic agents on schedules that will enhance the effectiveness of the drugs.

In summary, the elucidation of the hormone requirements of various cell types will allow cultivation of cells in defined media without serum and will permit a variety of elegant in vitro and in vivo studies analyzing integrative physiological processes. Furthermore, the very definition of a hormone may be re-evaluated to accommodate a more unifying concept of regulatory molecules and mechanisms.

REFERENCES

1. Aaronson, S. A., and Todaro, G. T. (1968). J. Cell. Physiol. 72:141.
2. Antoniades, H., Stathakos, D., and Scher, C. D. (1975). Proc. Natl. Acad. Sci. USA 72:2635.
3. Armelin, H. (1973). Proc. Natl. Acad. Sci. USA 70:2702.
4. Armelin, H. A. (1975). *In* G. Litwack (ed.), Biochemical Action of Hormones, Vol. III, pp. 1–21. Academic Press, New York.
5. Armelin, H. A., Nishikawa, K., and Sato, G. (1974). *In* B. Clarkson and R. Baserga (eds.), Control of Proliferation in Animal Cells, pp. 97–104. Cold Spring Harbor Press, Cold Spring Harbor, New York.
6. Armelin, H. A., and Sato, G. H. (1974). *In* P. Ts'o and J. DiPaolo (eds.), World Symposium on "Model Studies in Chemical Carcinogenesis," Part B, pp. 483–501. Marcel Dekker, Inc., New York.
7. Arnold, J. (1887). Arch. Mikr. Anat. 30:205.

8. Baker, L. E., and Carrel, A. (1928). J. Exp. Med. 47:371.
9. Baker, L. E., and Ebeling, A. H. (1939). J. Exp. Med. 69:365.
10. Biskind, M. S., and Biskind, G. S. (1944). Proc. Soc. Exp. Biol. Med. 55:176.
11. Bocchini, V., and Angeletti, P. U. (1969). Proc. Natl. Acad. Sci. USA 64: 787.
12. Brooks, R. F. (1975). *In* A. C. Allison (ed.), Structure and Function of Plasma Proteins, pp. 1–112. Plenum Press, New York.
13. Cameron, G., and Chambers, R. (1937). Am. J. Cancer 30:115.
14. Clark, J. L., Jones, K. L., Gospodarowicz, D., and Sato, G. H. (1972). Nature (Lond.) 236:180.
15. Cohen, K. L., and Nissley, S. P. (1974). Endocrinology 97:654.
16. Cohen, S., and Carpenter, G. (1975). Proc. Natl. Acad. Sci. USA 72:1317.
17. Cohen, S., and Savage, C. R., Jr. (1974). Recent Progr. Horm. Res. 30: 551.
18. Cohen, S., and Taylor, J. M. 1974. Recent Progr. Horm. Res. 30:533.
19. Cohn, E. J., Gurd, F. R. N., Surgenor, D. M., Barnes, B. A., Brown, R. K., Derouaux, G., Gilliespie, J. M., Kahnt, F. W., Lever, W. F., Liu, C. H., Mittelman, D., Mouton, R. F., Schmid, K., and Uroma, E. (1950). J. Am. Chem. Soc. 72:465.
20. Coon, H. G. (1968). J. Cell Biol. 39:29a.
21. Daughaday, W. H., Hall, K., Raben, M. S., Salmon, W. D., Van den Brande, J. L., and Van Wyk, J. J. (1972). Nature (Lond.) 235:107.
22. Daughaday, W. H., Phillips, L. S., and Herington, A. C. (1975). *In* R. Luft and K. Hall (eds.), Advances in Metabolic Disorders, 8:151–157. Academic Press, New York.
23. Davies, P., Allison, A. C., and Cardella, C. J. (1975). Philos. Trans. R. Soc. Lond. (Biol. Sci.) 271:363.
24. Desmond, W., Wolbers, S., and Sato, G. (1976). Cell 8:79.
25. Dulak, N. C., and Temin, H. M. (1973). J. Cell Physiol. 81:153.
26. Dulak, N. C., and Temin, H. M. (1973). J. Cell. Physiol. 81:161.
27. Earle, W. R., Schilling, E. L., Bryant, J. C., and Evans, V. I. (1954). J. Natl. Cancer Inst. 14:1159.
28. Eagle, H. (1955a). J. Exp. Med. 102:37.
29. Eagle, H. (1955b). Science 122:501.
30. Fentiman, I., Taylor-Papadimitriou, J., and Stoker, M. (1976). Nature (Lond.) 264:760.
31. Fischer, A. (1941). Acta Physiol. Scand. 2:145.
32. Fleischmajer, R., and Billingham, R. (1968). Epithelial-mesenchymal Interactions. The Williams and Wilkins Co. Baltimore, Maryland.
33. Fogh, J., and Trempe, G. (1975). *In* J. Fogh (ed.), Human Tumor Cells In Vitro, pp. 115–159. Plenum Press, New York.
34. Frazier, W. A., Angelletti, R. H., and Bradshaw, R. A. (1972). Science 176: 482.
35. Froesch, E. R., Bürgi, H., Ramseier, E. B., Bally, P., and Labhart, A. (1963). J. Clin. Invest. 42:1816.
36. Froesch, E. R., Schlumpf, U., Heimann, R., Zapf, J., Humbel, R. E., and Ritschard, W. J. (1975). Adv. Metab. Disord. 8:203.
37. Froesch, E. J., Zapf, J., Meuli, C., Mäder, M., Waldvogel, M., Kaufmann, U., and Morell, B. (1975). Adv. Metab. Disord. 8:211.
38. Froesch, E. R., Zapf, J., Audhya, T. K., Ben-Porath, E., Segen, B. J., and Gibson, K. D. (1975). Proc. Natl. Acad. Sci. USA 73:2904.
39. Fryklund, L., Uthne, K., and Sievertsson, H. (1974). Biochem. Biophys. Res. Commun. 61:950.

40. Gey, G. O., Coffman, W. D., and Kubicek, M. T. (1952). Cancer Res. 12:264.
41. Gey, G. O., and Thalhimer, W. (1924). J. Am. Med. Assoc. 82:1609.
42. Gospodarowicz, D. (1975). J. Biol. Chem. 250:2515.
43. Gospodarowicz, D., Jones, K. L., and Sato, G. (1974). Proc. Natl. Acad. Sci. USA 71:2295.
44. Green, H., Rheinwald, J. G., and Sun, T. J. Supramol. Struct. In press.
45. Grobstein, C. (1953). Science 118:52.
46. Hall, K., and Uthne, K. (1971). Acta Med. Scand. 190:137.
47. Hall, K., Takano, K., Fryklund, L., and Sievertsson, H. (1975). Adv. Metab. Disord. 8:19.
48. Ham, R. G. (1965). Proc. Natl. Acad. Sci. USA 53:288.
49. Harris, M. (1952). J. Cell. Comp. Physiol. 40:279.
50. Harris, M. (1954). J. Exp. Zool. 125:85.
51. Harrison, R. G. (1907). Proc. Soc. Exp. Biol. Med. 4:140.
52. Hayashi, I., Hutchings, S., Mather, J., and Sato, G. Excerpta Medica.
53. Hayashi, I., and Sato, G. H. (1976). Nature (Lond.) 259:132.
54. Hogue-Angelletti, R., Bradshaw, R. A., and Frazier, W. A. (1975). Adv. Metab. Disord. 8:285.
55. Jakob, E., Boon, T., Gaillard, J., Nicholas, J. F., and Jacob, F. (1973). Ann. Microbiol. Inst. Pasteur. 124:269.
56. Jones, K. L., and Gospodarowicz, D. (1974). Proc. Natl. Acad. Sci. USA 71:3372.
57. Kefalides, N. A. (1975). J. Invest. Dermatol. 65:85.
58. Kefalides, N. A. (1975). Dermatologica 150:4.
59. Klebe, R. J. (1974). Nature (Lond.) 250:248.
60. Kohler, N., and Lipton, A. (1974). Exp. Cell Res. 87:297.
61. Leibovich, S. J., and Ross, R. (1976). Amer. J. Pathol. 84:501.
62. Levi-Montalcini, R., and Angeletti, P. U. (1968). Physiol. Rev. 48:534.
63. Levi-Montalcini, R., Meyer, H., and Hamburger, V. (1954). Cancer Res. 14:49.
64. Lewis, M. R. (1916). Anat. Rec. 10:287.
65. Lewis, M. R., and Lewis, W. H. (1911a). Anat. Rec. 5:277.
66. Lewis, M. and Lewis, W. H. (1911b). Bull. Johns Hopkins Hosp. 22:126.
67. Lewis, M. R., and Lewis, W. H. (1912). Anat. Rec. 6:207.
68. Loewenstein, W. R. (1969). Can. Cancer Conf. 8:162.
69. Luft, R., and Hall, K. (ed.), (1975). Adv. Metab. Disord. 8
70. MacPherson, I., and Stoker, M. (1962). Virology 16:147.
71. Maciag, T., Gaudreau, J., Kelly, B., Hutchings, S., and Sato, G. (Unpublished.)
72. Masui, H., Ohasa, S., and Mackensen, S. Exp. Cell Res. In press.
73. Mather, J., and Sato, G. (Unpublished.)
74. McKeehan, W., Hamilton, W. G., and Ham, R. G. (1976). Proc. Natl. Acad. Sci. USA 73:2023.
75. McNutt, N. S., Hershberg, R. A., and Weinstein, R. S. (1971). J. Cell Biol. 51:805.
76. Morgan, J. F., Morton, H. J., and Parker, R. C. (1950). Proc. Soc. Exp. Biol. Med. 73:1.
77. Nakamura, T., and Sato, G. (Unpublished.)
78. Nishikawa, K., Armelin, H. A., and Sato, G. (1975). Proc. Natl. Acad. Sci. USA 72:483.
79. Oelz, O., Froesch, E. R., Bünzli, H. F., Humbel, R. E., and Ritschard, J. (1972). *In* I. D. Steiner and N. Freinkel (eds.), The Handbook of Physio-

logy, Section 7: Endocrinology, Vol. 1, p. 685. American Physiological Society, Washington, D.C.
80. Ohasa, S., Sato, G., and Masui, H. Cell. In press.
81. Ozzello, L., and Speer, F. D. (1958). Am. J. Pathol. 34:993.
82. Pitts, J. D., and Burk, R. R. (1976). Nature (Lond.) 264:762.
83. Pierson, R. W., Jr., and Temin, H. M. (1972). J. Cell. Physiol. 79:319.
84. Read, S. (1974). Guy's Hospital Reports. 123:53.
85. Reid, L. M., Stiles, C. D., Saier, M. H. Jr., and Rindler, M. J. Submitted for publication.
86. Rinderknecht, E., and Humbel, R. E. (1976). Proc. Natl. Acad. Sci. USA 73:4379.
87. Rinderknecht, E., and Humbel, R. E. (1976). Proc. Natl. Acad. Sci. USA 73:2365.
88. Rizzino, A., and Sato, G. In press.
89. Rizzino, A., Yeh, I., Friedman, H., and Sato, G. (Unpublished.)
90. Ross, R., Glemset, J., Kariya, B., and Harper, L. (1973). Proc. Natl. Acad. Sci. USA 71:1207.
91. Sakakura, T., Nishizuka, Y., and Dawer, C. (1976). Science 194:1439.
92. Salmon, N. D., and Daughaday, W. H. (1957). J. Lab. Clin. Med. 49:825.
93. Sanford, K. K., Waltz, H., Shannon, J. E., Jr., Earle, W. R., and Evans, V. J. (1952). J. Natl. Cancer Inst. 13:121.
94. Sato, G. H. (1975). *In* G. Litwack, (ed.), Biochemical Actions of Hormones, Vol. III, pp. 391–396. Academic Press, New York.
95. Sato, G. (1977). *In* C. Borek, C. M. Fenoglio, and D. W. King (eds.), Differentiation and Carcinogenesis. Advances in Pathobiology, No. 6, pp. 227–232. Stratton Intercontinental Medical Book Corporation, New York.
96. Saunders, J. W., Jr., Gasseling, M. T., and Gfeller, M. D. (1958). J. Exp. Zool. 137:39.
97. Smith, G. L., and Temin, H. M. (1974). J. Cell. Physiol. 84:181.
98. Stoker, M. G. P., Pigott, D., and Taylor-Papadimitriou, J. (1976). Nature (Lond.) 264:764.
99. Takeuchi, J. R., Tchao, and Leighton, J. (1974). Cancer Res. 34:161.
100. Takemoto, H., Yokoyo, K., Furth, J., and Cohen, A. I. (1962). Cancer Res. 22:917.
101. Temin, H. M. (1967). J. Cell. Physiol. 69:377.
102. Tischler, A. S., and Greene, L. A. (1975). Nature (Lond.) 258:341.
103. Tormey, D. C., Imrie, R. C., and Mueller, G. C. (1972). Exp. Cell Res. 74:163.
104. Uthne, K. (1973). Acta Endocrinol. 175:1.
105. Van Wyk, J. J., and Hintz, R. L. (1977). In press.
106. Van Wyk, J. J., Underwood, L. E., Baseman, J. B., Hintz, R. L., Clemmons, D. R., and Marshall, R. N. (1975). Adv. Metab. Disord. 8:127.
107. Van Wyk, J. J., Underwood, L. E., Hintz, R. L., Clemmons, D. R., Voina, S. J., and Weaver, R. P. (1975). Recent Progr. Horm. Res. 30:259.
108. Vaughan, F. L., and Bernstein, I. A. (1976). Mol. Cell. Biochem. 12:171.
109. Waymouth, C. (1972). *In* G. H. Rothblat and V. J. Cristofalo (eds.), Growth, Nutrition, and Metabolism of Cells in Culture, Academic Press, New York. Vol. 1, pp. 11–47.
110. White, P. R. (1946). Growth 10:231.
111. White, P. R. (1949). J. Cell. Comp. Physiol. 34:221.
112. White, P. R. (1955). J. Natl. Cancer Inst. 16:769.
113. Yasumura, Y., Tashjian, A. H., Jr., and Sato, G. (1966). Science 154:1186.
114. Zimmerman, J., Devlin, T., and Pruss, M. (1960). Nature (Lond.) 185:315.

Index